# Statistics and Computing

*Series Editors:*
J. Chambers
W. Eddy
W. Härdle
S. Sheather
L. Tierney

## Springer

*New York*
*Berlin*
*Heidelberg*
*Hong Kong*
*London*
*Milan*
*Paris*
*Tokyo*

# Statistics and Computing

James E. Gentle

# Random Number Generation and Monte Carlo Methods

## Second Edition

With 54 Illustrations

 Springer

James E. Gentle
School of Computational Sciences
George Mason University
Fairfax, VA 22030-4444
USA
jgentle@gmu.edu

*Series Editors:*

J. Chambers
Bell Labs, Lucent Technologies
600 Mountain Avenue
Murray Hill, NJ 07974
USA

W. Eddy
Department of Statistics
Carnegie Mellon University
Pittsburgh, PA 15213
USA

W. Härdle
Institut für Statistik und Ökonometrie
Humboldt-Universität zu Berlin
Spandauer Str. 1
D-10178 Berlin
Germany

S. Sheather
Australian Graduate School
  of Management
University of New South Wales
Sydney, NSW 2052
Australia

L. Tierney
School of Statistics and Actuarial Science
University of Iowa
Iowa City, IA 52242-1419
USA

Library of Congress Cataloging-in-Publication Data
Gentle, James E., 1943–
    Random number generation and Monte Carlo methods / James E. Gentle.
      p. cm. — (Statistics and computing)
    Includes bibliographical references and index.

    1. Monte Carlo method.    2. Random number generators.    I. Title.    II. Series.
    QA298 .G46 2003
    519.2'82—dc21                                    2003042437

ISBN  978-1-4419-1808-6          e-ISBN  978-0-387-21610-2

Printed in the United States of America.      (EB)

springeronline.com

*To María*

# Preface

The role of Monte Carlo methods and simulation in all of the sciences has increased in importance during the past several years. This edition incorporates discussion of many advances in the field of random number generation and Monte Carlo methods since the appearance of the first edition of this book in 1998.

These methods play a central role in the rapidly developing subdisciplines of the computational physical sciences, the computational life sciences, and the other computational sciences. The growing power of computers and the evolving simulation methodology have led to the recognition of computation as a third approach for advancing the natural sciences, together with theory and traditional experimentation. At the kernel of Monte Carlo simulation is random number generation.

Generation of random numbers is also at the heart of many standard statistical methods. The random sampling required in most analyses is usually done by the computer. The computations required in Bayesian analysis have become viable because of Monte Carlo methods. This has led to much wider applications of Bayesian statistics, which, in turn, has led to development of new Monte Carlo methods and to refinement of existing procedures for random number generation.

Various methods for generation of random numbers have been used. Sometimes, processes that are considered random are used, but for Monte Carlo methods, which depend on millions of random numbers, a physical process as a source of random numbers is generally cumbersome. Instead of "random" numbers, most applications use "pseudorandom" numbers, which are deterministic but "look like" they were generated randomly. Chapter 1 discusses methods for generation of sequences of pseudorandom numbers that simulate a uniform distribution over the unit interval $(0, 1)$. These are the basic sequences from which are derived pseudorandom numbers from other distributions, pseudorandom samples, and pseudostochastic processes.

In Chapter 1, as elsewhere in this book, the emphasis is on methods that *work*. Development of these methods often requires close attention to details. For example, whereas many texts on random number generation use the fact that the uniform distribution over $(0, 1)$ is the same as the uniform distribution over $(0, 1]$ or $[0, 1]$, I emphasize the fact that we are simulating this distribution with a discrete set of "computer numbers". In this case whether 0 and/or 1 is included does make a difference. A uniform random number generator should not yield a 0 or 1. Many authors ignore this fact. I learned it over twenty years ago, shortly after beginning to design industrial-strength software.

The Monte Carlo methods raise questions about the quality of the pseudorandom numbers that simulate physical processes and about the ability of those numbers to cover the range of a random variable adequately. In Chapter 2, I address some issues of the quality of pseudorandom generators.

Chapter 3 describes some of the basic issues in quasirandom sequences. These sequences are designed to be very regular in covering the support of the random process simulated.

Chapter 4 discusses general methods for transforming a uniform random deviate or a sequence of uniform random deviates into a deviate from a different distribution. Chapter 5 describes methods for some common specific distributions. The intent is not to provide a compendium in the manner of Devroye (1986a) but, for many standard distributions, to give at least a simple method or two, which may be the best method, but, if the better methods are quite complicated, to give references to those methods. Chapter 6 continues the developments of Chapters 4 and 5 to apply them to generation of samples and nonindependent sequences.

Chapter 7 considers some applications of random numbers. Some of these applications are to solve deterministic problems. This type of method is called Monte Carlo.

Chapter 8 provides information on computer software for generation of random variates. The discussion concentrates on the S-Plus, R, and IMSL software systems.

Monte Carlo methods are widely used in the research literature to evaluate properties of statistical methods. Chapter 9 addresses some of the considerations that apply to this kind of study. I emphasize that a Monte Carlo study uses an *experiment*, and the principles of scientific experimentation should be observed.

The literature on random number generation and Monte Carlo methods is vast and ever-growing. There is a rather extensive list of references beginning on page 336; however, I do not attempt to provide a comprehensive bibliography or to distinguish the highly-varying quality of the literature.

The main prerequisite for this text is some background in what is generally called "mathematical statistics". In the discussions and exercises involving multivariate distributions, some knowledge of matrices is assumed. Some scientific computer literacy is also necessary. I do not use any particular software system in the book, but I do assume the ability to program in either Fortran or C and the availability of either S-Plus, R, Matlab, or Maple. For some exercises, the required software can be obtained from either `statlib` or `netlib` (see the bibliography).

The book is intended to be both a reference and a textbook. It can be used as the primary text or a supplementary text for a variety of courses at the graduate or advanced undergraduate level.

A course in Monte Carlo methods could proceed quickly through Chapter 1, skip Chapter 2, cover Chapters 3 through 6 rather carefully, and then, in Chapter 7, depending on the backgrounds of the students, discuss Monte Carlo applications in specific fields of interest. Alternatively, a course in Monte Carlo methods could begin with discussions of software to generate random numbers, as in Chapter 8, and then go on to cover Chapters 7 and 9. Although the material in Chapters 1 through 6 provides the background for understanding the methods, in this case the details of the algorithms are not covered, and the material in the first six chapters would only be used for reference as necessary.

General courses in statistical computing or computational statistics could use the book as a supplemental text, emphasizing either the algorithms or the Monte Carlo applications as appropriate. The sections that address computer implementations, such as Section 1.2, can generally be skipped without affecting the students' preparation for later sections. (In any event, when computer implementations are discussed, note should be taken of my warnings about use of software for random number generation that has not been developed by software development professionals.)

In most classes that I teach in computational statistics, I give Exercise 9.3 in Chapter 9 (page 311) as a term project. It is to replicate and extend a Monte Carlo study reported in some recent journal article. In working on this exercise, the students learn the sad facts that many authors are irresponsible and many articles have been published without adequate review.

# Acknowledgments

I thank John Kimmel of Springer for his encouragement and advice on this book and other books on which he has worked with me. I thank Bruce McCullough for comments that corrected some errors and improved clarity in a number of spots. I thank the anonymous reviewers of this edition for their comments and suggestions. I also thank the many readers of the first edition who informed me of errors and who otherwise provided comments or suggestions for improving the exposition. I thank my wife María, to whom this book is dedicated, for everything.

I did all of the typing, programming, etc., myself, so all mistakes are mine. I would appreciate receiving suggestions for improvement and notice of errors. Notes on this book, including errata, are available at

   http://www.science.gmu.edu/~jgentle/rngbk/

Fairfax County, Virginia

James E. Gentle
April 8, 2003

# Contents

# Chapter 1

# Simulating Random Numbers from a Uniform Distribution

## Introduction

Because many statistical methods rely on random samples, applied statisticians often need a source of "random numbers". Older reference books for use in statistical applications contained tables of random numbers, which were intended to be used in selecting samples or in laying out a design for an experiment. Statisticians now rarely use printed tables of random numbers, but occasionally computer-accessed versions of such tables are used. Far more often, however, the computer is used to generate "random" numbers directly.

The use of random numbers in statistics has expanded beyond random sampling or random assignment of treatments to experimental units. More common uses now are in simulation studies of stochastic processes, analytically intractable mathematical expressions, or a population by resampling from a given sample from that population. Although we do not make precise distinctions among the terms, these three general areas of application are sometimes called "simulation", "Monte Carlo", and "resampling".

In engineering and the natural sciences, simulation is used extensively in studying physical and biological processes. Other common uses of random numbers are in cryptography. Applications in cryptography require somewhat different criteria for random numbers than those used in simulation. In this book, we consider the cryptographic criteria only in passing.

**Randomness and Pseudorandomness**

The digital computer cannot generate random numbers, and it is generally not convenient to connect the computer to some external source of random events. For most applications in statistics, engineering, and the natural sciences, this is not a disadvantage if there is some source of *pseudorandom* numbers, samples of which *seem* to be randomly drawn from some *known* distribution. There are many methods that have been suggested for generating such pseudorandom numbers.

It should be noted that there are *two* issues: randomness and knowledge of the distribution. Although, at least heuristically, there are many external physical processes that could perhaps be used as sources of *random* numbers— rather than pseudorandom numbers—there would still be the issue of *what is the distribution* of the realizations of that external random process. For random numbers to be useful in general applications, their distribution must be known. Other issues to consider for an external process are the independence of consecutive realizations and the constancy of the distribution. For random numbers to be useful, they usually must be identically and independently distributed (i.i.d.).

The most commonly used generator that is truly random according to generally accepted understandings of that concept is a substance undergoing atomic decay. The subatomic particles comprising the decaying substance transmute into other particles at random points in time. At a macro level (that is, given an amount of the substance that contains a very large number of atoms), both theory and empirical observations suggest that there are no dependencies among consecutive events and that the process is constant over sufficiently short time intervals. ("Sufficiently short" can be several years.) If we can measure the times between the events, and if the process is stationary, we can form a random variable with a known distribution. For observed intervals between events, $s_1, s_2, \ldots$, let

$$
\begin{aligned}
X &= 1 \quad \text{if} \quad s_{2i-1} < s_{2i}; \\
&= 0 \quad \text{otherwise.}
\end{aligned}
$$

Then, $X$ is a random variable with a Bernoulli distribution with probability parameter 0.5. This random variable can be transformed easily into other random variables. For example, a discrete uniform distribution over the set of all numbers between 0 and 1 that have a $d$-bit terminating binary representation can be generated by taking successive realizations of the Bernoulli.

The difficulty in using a generator based on atomic decay of course is measuring the time intervals and inputting those measurements into the computer. John Walker at his Fourmilab has assembled a radiation source (krypton-85), a sensor/timer, and a computer to obtain realizations of Bernoulli random variables. A file containing a random sample can be obtained at

    http://www.fourmilab.ch/hotbits/

Each sample is generated in response to the user's request, so the samples are unique.

Physical processes on the user's computer can also be used to generate random data. There are many ways to do this. For example, Davis, Ihaka, and Fenstermacher (1994) describe a method of using randomness in the air turbulence of disk drives. Toshiba produces a commercial product, Random Master, that uses thermal noise in a semiconductor to generate uniform random sequences. Random Master is packaged as a PCI board compatible with a range of computer architectures from personal computers to supercomputers.

There are many issues to consider in developing "truly random" generators. Our interest in this chapter will be in deterministic generators that can be implemented in ordinary computer programs. The output of such generators is pseudorandom.

## Multiple Recursion

The most useful type of generator of pseudorandom processes updates a current sequence of numbers in a manner that appears to be random. Such a deterministic generator, $f$, yields numbers recursively, in a fixed sequence. The previous $k$ numbers (often just the single previous number) determine(s) the next number:

$$x_i = f(x_{i-1}, \cdots, x_{i-k}). \qquad (1.1)$$

The number of previous numbers used, $k$, is called the "order" of the generator. Because the set of numbers directly representable in the computer is finite, the sequence will repeat.

The set of values at the start of the recursion is called the *seed*. Each time the recursion is begun with the same seed, the same sequence is generated.

The length of the sequence prior to beginning to repeat is called the *period* or cycle length. (Sometimes, it is necessary to be more precise in defining the period to account for the facts that, with some generators, different starting subsequences will yield different periods and that the repetition may begin without returning to the initial state.)

## Predictability

Random number generation has applications in cryptography, where the requirements for "randomness" are generally much more stringent than for ordinary applications in simulation. In cryptography, the objective is somewhat different, leading to a dynamic concept of randomness that is essentially one of predictability: a process is "random" if the known conditional probability of the next event, given the previous history (or any other information, for that matter), is no different from the known unconditional probability. (The condition of being "known" in such a definition is a primitive—undefined—concept.) This kind of definition leads to the concept of a "one-way function" (see Luby, 1996). A one-way function is a function $f$ such that, for any $x$ in its domain,

$f(x)$ can be computed in polynomial time and, given $f(x)$, $x$ cannot be computed in polynomial time. ("Polynomial time" means that the time required can be expressed as or bounded by a polynomial in some measure of the size of the problem.) In random number generation, the function of interest yields a stream of "unpredictable" numbers; that is, the function $f$ in equation (1.1) is easily computable but $x_{i-1}$, given $x_i$, ..., $x_{i-k}$, is not easily computable. The existence of a one-way function has not been proven, but the generator of Blum, Blum, and Shub (1986) (see page 37) is unpredictable under certain assumptions.

Boyar (1989) and Krawczyk (1992) consider the general problem of predicting the output of pseudorandom number generators. They define the problem as a game in which a predictor produces a guess of the next value to be generated and the generator then provides the value. If the period of the generator is $p$, then clearly a naive guessing scheme would become successful after $p$ guesses. For certain kinds of common generators, Boyar (1989) and Krawczyk (1992) give methods for predicting the output in which the number of guesses can be bounded by a polynomial in $\log p$. The reader is referred to those papers for the details.

A simple approach to unpredictability has been described and implemented by André Seznec and Nicolas Sendrier. They suggest use of computer system state information that is available to the user to generate starting values for pseudorandom number generators. They have developed a system-dependent method, HAVEGE (for "HArdware Volatile Entropy Gathering and Expansion"), to access the system state for a variety of types of computers. The number of possible states is very large and on current systems changes at a rate of over 100 megabits per second. More information on HAVEGE and programs implementing the method are available at

    http://www.irisa.fr/caps/projects/hipsor/HAVEGE.html

A survey of some uses of random numbers in cryptography is available in Lagarias (1993), and additional discussion of the applications of pseudorandomness in cryptography is provided by Luby (1996).

## Terminology Used in This Book

Although we understand that the generated stream of numbers is really only pseudorandom, in this book we usually use just the term "random", except when we want to emphasize the fact that the process is not really random, and then we use the term "pseudorandom". Pseudorandom numbers are meant to simulate random sampling. Generating pseudorandom numbers is the subject of this chapter. In Chapter 3, we consider an approach that seeks to ensure that, rather than appearing to be a random sample, the generated numbers are spread out more uniformly over their range. Such a sequence of numbers is called a quasirandom sequence.

We use the terms "random number generation" (or "generator") and "sampling" (or "sampler") interchangeably.

Another note on terminology: Some authors distinguish "random numbers" from "random variates". In their usage, the term "random numbers" applies to pseudorandom numbers that arise from a uniform distribution, and the term "random variates" applies to pseudorandom numbers from some other distribution. Some authors use the term "random variates" only when those numbers resulted from transformations of "random numbers" from a uniform distribution. I do not understand the motivation for these distinctions, so I do not make them. In this book, "random numbers" and "random variates", as well as the additional term "random deviates", are all used interchangeably. I generally use the term "random variable" with its usual meaning, which is different from the meaning of the other terms. *Random numbers* or *random variates* simulate realizations of *random variables*. I will also generally follow the notational convention of using capital Latin letters for random variables and corresponding lowercase letters for their realizations.

## 1.1 Uniform Integers and an Approximate Uniform Density

In most cases, we want the generated pseudorandom numbers to simulate a uniform distribution over the unit interval $(0, 1)$ (that is, the distribution with the probability density function),

$$
\begin{aligned}
p(x) &= 1 \quad \text{if} \ \ 0 < x < 1; \\
&= 0 \quad \text{otherwise.}
\end{aligned}
$$

We denote this distribution by $U(0, 1)$. More generally, we use the notation $U(a, b)$ to denote the absolutely continuous uniform distribution over the interval $(a, b)$. The uniform distribution is a convenient one to work with because there are many simple techniques to transform the uniform samples into samples from other distributions of interest.

### Computer Arithmetic

In generating pseudorandom numbers on the computer, we usually first generate pseudorandom integers over some fixed range and then scale them into the interval $(0, 1)$. The set of integers available is denoted by $\mathbb{II}$ (see page 315 for brief discussions of computer numbers). If the range of the integers is large enough, the resulting granularity is of little consequence in modeling a continuous distribution. ("Granularity" refers to the discrete nature of the set of numbers. The granularity is greater when the distances between successive numbers in the set are greater.) The granularity of pseudorandom numbers from good generators is no greater than the granularity of the numbers with which the computer ordinarily works.

In the standard model for floating-point representation of numbers with base or radix $b$ and $p$ positions in the significand, we approximate any real number by

$$\pm(d_1 b^{e-1} + d_2 b^{e-2} + \cdots + d_p b^{e-p}), \tag{1.2}$$

which we may write as

$$\pm 0.d_1 d_2 \cdots d_p \times b^e,$$

where each $d_j$ is a nonnegative integer less than $b$, and $e$ is an integer between the fixed numbers $e_{\min}$ and $e_{\max}$ inclusive (see Gentle, 1998, pages 7–11 for further discussion of this representation and the number system that it supports). In the most common computer systems, $b = 2$.

We denote the finite subset of the reals, $\mathbb{R}$, that is representable in this form as $\mathbb{F}$. This set, together with two operations that are similar to the addition and multiplication that determine the field $\mathbb{R}$, constitutes a rather complicated object that we also denote by $\mathbb{F}$. This object is similar to a field, but it is not one. (Note that we overload symbols to represent both a set and an object that consists of the set plus some operations.)

In this representation for the elements of a subset of the real numbers, the smallest and largest numbers in the interval $(0, 1)$ that can be represented are $b^{e_{\min}-p}$ and $1 - b^{-p}$, respectively. The computer numbers that approximate the real numbers in the open interval $(0, 1)$ are therefore a finite subset of the closed interval $[b^{e_{\min}-p}, 1 - b^{-p}]$, which is

$$S = \cup_{i=e_{\min}}^{p-1} [b^{i-p}, b^{i+1-p}] \setminus \{1\}. \tag{1.3}$$

The number of representable real numbers is finite, and they are not uniformly distributed over $S$.

For $e_{\min} \le j \le e_{\max} - 1$, the numbers in the interval $[b^j, b^{j+1}]$ are approximated in the computer by numbers from the discrete set

$$\{b^j, \ b^j + b^{j+1-p}, \ \ldots, \ (b-1)b^j + (b-1)b^{j-1} + \cdots + (b-1)b^{j+1-p}, \ b^{j+1}\}.$$

An ideal uniform generator would produce output such that, for a given value of $e \le 0$, the distribution of each digit in the representation (1.2) is independent of the other $p - 1$ digits and has a discrete uniform distribution over the set $\{0, \ldots, b - 1\}$. If $X$ is a random variable whose representation in the form (1.2) has $e = 0$ and has independent discrete uniform distributions for the digits, then

$$\mathrm{E}(X) \approx 1/2 \tag{1.4}$$

and

$$\mathrm{V}(X) \approx 1/12 \tag{1.5}$$

where $\mathrm{E}(X)$ represents the expectation of $X$, and $\mathrm{V}(X)$ represents the variance of $X$. The approximations (1.4) and (1.5) are slightly greater than the true values. If the restriction is added that $0.0 \cdots 0$ is not allowed, then $\mathrm{E}(X) = 1/2$. The expectation and variance of a random variable with a $\mathrm{U}(0, 1)$ distribution are $1/2$ and $1/12$, respectively.

In numerical analysis, although we may not be able to deal with the numbers in the interval $(1 - b^{-p}, 1)$, we do expect to be able to deal with numbers in the interval $(0, b^{-p})$, which is of the same length. (This is because, usually, relative differences are more important than absolute differences.) A random variable such as $X$ above, defined on the computer numbers with $e = 0$, is not adequate for simulating a $U(0, 1)$ random variable. Although the density of computer numbers near 0 is greater than that of the numbers near 1, a good random number generator will yield essentially the same proportion of numbers in the interval $(0, k\epsilon)$ as in the interval $(1 - k\epsilon, 1)$, where $k$ is some small number such as 3 or 4, and $\epsilon = b^{-p}$, which is a machine epsilon. (The phrase "machine epsilon" is used in at least two different ways. In general, the machine epsilon is a measure of the relative spacing of computer numbers. This is just the difference between 1 and the two numbers on either side of 1 in the set of computer numbers. The difference between 1 and the next smallest representable number is the machine epsilon used above. It is also called the smallest relative spacing. The difference between 1 and the next largest representable number is another machine epsilon. It is also called the largest relative spacing.) If random numbers on the computer were to be generated by generating the components of equation (1.2) directly, we would have to use a rather complicated joint distribution on $e$ and the $d$s.

## Modular Arithmetic

The standard methods of generating pseudorandom numbers use modular reduction in congruential relationships. There are currently two basic techniques in common use for generating uniform random numbers: congruential methods and feedback shift register methods. For each basic technique there are many variations. (Both classes of methods use congruential relationships, but for historical reasons only one of the classes of methods is referred to as "congruential methods".)

Both the congruential and feedback shift register methods use modular arithmetic, so we now describe a few of the properties of this arithmetic. For more details on general properties, the reader is referred to Ireland and Rosen (1991), Fang and Wang (1994), or some other text on number theory. Zaremba (1972) and Fang and Wang (1994) discuss several specific applications of number theory in random number generation and other areas of numerical analysis.

The basic relation of modular arithmetic is *equivalence modulo m*, where $m$ is some integer. This is also called *congruence modulo m*. Two numbers are said to be equivalent, or congruent, modulo $m$ if their difference is an integer evenly divisible by $m$. For $a$ and $b$, this relation is written as

$$a \equiv b \bmod m.$$

For example, 5 and 14 are congruent modulo 3 (or just "mod 3"); 5 and $-1$ are also congruent mod 3. Likewise, 1.33 and 0.33 are congruent mod 1. It is clear from the definition that congruence is

- symmetric:
  $a \equiv b \bmod m$   implies   $b \equiv a \bmod m$

- reflexive:
  $a \equiv a \bmod m$   for any $a$

- transitive:
  $a \equiv b \bmod m$   and   $b \equiv c \bmod m$   implies   $a \equiv c \bmod m$;

that is, congruence is an *equivalence relationship*.

A basic operation of modular arithmetic is *reduction modulo m*; that is, for a given number $b$, find $a$ such that $a \equiv b \bmod m$ and $0 \le a < m$. If $a$ satisfies these two conditions, then $a$ is called the *residue* of $b$ modulo $m$. The residues form equivalence classes.

Reduction of $b$ modulo $m$ can also be defined as

$$a = b - \lfloor b/m \rfloor m,$$

where the floor function $\lfloor \cdot \rfloor$ is the greatest integer less than or equal to the argument.

From the definition of congruence, we see that the numbers $a$ and $b$ are congruent modulo $m$ if and only if there exists an integer $k$ such that

$$km = a - b.$$

(In this expression, $a$ and $b$ are not necessarily integers, but $m$ and $k$ are.) This consequence of congruence is very useful in determining equivalence relationships. For example, using this property, it is easy to see that modular reduction distributes over both addition and multiplication:

$$(a + b) \bmod m \equiv a \bmod m + b \bmod m$$

and

$$ab \bmod m \equiv (a \bmod m)(b \bmod m).$$

For a given modulus $m$, each set of integers that are equivalent modulo $m$ forms a *residue class* modulo $m$. For $m = 5$, there are five residue classes:

$$
\begin{aligned}
&\{\cdots, \quad -10, \quad -5, \quad 0, \quad 5, \quad 10, \quad \cdots\} \\
&\{\cdots, \quad -9, \quad -4, \quad 1, \quad 6, \quad 11, \quad \cdots\} \\
&\{\cdots, \quad -8, \quad -3, \quad 2, \quad 7, \quad 12, \quad \cdots\} \\
&\{\cdots, \quad -7, \quad -2, \quad 3, \quad 8, \quad 13, \quad \cdots\} \\
&\{\cdots, \quad -6, \quad -1, \quad 4, \quad 9, \quad 14, \quad \cdots\}
\end{aligned}
$$

For any integer $m \ne 0$, there are $|m|$ residue classes, and the union of all $|m|$ residue classes is the set of all integers. (In the following, we will deal only with moduli that are positive.) In applications, the residue classes whose members are relatively prime to $m$ are important. The number of such residue classes is of course just the number of positive integers less than $m$ that are relatively prime to $m$. The function that assigns to $m$ the number of residue classes (mod $m$) that are relatively prime to $m$ is called Euler's totient function and is denoted by $\phi(m)$. Of the residue classes mod 5, four of them are relatively prime to 5, and for any prime $m$, we have $\phi(m) = m - 1$. The totient function plays an important role in determining the period of some random number generators.

A useful fact that is easy to see is that if $p$ is a prime and $e$ is a positive integer, then

$$\phi(p^e) = p^{e-1}(p-1). \tag{1.6}$$

Another useful fact that is a little more difficult to show (see Ireland and Rosen, 1991) is that if $n$ and $m$ are relatively prime, then

$$\phi(nm) = \phi(n)\phi(m). \tag{1.7}$$

With these two facts, we can evaluate $\phi$ at any positive integer.

## Finite Fields

A system of modular arithmetic is usually defined on nonnegative integers. Modular reduction together with the two operations of the ring results in a finite field (or *Galois field*) on a set of integers. The cardinality of the field is less than or equal to $m$ and is equal to $m$ if and only if $m$ is a prime. We will denote a Galois field over a set with $m$ elements as $\mathbb{G}(m)$. If $m = 5$, for example, a finite field is defined on the set $\{0, 1, 2, 3, 4\}$ with the addition and multiplication of the field being defined in the usual way followed by a reduction modulo 5. If $m = 6$, however, a finite field can be defined on the set $\{0, 2, 4\}$, the set $\{0, 3\}$, or the set $\{0, 1, 5\}$, again with addition and multiplication being defined in the usual way followed by a reduction modulo 6.

Simple random number generators based on congruential methods commonly use a finite field of integers consisting of the nonnegative integers that are directly representable in the computer (that is, of about $2^{31}$ integers).

## Modular Reduction in the Computer

Modular reduction is a binary operation, or a function with two arguments. In the C programming language, the operation is represented as "b%m". (There is no obvious relation of the symbolic value of "%" to the modular operation. No committee passed judgment on this choice before it became a standard part of the language. Sometimes, design by committee helps.) In Fortran, the operation is specified by the function "mod(b,m)", in Matlab by the function "rem(b,m)", and in Maple by "b mod m". There is no modulo function in S-Plus, but the operation can be implemented using the "floor" function, as was shown above.

Modular reduction can be performed by using the lower-order digits of the representation of a number in a given base. For example, taking the two lower-order digits of the ordinary base-ten representation of a nonnegative integer yields the decimal representation of the number reduced modulo 100. When numbers represented in a fixed-point scheme in the computer are multiplied, except for consideration of a sign bit, the product when stored in the same fixed-point scheme is the residue of the product modulo the largest representable number. In a twos-complement representation, if the sign bit is changed, the meaning of the remaining bits is changed. For positive integers $x$ and $y$ represented in the fixed-point variables ix and iy in 32-bit twos-complement, the product

$$\texttt{iz = ix*iy}$$

contains either $xy \bmod 2^{31}$ or $xy \bmod 2^{31} - 2^{31}$, which is negative.

Because the pseudorandom numbers that we wish to generate are between 0 and 1, in some algorithms reduction modulo 1 is used. The resultants are the fractional parts of real numbers.

### Modular Arithmetic with Uniform Random Variables

Modular arithmetic has some useful applications with true random variables also. An interesting fact, for example, is that if $Y$ is a random variable distributed as $U(0,1)$ and

$$X \equiv (kY + c) \bmod 1, \tag{1.8}$$

where $k$ is an integer constant not equal to 0, and $c$ is a real constant, then $X$ has a $U(0,1)$ distribution. (You are asked to show this in Exercise 1.4a, page 57.) Modular arithmetic can also be used to generate two independent random numbers from a single random number. If

$$\pm 0.d_1 d_2 d_3 \cdots$$

is the representation, in a given base, of a uniform random number $Y$, then any subsequence of the digits $d_1, d_2, \ldots$ can be used to form other uniform numbers. (If the subsequence is finite, as of course it is in computer applications, the numbers are discrete uniform, but if the subsequence is long enough, the result is considered continuous uniform.) Furthermore, any two disjoint subsequences can be used to form *independent* random numbers.

The sequence of digits $d_1, d_2, \ldots$ can be rearranged to form more than one uniform variate; for example,

$$\pm 0.d_1 d_3 d_5 \cdots$$

and

$$\pm 0.d_2 d_4 d_6 \cdots.$$

The use of subsequences of bits in a fixed-point binary representation of pseudorandom numbers to form other pseudorandom numbers is called *bit stripping*.

# 1.2 Simple Linear Congruential Generators

D. H. Lehmer in 1948 (see Lehmer, 1951) proposed a simple *linear congruential generator* as a source of random numbers. In this generator, each single number determines its successor by means of a simple linear function followed by a modular reduction. Although this generator is limited in its ability to produce very long streams of numbers that appear to be independent realizations of a uniform process, it is a basic element in other, more adequate generators. Understanding its properties is necessary in order to use it to build better generators.

The form of the linear congruential generator is

$$x_i \equiv (ax_{i-1} + c) \bmod m, \quad \text{with} \quad 0 \le x_i < m; \tag{1.9}$$

$a$ is called the "multiplier", $c$ is called the "increment", and $m$ is called the "modulus" of the generator. Often, $c$ in equation (1.9) is taken to be 0, and, in this case, the generator is called a "multiplicative congruential generator":

$$x_i \equiv ax_{i-1} \bmod m, \quad \text{with} \quad 0 < x_i < m. \tag{1.10}$$

For $c \ne 0$, the generator is sometimes called a "mixed congruential generator". The seed for this generator is just the single starting value in the recursion, $x_0$. A sequence resulting from the recursion (1.9) is called a *Lehmer sequence*. Each $x_i$ is scaled into the unit interval $(0,1)$ by division by $m$, that is,

$$u_i = x_i/m.$$

If $a$ and $m$ are properly chosen, the $u_i$s will "look like" they are randomly and uniformly distributed between 0 and 1.

The recurrence in (1.10) for the integers is equivalent to the recurrence

$$u_i \equiv au_{i-1} \bmod 1, \quad \text{with} \quad 0 < u_i < 1.$$

This recurrence has some interesting relationships to the first-order linear autoregressive model

$$U_i = \rho u_{i-1} + E_i,$$

where $\rho$ is the autoregressive coefficient and $E_i$ is a random variable with a $U(0,1)$ distribution (see Lawrance, 1992).

## Period

Because $x_i$ is determined by $x_{i-1}$ and since there are only $m$ possible different values of the $x$s, the maximum period or cycle length of the linear congruential generator is $m$. Also, since $x_{i-1} = 0$ cannot be allowed in a multiplicative generator, the maximum period of the multiplicative congruential generator is $m - 1$.

When computing was expensive, values of $m$ used in computer programs were often powers of 2. Such values could result in faster computer arithmetic. The maximum period of multiplicative generators with such moduli is $m/4$, and, interestingly, this period is achieved for any multiplier that is $\pm 3 \bmod 8$ (see Knuth, 1998).

The period of a multiplicative congruential generator with multiplier $a$ and modulus $m$ depends on the smallest positive value of $k$ for which

$$a^k \equiv 1 \bmod m. \tag{1.11}$$

This is because when that relationship is satisfied, the sequence begins to repeat. The period, therefore, can be no greater than $k$. The Euler–Fermat Theorem (see Ireland and Rosen, 1991) states that if $a$ and $m$ are relatively prime, then $a^{\phi(m)} \equiv 1 \bmod m$, where $\phi(m)$ is the Euler totient function. The period, therefore, can be no greater than $\phi(m)$. For a given value of $m$, we seek $a$ such that $k$ in equation (1.11) is $\phi(m)$. Such a number $a$ is called a *primitive root modulo m*. (See Ireland and Rosen, 1991, or other texts on number theory for general discussions of primitive roots; see Fuller, 1976, for methods to determine whether a number is a primitive root; and see Exercise 1.14, page 59, for some computations.) If $m$ is a prime, the number of primitive roots modulo $m$ is $\phi(m-1)$.

For example, consider $m = 31$ and $a = 7$, that is,

$$x_i \equiv 7x_{i-1} \bmod 31,$$

and begin with $x_0 = 19$. The next integers in the sequence are

$$9, 1, 7, 18, 2, 14, 5, 4, 28, 10, 8, 25, 20, 16, 19,$$

so, of course, at this point the sequence begins to repeat. The period is 15. We have

$$7^{15} \equiv 1 \bmod 31,$$

that is, 7 is not a primitive root modulo 31.

Now consider $m = 31$ and $a = 3$, and again begin with $x_0 = 19$. We go through 30 numbers before we get back to 19. This is because 3 is a primitive root modulo 31. There are $\phi(30) = 8$ primitive roots modulo 31.

It turns out that $m$ has a primitive root if and only if $m$ is of the form $2^{e_0}p^{e_1}$, where $p$ is an odd prime, $e_0 = 0$ or 1, and $e_1 \geq 1$.

For any $m$, it is of interest to determine an $a$ such that $k$ is as large as possible for that $m$. Such a number $a$ is called a *primitive element modulo m*. For $m$ of the general form $2^{e_0}p_1^{e_1} \cdots p_t^{e_t}$, where each $p_i$ is an odd prime and each $e_i \geq 0$, Knuth (1998) gives minimum values of $k$ in equation (1.11) and various conditions for $a$ to be a primitive element modulo $m$. (We quoted one of those results above: for $m = 2^{e_0}$, with $e_0 \geq 4$, $k = 2^{e_0-2}$, and $a$ must be of the form $\pm 3 \bmod 8$.)

For a random number generator to be useful in most practical simple applications, the period must be of the order of at least $10^9$ or so, which means that the modulus in a linear congruential generator must be at least that large. The values of the moduli in common use range in order from about $10^9$ to $10^{15}$. Even so, the period of such generators is relatively short because of the speed with which computers can cycle through the full period and in view of the very large sizes of some simulation experiments.

## Composite Modulus

The bits in the binary representations of the sequences from generators whose modulus is a power of 2 have very regular patterns. The period of the lowest-order bit is at most 1 (that is, it is always the same), the period of the next lowest-order bit is at most 2, the period of the next lowest-order bit is at most 4, and so on. In general, low-order bits in the streams resulting from a composite modulus will have periods corresponding to the factors. The small periods can render a bit-stripping technique completely invalid with random number generators with a modulus that is a power of 2.

Similar regular patterns occur anytime the modulus has a factor that is a small prime. If the modulus is even (so the multiplier must be odd), we get the first two patterns that we mentioned above: the period of the lowest-order bit is at most 1, and the period of the next lowest-order bit is at most 2. If the modulus is divisible by 4, the output exhibits these two patterns, plus the pattern mentioned above for the third-lowest bit.

It is easy to identify other patterns for other composite moduli. If the modulus is divisible by 3, for example, the lower-order pair of bits will have a period of at most 3.

Currently, the numbers used as moduli in production random number generators are usually primes, often Mersenne primes, which have the form $2^p - 1$. (For any prime $p \leq 31$, numbers of that form are prime except for the three values $p = 11$, 23, and 29. Most larger values of $p$ do not yield primes. A large one that does yield a prime is $p = 13\,466\,917$.)

## Moduli and Multipliers

A commonly used modulus is the Mersenne prime $2^{31} - 1$, and for that modulus, a common multiplier is $7^5$ (see the discussion of the "minimal standard" on page 20). The Mersenne prime $2^{61} - 1$ is also used occasionally. The primitive roots for these two moduli have been extensively studied.

Wu (1997) suggests multipliers of the form $\pm 2^{q_1} \pm 2^{q_2}$ because they result in particularly simple computations yet seem generally to have good properties. Wu suggests $2^{15} - 2^{10}$ and $2^{16} - 2^{21}$ for a modulus of $2^{31} - 1$, and $2^{30} - 2^{19}$ and $2^{42} - 2^{31}$ for a modulus of $2^{61} - 1$. The computational efficiency of such multipliers results from the fact that multiplication by a multiplier of the form $2^q$ and followed by a modular reduction with a modulus of the form $2^p - 1$ results in an exchange of the block of the $q$ most significant bits and the block of the $p - q$ least significant bits. Multipliers of the form suggested by Wu effectively do this kind of exchange twice and then add the results. L'Ecuyer and Simard (1999) point out, however, that an operation consisting of two exchanges of two blocks of bits followed by addition of the results tends to yield a value whose binary representation has a number of 1s similar to the number of 1s in the original value. The number of 1s in the binary representation of a value is called its *Hamming weight*. L'Ecuyer and Simard (1999) define a test of independence of Hamming weights of successive values in the output streams of random number generators and, in applying the test to generators with multipliers of the form $\pm 2^{q_1} \pm 2^{q_2}$, find that such generators perform poorly with respect to this criterion.

## 1.2.1   Structure in the Generated Numbers

In addition to concern about the length of the period, there are several other considerations. It is clear that if the period is $m$, then the output of the generator over a full cycle will be evenly distributed over the unit interval. If we ignore the sequential order of a full-period sequence from a congruential generator, it will appear to be U$(0, 1)$; in fact, the sample would appear *too much* like a sample from U$(0, 1)$.

A useful generator, however, must generate *subsamples* of the full cycle that appear to be uniformly distributed over the unit interval. Furthermore, the numbers should appear to be distributionally independent of each other; that is, the serial correlations should be small.

Unfortunately, the structure of a sequence resulting from a linear congruential generator is very regular. Marsaglia (1968) pointed out that the output of any congruential generator lies on a simple lattice in a $k$-space with axes representing successive numbers in the output. This is fairly obvious upon inspection of the algebra of the generator. How bad this is (that is, how much this situation causes the output to appear nonrandom) depends on the structure of the lattice. A lattice is defined in terms of integer combinations of a set of "basis vectors". Given a set of linearly independent vectors $\{v_1, v_2, \ldots, v_d\}$ in $\mathbb{R}^d$, a lattice is the set of vectors $w$ of the form $\sum_{i=1}^{d} z_i v_i$, where $z_i$ are integers. The set of vectors $\{v_i\}$ is a basis for the lattice. Figure 1.1 shows a lattice in two dimensions with basis $\{v_1, v_2\}$.

For an example of the structure in a stream of pseudorandom numbers produced by a linear congruential generator, consider the output of the generator (1.10) with $m = 31$ and $a = 3$ that begins with $x_0 = 9$. The next integers in the sequence are shown in Figure 1.2. That sequence then repeats. The period is 30; we know that 3 is a primitive root modulo 31.

A visual assessment or even computation of a few descriptive statistics does not raise serious concerns about whether this represents a sample from a discrete uniform distribution over the integers from 1 to 30 except for the fact that there are no repeats in the sample. The scaled numbers (the integers divided by 30) have a sample mean of 0.517 and a sample variance of 0.86. Both of these values are consistent with the expected values from a U$(0, 1)$ distribution. The

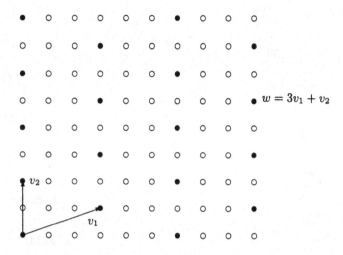

Figure 1.1: A Lattice in 2-D

27, 19, 26, 16, 17, 20, 29, 25, 13, 8, 24, 10, 30, 28, 22, 4, 12, 5, 15, 14, 11, 2, 6, 18, 23, 7, 21, 1, 3, 9,

Figure 1.2: An Output Stream from $x_i \equiv 3x_{i-1} \bmod 31$

autocorrelations for lags 1 through 5 are

$$0.27, 0.16, -0.10, 0.06, 0.07.$$

Although the lag 1 correlation is somewhat large for a sample of this size, these values are not inconsistent with a hypothesis of independence.

However, when we plot the successive (overlapping) pairs

$$(27, 19), (19, 26), (26, 16), \ldots$$

as in Figure 1.3, a disturbing picture emerges. All points in the lattice of pairs lie along just three lines, each with a slope of 3. (There are also ten lines with slope $-\frac{1}{10}$ and ten lines with slope $\frac{2}{11}$ if we count as lines the continuation by "wrapping" modulo 31.) In most applications, we probably would not use overlapping pairs. Successive nonoverlapping pairs are shown as solid (or open) circles in Figure 1.3. That pattern appears even less random, as the nonoverlapping pairs cluster on one line or another.

This pattern is in fact related to the relatively large correlation at lag 1. Although the correlation may not appear so large for the small sample size, that large value of the correlation would persist even if we were to increase the sample size by generating more random numbers because the random numbers would just repeat themselves. It is easy to see that this kind of pattern results from the small value of the multiplier. The same kind of problem would also

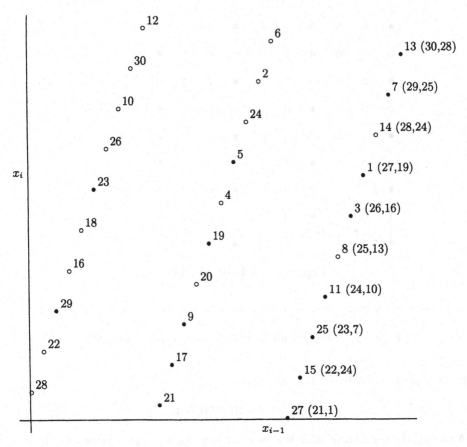

Figure 1.3: Pairs of Successive Numbers from $x_i \equiv 3x_{i-1} \bmod 31$

result from a multiplier that is too close to the modulus, such as $a = 27$, for example.

There are eight primitive roots modulo 31, so we might try another one, say 12. Let $a = 12$ and again begin with $x_0 = 9$. The next integers in the sequence are shown in Figure 1.4.

15, 25, 21, 4, 17, 18, 30, 19, 11, 8, 3, 5, 29, 7, 22, 16, 6, 10, 27, 14, 13, 1, 12, 20, 23, 28, 26, 2, 24, 9.

Figure 1.4: An Output Stream from $x_i \equiv 12x_{i-1} \bmod 31$

A visual assessment does not show much difference between this sequence and the sequence of numbers generated by $a = 3$ shown in Figure 1.2. The numbers themselves are exactly the same as those before, so the static properties of mean and variance are the same. The autocorrelations are different, however.

For lags 1 through 5, they are

$$-0.01, -0.07, -0.17, -0.15, 0.03, 0.35.$$

The smaller value for lag 1 indicates that the structure of successive pairs may be better, and, in fact, the points do appear better distributed, as we see in Figure 1.5. There are six lines with slope $-\frac{2}{5}$ and seven lines with slope $\frac{5}{3}$.

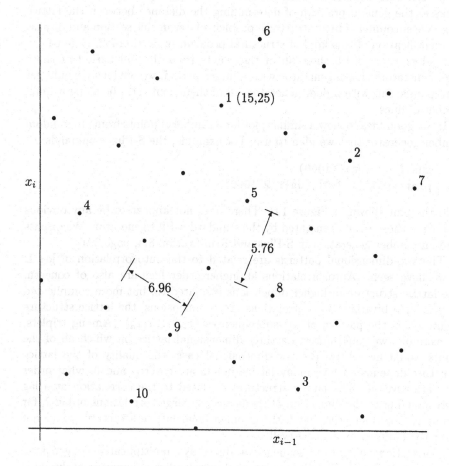

Figure 1.5: Pairs of Successive Numbers from $x_i \equiv 12x_{i-1} \bmod 31$

From a visual inspection, we can conclude that a generator with a small number of lines in any direction does not cover the space well. The generator with output shown in Figure 1.3 has ten lines with slope $-\frac{1}{10}$ but only three lines with slope 3. Marsaglia (1968) showed that when $d$ overlapping sequences of the output of a congruential generator with modulus $m$ are plotted as we have done for $d = 2$, the points will lie on parallel hyperplanes (parallel lines in Figures 1.3 and 1.5), and there will be a direction of some hyperplane in which there will be no more than $(d!m)^{1/k}$ parallel hyperplanes. In the simple example above, for one modulus there were only three parallel lines, but for the other modulus there were six, which is close to the bound of seven.

Another quantitative measure of the severity of the lattice structure is the distance between the lines—specifically, the shortest distance between two sides of the maximal volume parallelogram formed by four points and not enclosing any points. The distance between the lines with slope $\frac{5}{3}$ is 6.96, as shown in Figure 1.5. The distance between the lines with slope $-\frac{2}{5}$ is 5.76. Dieter (1975) discusses the general problem of determining the distance between the lattice lines. We encounter similar structural problems later in this section and discuss the identification of this kind of structural problem in Section 2.2, page 64.

Figures 1.3 and 1.5 show all of the points from the full period of those small generators. For a generator with a larger period, we obviously would get more points; but with a poor generator, all of them could still lie along a small number of lines.

It is a good idea to view a similar plot for a sample of points from any random number generator that we plan to use. For example, the S-Plus commands

```
xunif <- runif(1000)
plot(xunif[1:999],xunif[2:1000])
```

yield the plot shown in Figure 1.6. There does not appear to be any obvious pattern in those points generated by the standard S-Plus generator. We discuss random number generation in S-Plus and R in Section 8.5, page 291.

The two-dimensional patterns are related to the autocorrelation of lag 1, as we have seen. Autocorrelations at higher-order lags are also of concern. The lattice structure in higher dimensions is related to, but more complicated than, simple bivariate autocorrelations. In $d$ dimensions, the lattice structure of interest is the pattern of the subsequences $(x_i, \ldots, x_{i+d})$. Among triplets, for example, we could observe a three-dimensional lattice on which all of the points would lie. As in the two-dimensional case, the quality of the lattice structure depends on how many lattice points are covered and in what order they are covered. The lattice structure is related to the correlation at a lag corresponding to the dimension of the lattice, so large correlations of lag 2, for example, would suggest that a three-dimensional lattice structure would not cover three-dimensional space well.

Correlations of lag 1 in a sequence produced by a multiplicative congruential generator will be small if the square of the multiplier is approximately equal to the modulus. In this case, however, the correlations of lag 2 are likely to be large, and consequently the three-dimensional lattice structure will be poor. (In our examples in Figures 1.3 and 1.5, we had $a^2 \equiv 9 \bmod 31$ and $a^2 \equiv 28 \bmod 31$, respectively, so the generator used in Figure 1.5 would have poor lattice structure in three dimensions.) An example of a generator with very good two-dimensional lattice structure yet very poor three-dimensional lattice structure is implemented in the program RANDU, which for many years was the most widely used random number generator in the world.

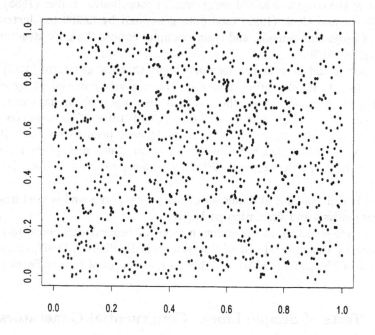

Figure 1.6: Pairs of Successive Numbers from the S-Plus Function `runif`

The generator in `RANDU` is essentially

$$x_i \equiv 65539 x_{i-1} \bmod 2^{31}. \tag{1.12}$$

(`RANDU` was coded to take advantage of integer overflow and made an adjustment when overflow occurred, so the generator is not exactly the same as equation (1.12).)

The generator in equation (1.12) can be written to express the relationship among three successive members of the output sequence:

$$
\begin{aligned}
x_i &\equiv (65\,539)^2 x_{i-2} \bmod 2^{31} \\
&\equiv (2^{16} + 3)^2 x_{i-2} \bmod 2^{31} \\
&\equiv (6 x_{i-1} - 9 x_{i-2}) \bmod 2^{31};
\end{aligned}
$$

that is,

$$x_i - 6 x_{i-1} + 9 x_{i-2} = c2^{31}, \tag{1.13}$$

where $c$ is an integer. Since $0 < x_i < 2^{31}$, all such triplets must lie on no more than 15 planes in $\mathbb{R}^3$. If a thin slice of a cube is projected onto one face of the cube, all of the points in the slice would appear in only a small number of linear regions. (See Exercise 1.7, page 58.) Plots of triples or other tuples in parallel coordinates can also reveal linear dependencies. See Gentle (2002, page 181) for a plot of the output of RANDU using parallel coordinates. Huber (1985) and later Cabrera and Cook (1992) used data generated by RANDU to illustrate a method of projection pursuit, and their examples expose the poor structure of the output of RANDU.

Although it had been known for many years that the generator (1.12) had problems (see Coldwell, 1974), and even the analysis represented by equation (1.13) had been performed, the exact nature of the problem has sometimes been misunderstood. For example, as James (1990) states: "We now know that any multiplier congruent to 5 mod 8 ... would have been better ...." Being congruent to 5 mod 8 does not solve the problem. Such multipliers have the same problem if they are close to $2^{16}$ for the same reason (see Exercise 1.8, page 58).

RANDU is still available at a number of computer centers and is used in some statistical analysis and simulation packages.

The lattice structure of the common types of congruential generators can be assessed by the spectral test of Coveyou and MacPherson (1967) or by the lattice test of Marsaglia (1972a). We discuss these types of tests in Section 2.2, page 64.

## 1.2.2   Tests of Simple Linear Congruential Generators

Any number of statistical tests can be developed to be applied to the output of a given generator. The simple underlying idea is to form any transformation on the subsequence, determine the distribution of the transformation under the null hypothesis of independent uniformity of the sequence, and perform a goodness-of-fit test of that distribution. A simple transformation is just to add successive terms. (Adding two successive terms should yield a triangular distribution.) We discuss goodness-of-fit tests in Chapter 2, but the reader familiar with such tests should be able, with a little imagination, to devise chi-squared tests for uniformity in any dimension, chi-squared tests for triangularity of sums of two successive numbers, and so on. Tests for serial correlation of various lags and various sign tests are other possibilities that should come to mind to anyone with some training in statistics.

For some specific generators or families of generators, there are extensive empirical studies reported in the literature. For $m = 2^{31} - 1$, for example, empirical studies by Fishman and Moore (1982, 1986) indicate that different values of multipliers, all of which perform well under the lattice test and the spectral test (see Section 2.2, page 64), may yield samples statistically distinguishable from samples from a true uniform distribution.

Park and Miller (1988) summarize some problems with random number generators commonly available and propose a "minimal standard" for a linear congruential generator. The generator must perform "at least as well as" one with $m = 2^{31} - 1$ and $a = 16807$, which is a primitive root.

This choice of $m$ and $a$ was made by Lewis, Goodman, and Miller (1969) and is very widely used. (The smallest primitive root of $2^{31}-1$ is 7, and $7^5 = 16807$ is the largest power of 7 such that $7^p x$, for the largest value of $x$ (which is $2^{31}-2$), can be represented in the common 64-bit floating-point format.) Results of extensive tests by Learmonth and Lewis (1973) are available for it. It is provided as one option in the IMSL Libraries. Fishman and Moore (1986) found the value of 16807 to be marginally acceptable as the multiplier, but there were several other multipliers that performed better in their battery of tests.

The article by Park and Miller generated extensive discussion; see the "Technical Correspondence" in the July 1993 issue of *Communications of the ACM*, pages 105 through 110. It is relatively easy to program the minimal standard, as we see in the next section, but the algorithm given by Carta (1990) ostensibly to implement the minimal standard should be avoided.

Ferrenberg, Landau, and Wong (1992) used some of the generators that meet the Park and Miller minimal standard to perform several simulation studies in which the correct answer was known. Their simulation results suggested that even some of the "good" generators could not be relied on in some simulations. Vattulainen, Ala-Nissila, and Kankaala (1994) likewise used some of these generators as well as generators of other types and found that their simulations often did not correspond to the processes they were modeling. The point is that the "minimal standard" is *minimal*.

Sets of random numbers sequentially produced by linear congruential generators exhibit a certain type of departure from what would be expected in a random sample if the number of random numbers in the set exceeds approximately the square root of the period of the generator. (The apparent nonrandomness is in the distribution of the interpoint distances in lattices of various dimensions formed by the random numbers, as in Figures 1.3 and 1.5 for two dimensions. See L'Ecuyer and Hellekalek, 1998, and L'Ecuyer, Cordeau, and Simard, 2000, for discussions of results of empirical tests on linear congruential generators.) The useful, "safe" period of the "minimal standard", therefore, is less than 50,000. In Section 2.3 beginning on page 71, we further discuss general empirical tests of random number generators.

Deng and Lin (2000) contend that because of its relatively short period (even at the full period of $2^{31}-1$, rather than the "safe" period) and its lattice structure, the "minimal standard" is not acceptable for serious work. They suggest use of matrix congruential generators (see generators (1.31) and (1.32) in Section 1.4.2).

## 1.2.3 Shuffling the Output Stream

MacLaren and Marsaglia (1965) suggest that the output stream of a linear congruential random number generator be *shuffled* by using another, perhaps simpler, generator to permute subsequences from the original generator. This shuffling can increase the period (because it is no longer necessary for the same value to follow a given value every time it occurs) and can also break up the lattice structure. (There will still be a lattice, of course; it will just have a different number of planes.)

Because a single random number can be used to generate independent random numbers ("bit stripping", see page 10), a single generator can be used to shuffle itself.

Bays and Durham (1976) describe a method of using a single generator to fill a table of length $k$ and then using a single stream to select a number from the table and to replenish the table. After initializing a table $T$ to contain $x_1, x_2, \ldots, x_k$, set $i = k + 1$ and generate $x_i$ to use as an index to the table. Then, update the table with $x_{i+1}$. The method is shown in Algorithm 1.1 to generate the stream $y_i$ for $i = 1, 2, \ldots$.

**Algorithm 1.1 Bays–Durham Shuffling of Uniform Deviates**

0. Initialize the table $T$ with $x_1, x_2, \ldots, x_k$, $i = 1$, generate $x_{k+i}$, and set $y_i = x_{k+i}$.

1. Generate $j$ from $y_i$ (use bit stripping or mod $k$).

2. Set $i = i + 1$.

3. Set $y_i = T(j)$.

4. Generate $x_{k+i}$, and refresh $T(j)$ with $x_{k+i}$.                                   ∎

The period of the generator may be increased by this shuffling. Bays and Durham (1976) show that the period under this shuffling is $O(k!c)^{\frac{1}{2}}$, where $c$ is the cycle length of the original, unshuffled generator. If $k$ is chosen so that $k! > c$, then the period is increased.

For example, with the generator used in Figure 1.3 ($m = 31$, $a = 3$, and beginning with $x_0 = 9$), which yielded the sequence

27, 19, 26, 16, 17, 20, 29, 25, 13, 8, 24, 10, 30, 28, 22, 4, 12, 5, 15, 14, 11, 2, 6, 18, 23, 7, 21, 1, 3, 9,

we select $k = 8$ and initialize the table as

$$27, 19, 26, 16, 17, 20, 29, 25.$$

We then use the next number, 13, as the first value in the output stream and also to form a random index into the table. If we form the index as $13 \bmod 8 + 1$, we get the sixth tabular value, 20, as the second number in the output stream. We generate the next number in the original stream, 8, and put it in the table, so we now have the table

$$27, 19, 26, 16, 17, 8, 29, 25.$$

Now, we use 20 as the index to the table and get the fifth tabular value, 17, as the third number in the output stream. By continuing in this manner to yield 10,000 deviates and plotting the successive pairs, we get Figure 1.7. The very bad lattice structure shown in Figure 1.3 has diminished. (Remember that there are only 30 different values, however.)

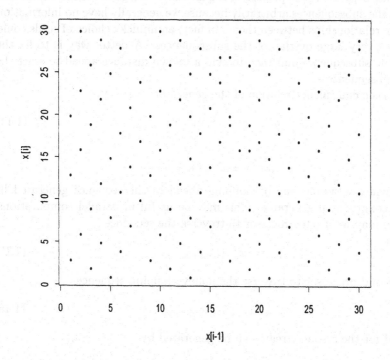

Figure 1.7: Pairs of Successive Numbers from a Shuffled Version of $x_i \equiv 3x_{i-1} \bmod 31$ (Compare with Figure 1.3)

## 1.2.4   Generation of Substreams in Simple Linear Congruential Generators

Sometimes, it is useful to generate separate, independent subsequences with the same generator. The reason may be that the Monte Carlo experiment is being run in blocks or it may be because the computations are being performed on a parallel processing computer system.

    In this subsection, we discuss the basic methods for generating separate substreams using a simple linear congruential generator and some of the properties of such substreams. As we indicated at the beginning of this section, although the simple linear congruential generator forms the basis for many good random number generators, by itself it is generally not adequate for serious applications. In later sections, we discuss various generators based on the simple linear congruential generator, and in Section 1.10, beginning on page 51, we return to the topic of the present subsection and use these basic methods for other generators that have longer periods and better properties.

**Nonoverlapping Blocks**

To generate separate subsequences, it is generally not a good idea to choose two seeds for the subsequences arbitrarily because we generally have no information about the relationships between them. In fact, an unlucky choice of seeds could result in a very large overlap of the subsequences. A better way is to fix the seed of one subsequence and then to skip a known distance ahead to start the second subsequence.

The basic equivalence relation of the generator

$$x_{i+1} \equiv ax_i \bmod m \qquad (1.14)$$

implies

$$x_{i+k} \equiv a^k x_i \bmod m.$$

This provides a simple way of skipping ahead in the sequence generated by a linear congruential generator. This may be useful in parallel computations, where we may want one processor to traverse the sequence

$$x_s, \; x_{s+1}, \; x_{s+2}, \; \cdots \qquad (1.15)$$

and a second processor to traverse the nonoverlapping sequence

$$x_{s+k}, \; x_{s+k+1}, \; x_{s+k+2}, \; \cdots . \qquad (1.16)$$

The seed for the second stream can be generated by

$$x_{s+k-1} \equiv bx_0 \bmod m,$$

where

$$b \equiv a^k \bmod m. \qquad (1.17)$$

Note that any element in the second sequence (1.16) can be obtained by multiplication by $b$ of the corresponding element in the first sequence (1.15) followed by reduction modulo $m$.

**Leapfrogging**

Another interesting subsequence that is "independent" of the first sequence is

$$x_s, \; x_{s+k}, \; x_{s+2k}, \; \cdots . \qquad (1.18)$$

This sequence is generated by

$$x_{i+1} \equiv bx_i \bmod m, \qquad (1.19)$$

where $b$ is as before. This method of generating independent streams is called *leapfrogging* by $k$. Different "independent" subsequences can be formed using various leapfrog distances. (The distances must be chosen carefully, of course. A minimum requirement is that the distances be relatively prime.) We could let one processor leapfrog beginning at $x_s$ and let a second processor leapfrog beginning at $x_{s+1}$ and using the same leapfrog distance.

If $a$ in equation (1.14) is a primitive root modulo $m$, then any $b$ in equation (1.19) can be obtained by some integer $k$ in equation (1.17). For example, consider the simple generators discussed in Section 1.2.1,

$$x_i \equiv 3x_{i-1} \bmod 31$$

and

$$x_i \equiv 12x_{i-1} \bmod 31.$$

Because

$$12 \equiv 3^{19}x_{i-1} \bmod 31, \qquad (1.20)$$

the sequence generated by the second generator (shown on page 16) can be obtained from the sequence generated by the first generator by repeating the sequence in Figure 1.2 on page 15 and then taking every nineteenth number.

If the skip distance is relatively prime to the period, then the sequence formed by leapfrogging will have the same period of the original sequence. If this is not the case, a sequence formed by leapfrogging is not of full period. Suppose, for example, that in one of the sequences above, each with period 30, we choose to take every fifth element. From the sequence in Figure 1.2, we could form the subsequence

$$27, 20, 24, 4, 11, 7, 27, \ldots.$$

Any other starting point would likewise yield a subsequence with a period of 6. From equation (1.17), the generator for skipping five elements ahead is

$$x_i \equiv 3^5 x_{i-1} \bmod 31,$$

and 26 is not a primitive root modulo 31. Any of the 30 possible seeds will generate a subsequence with a period of 6, and there will be five subsequences that do not overlap.

Given a basic stream from the generator (1.14), Figure 1.8 shows two nonoverlapping blocks of substreams and two leapfrogged substreams.

| | | | | |
|---|---|---|---|---|
| Basic stream | $x_1, \ldots, x_t, \ldots, x_k, \ldots, x_{(s-1)k+1}, \ldots, x_{(s-1)k+t}, \ldots, x_{sk}, \ldots,$ | | | |
| Block 1 | $x_1, \ldots, x_t, \ldots, x_k$ | | | |
| Block $s$ | | $x_{(s-1)k+1}, \ldots, x_{(s-1)k+t}, \ldots, x_{sk}$ | | |
| Leapfrogged stream 1 | $x_1,$ | $x_{(s-1)k+1},$ | | $\ldots$ |
| Leapfrogged stream $t$ | $x_t,$ | | $x_{(s-1)k+t},$ | $\ldots$ |

Figure 1.8: Nonoverlapping Blocks of Substreams and Leapfrogged Substreams

## Lehmer Trees

Frederickson et al. (1984) describe a way of combining linear congruential generators to form what they call a Lehmer tree, which is a binary tree in which all right branches or all left branches form a sequence from a Lehmer linear congruential generator. The tree is defined by the two recursions, both of which are the basic recursion (1.9)

$$x_i^{(L)} \equiv (a_L x_{i-1} + c_L) \bmod m$$

and

$$x_i^{(R)} \equiv (a_R x_{i-1} + c_R) \bmod m.$$

At the $(i-1)^{\text{th}}$ node in the tree, an ordinary Lehmer sequence with seed $x_{i-1}$ is generated by all right-branch nodes below it. A new sequence is initiated by taking a left branch to the first node below it and then all right branches from then on. The question is whether a (finite) sequence of all right-branch nodes below a given node is independent from the (finite) sequence of all right-branch nodes below the left-branch node immediately below the given node. ("Independent" for these finite subsequences can be interpreted strictly as having no elements in common.) Frederickson et al. gave conditions on $a_L$, $c_L$, $a_R$, $c_R$, and $m$ that would guarantee the independence for a fixed length of the sequences.

## Correlations Between Substreams

Care is necessary in choosing the seed and the distance to skip ahead. Anderson and Titterington (1993) show that, for a multiplicative congruential generator, the correlation between the sequences $(x_i, \ldots)$ and $(x_{i+k}, \ldots)$ is approximately asymptotically $(n_1 n_2)^{-1}$, where $x_{i+k} = (n_1/n_2)x_i$, and $n_1$ and $n_2$ are relatively prime. Figure 1.9 shows a plot of two subsequences, each of length 100, having seeds that differ by a factor of 5. The correlation of these two subsequences is 0.375. (See also Exercise 1.15, page 59.)

This would indicate that when choosing to skip ahead $k$ steps, the seed should not be a small multiple of, or a small fraction of, $k$.

Sometimes, instead of "independent" sequences, we may want a sequence that is strongly negatively correlated with the first sequence. (We call such sequences "antithetic". They may be useful in variance reduction; we discuss them further in Section 7.5, page 239.) If one sequence begins with the seed $x_0$, and another sequence begins with $m - x_0$, then the $i^{\text{th}}$ term in the two sequences will be $x_i$ and $m - x_i$. The sequences are perfectly negatively correlated.

Another pair of sequences that may be of interest are ones that go in reverse order; that is, if one sequence is

$$x_1, \ x_2, \ x_3, \ \ldots, \ x_k, \ \ldots, \tag{1.21}$$

then the other sequence is

$$x_k, \ x_{k-1}, \ x_{k-2}, \ \ldots, \ x_3, \ x_2, \ x_1, \ \ldots. \tag{1.22}$$

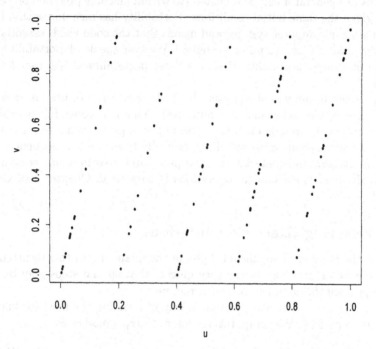

Figure 1.9: The Sequences $u_i$ and $v_i$ Generated by a Good Multiplicative Congruential Generator, Starting at Seeds 999 and 4995, Respectively

If the first sequence is generated by $x_{i+1} \equiv ax_i \bmod m$, then the second sequence is generated by $y_{i+1} \equiv by_i \bmod m$, where

$$b \equiv a^{p-1} \bmod m,$$

and $p$ is the period.

## 1.3 Computer Implementation of Simple Linear Congruential Generators

There are many important issues to consider in writing computer software to generate random numbers. Before proceeding to discuss other types of random number generators, we consider the computer implementation of simple linear congruential generators of the form (1.10),

$$x_i \equiv ax_{i-1} \bmod m.$$

These same kinds of considerations apply to other generators as well.

Requirements for a random number generator computer code include *strict reproducibility* and *portability*. Strict reproducibility means that any stream produced by the generator can be repeated (to within machine precision on each element), given the same initial conditions. Portability has meaning within the context of a set of computer systems and means that the code yields essentially results over that set of computer systems. We can speak of portability of executable programs, portability of source code, portability at the user level, and so on.

As a problem in numerical analysis, the basic recurrence of the linear congruential generator is somewhat ill-conditioned. This is because full precision must be maintained throughout; if any term in the sequence is not exact in the last place, all subsequent terms will differ radically from the true sequence.

Relevant background material on fixed-point and floating-point computations that simulate mathematical operations is covered in Chapter 1 of Gentle (1998).

### 1.3.1  Ensuring Exact Computations

Because of the number of significant digits in the quantities in this recurrence, even the novice programmer learns very quickly that special steps may be required to provide the complete precision required.

If the multiplier $a$ is relatively large, a way of avoiding the need for higher precision is to perform the computations for $x_i \equiv ax_{i-1} \bmod m$ as

$$x_i \equiv (c(dx_{i-1} \bmod m) + ex_{i-1} \bmod m)\bmod m, \qquad (1.23)$$

where $a = cd + e$. The values of $c$, $d$, and $e$ are chosen so that all products are exactly representable. This is possible as long as the available precision is at least $1.5\log_b m$ places in the arithmetic base $b$. The origins of the idea of using the decomposition (1.23), like so many of the ideas useful in numerical computations, are lost in antiquity. The idea has been rediscovered many times.

Even if $a$ is relatively small, as in the Park and Miller (1988) "minimal standard", the computations cannot be performed directly in ordinary floating-point words on computers with 32-bit words. In C, we can use double (fmod), and in Fortran we can use double precision, as shown in Figure 1.10.

Another way of avoiding the need for higher precision for relatively small multipliers—that is, multipliers smaller than $\sqrt{m}$ (as in the case shown in Figure 1.10)—is to perform the computations in the following steps. Let

$$q = \lfloor m/a \rfloor$$

and

$$r \equiv m \bmod a.$$

```
subroutine rnlcm (dseed, nr, u)
double precision dseed, dm, dmp
real u(*)
data dm/2147483647.d0/, dmp/2147483655.d0/
               !(dmp is computer specific)
do 10 i = 1, nr
   dseed = dmod (16807.d0*dseed, dm)
   u(i) = dseed/dmp
10 continue
end
```

Figure 1.10: A Fortran Program Illustrating a Congruential Generator for a Machine with 32-Bit Words

Then,

$$
\begin{aligned}
ax_{i-1} \bmod m &= (ax_{i-1} - \lfloor x_{i-1}/q \rfloor m) \bmod m \\
&= (ax_{i-1} - \lfloor x_{i-1}/q \rfloor (aq + r)) \bmod m \\
&= (a(x_{i-1} - \lfloor x_{i-1}/q \rfloor)q - \lfloor x_{i-1}/q \rfloor r) \bmod m \\
&= (a(x_{i-1} \bmod q) - \lfloor x_{i-1}/q \rfloor r) \bmod m.
\end{aligned}
\tag{1.24}
$$

For the operations of equation (1.24), L'Ecuyer (1988) gives coding of the following form:

```
k = x/q
x = a*(x - k*q) - k*r
if (x < 0) x = x + m
```

As above, it is often convenient to use the fact that the modular reduction is equivalent to

$$x_i = ax_{i-1} - m\lfloor ax_{i-1}/m \rfloor.$$

But in this equivalent formulation, the floor operation must be performed simultaneously with the division operation (that is, there can be no rounding between the operations). This may not be the case in some computing systems, and this is just another numerical "gotcha" for this problem.

## 1.3.2   Restriction that the Output Be in the Open Interval $(0, 1)$

There is another point that is often overlooked in general discussions of random number generation. We casually mentioned earlier in the chapter that we want to simulate a uniform distribution over $(0, 1)$. Mathematically, that is the same as a uniform distribution over $[0, 1]$; it is not the same on the computer, however. Using the methods that we have discussed, we must sometimes be careful to exclude the endpoints. Whenever the mixed congruential generator is used (i.e., $c \neq 0$ in the recurrence (1.9)), we may have to take special steps to handle the case where $x_i = 0$. For the multiplicative congruential generator (1.10), we do not have to worry about a 0. In any standard fixed-point and floating-point computer arithmetic system, if $x_i \geq 1$, the normalization $u_i = x_i/m$ will not yield a 0.

The normalization, however, can yield a 1 whether the generator is mixed or multiplicative. To avoid that we choose a different normalizer, $\tilde{m}$ ($> m$). See the code in Figure 1.10, and see Exercise 1.13, page 59.

### 1.3.3  Efficiency Considerations

In some computer architectures, operations on fixed-point numbers are faster than those on floating-point numbers. If the modulus is a power of 2, it may be possible to perform the modular reduction by simply retaining only the low-order bits of the product. Furthermore, if the power of 2 corresponds to the number of numeric bits, it may be possible to use the fixed-point overflow to do the modular reduction. This is an old trick, which was implemented in many of the early generators, including RANDU. This trick can also be implemented using a bitwise "and" operation. (The Fortran intrinsic iand does this.) This can have the same effect but without causing the overflow. Overflow may be considered an arithmetic error. A power of 2 is generally not a good modulus, however, as we have already pointed out.

The fixed-point overflow trick can be modified for doing a modular reduction for $m = 2^p - 1$. Let

$$\tilde{x}_i \equiv (ax_{i-1} + c) \bmod (m+1),$$

and if $\tilde{x}_i < m + 1$, then

$$x_i = \tilde{x}_i;$$

otherwise,

$$x_i = \tilde{x}_i - m.$$

This trick can be implemented three times if the multiplier $a$ is large, and the representation of equation (1.23) is used.

Some multiplications, especially in fixed-point, are performed more rapidly if one of the multiplicands is a power of 2. This fact provides another opportunity for accelerating computations that make use of equation (1.23). The decomposition of the multiplier $a$ can have two components that are powers of 2.

### 1.3.4  Vector Processors

Because a random number generator may be invoked millions of times in a single program, it is important to perform the operations efficiently. Brophy et al. (1989) describe an implementation of the linear congruential generator for a vector processor. If the multiplier is $a$, the modulus is $m$, and the vector register length is $k$, the quantities

$$a_j \equiv a^j \bmod m, \quad \text{for } j = 1, 2, \ldots, k,$$

are precomputed and stored in a vector register. The modulo reduction operation is not (usually) a vector operation, however. For the particular generator that they considered, $m = 2^{31} - 1$, and because

$$\frac{x}{2^{31} - 1} = x2^{-31} + x2^{-62} + \cdots,$$

they did the modular reduction as $a_j x - (2^{31} - 1) \lfloor a_j 2^{-31} x \rfloor$. For a given $x_i$, the $k$ subsequent deviates are obtained with a few operations in a vector register.

How to implement a random number generator in a given environment depends on such things as

- the type of architecture (vector, parallel, etc.),

- the precision available,

- whether integer arithmetic is available,

- whether integer overflow is equivalent to modular reduction,

- the base of the arithmetic,

- the relative speeds of multiplication and division,

- the relative speeds of MOD and INT.

The quality of the generator should never be sacrificed in the interest of efficiency, no matter what type of computer architecture is being used.

## 1.4 Other Linear Congruential Generators

There are many variations on the basic computer algorithm (1.1),

$$x_i = f(x_{i-1}, \cdots, x_{i-k}), \tag{1.25}$$

for producing pseudorandom numbers. In the simple linear congruential generator of Lehmer (1951), equation (1.9), $f$ is a simple linear function (that is, $k = 1$) combined with a modular reduction. A variation on the simple linear congruential generator involves shuffling of the output stream. Other variations of the linear congruential form involve use of $k$ greater than 1:

$$x_i \equiv (a^{\mathrm{T}} v_{i-1} + c) \bmod m, \quad \text{with} \quad 0 \le x_i < m, \tag{1.26}$$

where $a$ is a $k$-vector of constants, and $v_{i-1}$ is the $k$-vector $(x_{i-1}, \cdots, x_{i-k})$. Such a generator obviously requires a $k$-vector for a seed. Two widely used linear congruential generators of the form (1.26) with $k > 1$ are the "multiple recursive" generators, discussed in Section 1.4.1, in which $c = 0$ and the lagged Fibonacci generators, discussed on page 33, in which $c = 0$, and $a = (0, \ldots, 0, 1, 0, \cdots, 0, 1)$.

There are several other variations of the basic congruential generator. Some maintain linearity but include conditional values of $c$, such as the add-with-carry generators discussed in Section 1.4.3. Other variations involve nonlinear operations, such as the generators discussed in Sections 1.5.1 and 1.5.2. All of these generators share the fundamental operation of modular reduction that provides the "randomness" or the "unpredictability".

In general, the variations that have been proposed increase the complexity of the generator. For a generator to be useful, however, we must be able to analyze and understand its properties at least to the point of having some assurance that there is no deleterious nonrandomness lurking in the output of the generator. We should also be able to skip ahead in the generator a fixed distance.

## 1.4.1   Multiple Recursive Generators

A simple extension of the multiplicative congruential generator is to use multiples of the previous $k$ values to generate the next one:

$$x_i \equiv (a_1 x_{i-1} + a_2 x_{i-2} + \cdots a_k x_{i-k}) \bmod m. \tag{1.27}$$

When $k > 1$, this is sometimes called a "multiple recursive" multiplicative congruential generator. The number of previous numbers used, $k$, is called the "order" of the generator. (If $k = 1$, it is just a multiplicative congruential generator.)

The period of a multiple recursive generator can be much longer than that of a simple multiplicative generator. For $m$ a prime, Knuth (1998) has shown that the maximum period is $m^k - 1$ and has given conditions for achieving that period in terms of a polynomial

$$f(z) = z^k - (a_1 z^{k-1} + \cdots + a_{k-1} z + a_k) \tag{1.28}$$

over $\mathbb{G}(m)$. The reader is referred to Knuth (1998, page 29 and following) for a discussion of the conditions.

Computational efficiency can be enhanced by choosing some of the $a_j$ as 0 or $\pm 1$. Deng and Lin (2000) propose a "fast multiple recursive generator" in which only $a_1 = 1$ and all other $a_j$ are 0 except for $a_k$. For a given modulus, they use the criteria of Knuth (1998) under the restriction that $a_1 = 1$ to determine $a_k$ that yields the maximum period. For $m = 2^{31} - 1$ and $k = 2, 3, 4$, they give several such values of $a_k$. They also discuss coding issues for efficiency and portability.

L'Ecuyer, Blouin, and Couture (1993) study some of the statistical properties of multiple recursive generators and make recommendations for the multipliers for some specific moduli, including the common one, $2^{31} - 1$. Their assessment of the quality of the generators is based on the Beyer ratio (see page 66) and applied for dimensions from 1 to 20. They discuss programming issues for such generators and give code for a short C program for a fifth-order multiple recursive generator,

$$x_i \equiv (107374182 x_{i-1} + 104480 x_{i-5}) \bmod 2^{31} - 1. \qquad (1.29)$$

Their code is portable and will not return 0 or 1. They also describe how to skip ahead in multiple recursive generators to form substreams using methods similar to those discussed in Section 1.2.4.

Kao and Tang (1997b) also tested various full-period multipliers for multiple recursive generators. They evaluated the lattice structure through dimension 8. In the tests of Kao and Tang (1997a), multiple recursive generators with fewer terms than the order of the generator, such as the lagged Fibonacci generator discussed below, did not perform well. This appears to be a consequence of the particular generators that they studied rather than a general conclusion. Other tests of some such generators have not shown this to be a problem.

**Lagged Fibonacci**

A simple Fibonacci sequence has $x_{i+2} = x_{i+1} + x_i$. Reducing the numbers mod $m$ produces a sequence that looks somewhat random but actually does not have satisfactory randomness properties.

We can generalize the Fibonacci recursion in two ways. First, instead of combining successive terms, we combine terms at some greater distance apart, so the *lagged Fibonacci congruential generator* is

$$x_i \equiv (x_{i-j} + x_{i-k}) \bmod m. \qquad (1.30)$$

The lagged Fibonacci generator is a simple multiple recursive generator. If $j$, $k$, and $m$ are chosen properly, the lagged Fibonacci generator can perform well. If $m$ is a prime and $k > j$, then the period can be as large as $m^k - 1$. More important, this generator is the basis for other generators such as those discussed in Section 1.7, page 43.

If the modulus in the lagged Fibonacci generator is a power of 2, say $2^p$, the maximum period possible is $(2^k - 1)2^{p-1}$. Mascagni et al. (1995) describe a method of identifying and using exclusive substreams, each with the maximal possible period.

Another way of generalizing the Fibonacci recursion is to use more general binary operators instead of addition modulo $m$. In a *general lagged Fibonacci generator*, we start with $x_1, x_2, \ldots, x_k$ ("random" numbers) and let

$$x_i = (x_{i-j} \circ x_{i-k}),$$

where ∘ is some binary operator, with

$$0 \le x_i \le m - 1 \quad \text{and} \quad 0 < j < k < i.$$

**Moduli in Multiple Recursive Generators**

Composite moduli in multiple recursive generators do not necessarily yield output with the same kinds of patterns described on page 13 for singly recursive generators. A modulus of the form $2^e$ is often acceptable for a multiple recursive generator, but it is generally advisable to use a prime modulus.

**Seeding Multiple Recursive Generators**

A multiple recursive generator may require an initial sequence rather than just a single number for the seed. Altman (1989) has shown that care must be exercised in selecting the initial sequence for a lagged Fibonacci congruential generator, but for a carefully chosen initial sequence, the bits in the output sequence seem to have better randomness properties than those from congruential and Tausworthe generators (page 38). This is an indictment of lagged Fibonacci generators; the seed should not matter in good generators (although, in applications, we advocate in Chapter 2 that ad hoc tests be conducted on the generator with the seed to be used in the application).

## 1.4.2   Matrix Congruential Generators

A generalization of the scalar linear congruential generator to a generator of pseudorandom vectors is straightforward:

$$x_i \equiv (Ax_{i-1} + c) \bmod m, \qquad (1.31)$$

where the $x_i$, $x_{i-1}$, and $c$ are vectors of length $d$, and $A$ is a $d \times d$ matrix. The elements of the vectors and matrices are integers between 1 and $m - 1$. The vector elements are then scaled into the interval $(0, 1)$ to simulate $U(0, 1)$ deviates. Such a generator is called a *matrix congruential generator*. As with the scalar generators, $c$ is often chosen as 0 (a vector with all elements equal to 0).

Reasons for using a matrix generator are to generate parallel streams of pseudorandom deviates or to induce a correlational structure in the random vectors.

Deng and Lin (2000) suggest a fast implementation of the matrix multiplicative congruential generator with

$$A = \begin{bmatrix} a_1 & -1 & 0 & \cdots & 0 \\ 0 & a_2 & -1 & \cdots & 0 \\ \vdots & \vdots & \vdots & \cdots & \vdots \\ 0 & 0 & 0 & \cdots & -1 \\ -1 & 0 & 0 & \cdots & a_d \end{bmatrix}.$$

This choice involves computations similar to those in their fast multiple recursive generator.

If $A$ is a diagonal matrix, the matrix generator is essentially a set of scalar generators with different multipliers. For more general $A$, the elements of the vectors are correlated. Rather than concentrating directly on the correlations, most studies of the matrix generators (e.g., Afflerbach and Grothe, 1988) have focused on the lattice structure. Choosing $A$ as the Cholesky factor of a target variance-covariance matrix may make some sense, and there may be other situations in which a matrix generator would be of some value. In Exercise 1.6, page 58, you are asked to experiment with a matrix generator to study the correlations. Generally, however, any desired correlational structure should be simulated by transformations on an independent stream of uniforms, as we discuss in Section 5.3, page 197, rather than trying to induce it in the congruential generator.

By analogy with the multiple recursive generator with prime modulus, it could be guessed that the maximal period of the matrix multiplicative generator (with $c = 0$) is $m^d - 1$, and indeed that is the case, although the analysis of the period of the matrix congruential generator is somewhat more complicated than that of the scalar generator (see Grothe, 1987). It is clear that $A$ should be nonsingular (within the finite field generated by the modulus), because otherwise it is possible to generate a zero vector at which point all subsequent vectors are zero.

A straightforward extension of the matrix congruential generator is the multiple recursive matrix random number generator:

$$x_i \equiv (A_1 x_{i-1} + \cdots A_k x_{i-k}) \bmod m. \tag{1.32}$$

This is the same idea as in the multiple recursive generator for scalars, as considered above. Also, as mentioned above, computational efficiency can be enhanced by choosing some of the $A_j$ as 0 matrices, identity (or negative identity) matrices, or simply scalars (that is, scalar multiples of the identity matrix). Niederreiter (1993, 1995a, 1995b, 1995d) discusses some of the properties of the multiple recursive matrix generator.

## 1.4.3 Add-with-Carry, Subtract-with-Borrow, and Multiply-with-Carry Generators

Marsaglia and Zaman (1991) describe two variants of a generator that they called "add-with-carry" (AWC) and "subtract-with-borrow" (SWB). The add-with-carry form is

$$x_i \equiv (x_{i-s} + x_{i-r} + c_i) \bmod m,$$

where $c_1 = 0$, and $c_{i+1} = 0$ if $x_{i-s} + x_{i-r} + c_i < m$ and $c_{i+1} = 1$ otherwise. The $c$ is the "carry".

Marsaglia and Zaman investigated various values of $s$, $r$, and $m$. For some choices that they recommended, the period of the generator is of the order of $10^{43}$. This generator can also be implemented in modulo 1 arithmetic. Although this at first appears to be a nonlinear generator (because of the branch), Tezuka and L'Ecuyer (1992) have shown that sequences resulting from this generator are essentially equivalent to sequences from linear congruential generators with very large prime moduli. (In fact, the AWC/SWB generators can be viewed as efficient ways of implementing such large linear congruential generators.) The work of Couture and L'Ecuyer (1994) and Tezuka, L'Ecuyer, and Couture (1994) indicates that the lattice structure in high dimensions may be very poor.

Marsaglia also described a multiply-with-carry random number generator that is a generalization of the add-with-carry random number generator. The multiply-with-carry generator is

$$x_i \equiv (ax_{i-1} + c_i) \bmod m.$$

Marsaglia suggests $m = 2^{32}$ and an implementation in 64-bit integers. The use of the lower-order 32 bits results in the modular reduction, as we have seen. The higher-order 32 bits determine the carry. Couture and L'Ecuyer (1995, 1997) suggest methods for finding parameters for the multiply-with-carry generator that yield good sequences.

## 1.5    Nonlinear Congruential Generators

If the function in the basic congruential recursion (1.1),

$$x_i = f(x_{i-1}, \cdots, x_{i-k}),$$

is linear, not only are the computations simpler and generally faster, but the analysis of the output is more tractable. Nevertheless, various nonlinear congruential generators have been proposed. More study is needed for most of the generators mentioned in this section.

### 1.5.1    Inversive Congruential Generators

Inversive congruential generators, introduced by Eichenauer and Lehn (1986), use the modular multiplicative inverse (if it exists) to generate the next variate in a sequence. The inversive congruential generator is

$$x_i \equiv (ax_{i-1}^- + c) \bmod m, \quad \text{with} \quad 0 \le x_i < m, \tag{1.33}$$

where $x^-$ denotes the multiplicative inverse of $x$ modulo $m$, if it exists, or else it denotes 0. Compare this generator with equation (1.9), page 11.

The multiplicative inverse, $x^-$, of $x$ modulo $m$ is defined for all nonzero $x$ relatively prime to $m$ by

$$1 \equiv x^- x \bmod m.$$

Eichenauer and Lehn (1986), Eichenauer, Grothe, and Lehn (1988), and Niederreiter (1988, 1989) show that the inversive congruential generators have good uniformity properties, in particular with regard to lattice structure and serial correlations.

Eichenauer-Herrmann and Ickstadt (1994) introduced the *explicit inversive congruential* method that yields odd integers between 0 and $m - 1$, and Eichenauer-Herrmann (1996) gives a modification of it that can yield all nonnegative integers up to $m - 1$. The modified explicit inversive congruential generator is given by

$$x_i \equiv i(ai + b)^- \bmod 2^p, \quad i = 0, 1, 2, \ldots,$$

where $a \equiv 2 \bmod 4$, $b \equiv 1 \bmod 2$, and $p \geq 4$. Eichenauer-Herrmann (1996) shows that the period is $m$ and the two-dimensional discrepancy (see page 69) is $O(m^{-\frac{1}{2}}(\log m)^2)$ for the modified explicit inversive generator.

Although there are some ways of speeding up the computations (see Gordon, 1989), the computational difficulties of the inversive congruential generators have prevented their widespread use. There have been mixed reports of the statistical properties of the pseudorandom streams produced by inversive congruential generators. Chou and Niederreiter (1995) describe a lattice test and some results of it for inversive congruential generators. An inversive congruential generator does not yield regular planes like a linear congruential generator. Leeb and Wegenkittl (1997) report on several tests of inversive congruential generators. The apparent randomness was better than that of the linear congruential generators that they tested. The inversive congruential generators did not perform so well on tests based on spacings reported by L'Ecuyer (1997), however. The bulk of the evidence currently does not recommend inversive congruential generators.

## 1.5.2 Other Nonlinear Congruential Generators

A simple generalization of the linear congruential generator (1.9) is suggested by Knuth (1998):

$$x_i \equiv (dx_{i-1}^2 + ax_{i-1} + c) \bmod m, \quad \text{with} \quad 0 \leq x_i < m. \tag{1.34}$$

Higher-degree polynomials could be used also. Whether there is any advantage in using higher-degree polynomials is not clear. They do have the apparent advantage of making the generator less obvious.

Blum, Blum, and Shub (1986) proposed a generator based on

$$x_i \equiv x_{i-1}^2 \bmod m. \tag{1.35}$$

Their method, however, goes beyond the introduction of just more complicated transformations in the sequence of integers. They form the output sequence as the bits $b_1 b_2 b_3 \cdots$, where $b_i = 0$, if $x_i$ is even, and otherwise $b_i = 1$. Blum, Blum, and Shub showed that if $m = p_1 p_2$, where $p_1$ and $p_2$ are distinct primes, each congruent to 3 mod 4, then the output of this generator is not predictable in polynomial time without knowledge of $p_1$ and $p_2$. Because of this unpredictability, this generator has important possible uses in cryptography.

A more general congruential generator uses a function $g$ of the previous value to generate the next one:

$$x_i \equiv g(x_{i-1}) \bmod m, \quad \text{with} \quad 0 \le x_i < m. \tag{1.36}$$

Such generators were studied by Eichenauer, Grothe, and Lehn (1988). The wide choice of functions can yield generators with very good uniformity properties, but there is none other than the simple linear or inversive ones for which a body of theory and experience would suggest its use in serious work.

Kato, Wu, and Yanagihara (1996a) consider a function $g$ in equation (1.36) that combines a simple linear function with the multiplicative inverse in equation (1.33). Their generator is

$$x_i \equiv (ax_{i-1}^- + bx_{i-1} + c) \bmod m, \quad \text{with} \quad 0 \le x_i < m.$$

They suggest a modulus that is a power of 2 and in Kato, Wu, and Yanagihara (1996b) derive properties for this generator similar to those derived by Niederreiter (1989) for the inversive generator in equation (1.33).

Eichenauer-Herrmann (1995) and Eichenauer-Herrmann, Herrmann, and Wegenkittl (1998) provide surveys of work done on various nonlinear generators, including the inversive congruential generators.

## 1.6   Feedback Shift Register Generators

Tausworthe (1965) introduced a generator based on a sequence of 0s and 1s generated by a recurrence of the form

$$b_i \equiv (a_p b_{i-p} + a_{p-1} b_{i-p+1} + \cdots + a_1 b_{i-1}) \bmod 2, \tag{1.37}$$

where all variables take on values of either 0 or 1. This is the multiple recursive generator of equation (1.27). The difference is that we will interpret the $b$s as bits and will form them into binary representations of integers.

Because the modulus is a prime, the generator can be related to a polynomial

$$f(z) = z^p - (a_1 z^{p-1} + \cdots + a_{p-1} z + a_p) \tag{1.38}$$

over the Galois field $\mathbb{G}(2)$ defined over the integers 0 and 1 with the addition and multiplication being defined in the usual way followed by a reduction modulo 2. An important result from the theory developed for such polynomials is that, as long as the initial vector of $b$s is not all 0s, the period of the recurrence (1.37) is $2^p - 1$ if and only if the polynomial (1.38) is irreducible over $\mathbb{G}(2)$. (An irreducible polynomial is also sometimes called a primitive polynomial.) It is easy to see that the maximal period is $2^p - 1$ because if any $p$-vector of $b$s repeats, all subsequent values are repeated values.

For computational efficiency, most of the $a$s in equation (1.37) should be zero. For a modulus equal to 2, there is only one binomial that is irreducible, and it is $z + 1$, which would yield an unacceptable period. There are, however, many trinomials and higher-degree polynomials. See Zierler and Brillhart (1968, 1969) for a list of all primitive trinomials modulo 2 up to degree 1000. Kumada et al. (2000) give additional trinomials.

The recurrence (1.37) often has the form

$$b_i \equiv (b_{i-p} + b_{i-p+q}) \bmod 2 \tag{1.39}$$

resulting from a trinomial. Addition of 0s and 1s modulo 2 is the binary exclusive-or operation, denoted by $\oplus$; thus, we may write the recurrence as

$$b_i = b_{i-p} \oplus b_{i-p+q}. \tag{1.40}$$

After this recurrence has been evaluated a sufficient number of times, say $l$ (with $l \leq p$), the $l$-tuple of $b$s is interpreted as a number in base 2. This is referred to as an $l$-wise decimation of the sequence of $b$s. If $l$ is relatively prime to $2^p - 1$ (the period of the sequence of $b$s), the period of the $l$-tuples will also be $2^p - 1$. In this case, the decimation is said to be proper. Note that the recurrence of bits is the same recurrence of $l$-tuples,

$$x_i = x_{i-p} \oplus x_{i-p+q}, \tag{1.41}$$

where the $x$s are the numbers represented by interpreting the $l$-tuples as binary notation, and the exclusive-or operation is performed bitwise. The underlying trinomial is

$$x^p + x^r + 1,$$

where $r = p - q$. A random number generator built on this recurrence is sometimes denoted by $\mathrm{R}(r, p)$.

As an example, consider the trinomial

$$x^4 + x + 1, \tag{1.42}$$

and begin with the bit sequence

$$1, 0, 1, 0.$$

For this polynomial, $p = 4$ and $q = 3$ in the recurrence (1.40). Operating the generator, we obtain

$$1, 1, 0, 0, 1, 0, 0, 0, 1, 1, 1, 1, 0, 1, 0,$$

at which point the sequence repeats; its period is $2^4 - 1$. A 4-wise decimation using the recurrence (1.41) yields the numbers

$$12, 8, 15, 5, \ldots$$

(in which the 5 required an additional bit in the sequence above). We could continue in this way to get 15 (that is, $2^4 - 1$) integers between 1 and 15 before the sequence began to repeat.

As with the linear congruential generators, different values of the *as* and even of the starting values of the *bs* can yield either good generators (i.e., ones with outputs that seem to be random samples of a uniform distribution) or bad generators.

This recurrence operation in equation (1.37) can be performed in a *feedback shift register*, which is a vector of bits that is shifted, say, to the left, one bit at a time, and the bit shifted out is combined with other bits in the register to form the rightmost bit. The operation can be pictured as shown in Figure 1.11, where the bits are combined using $\oplus$.

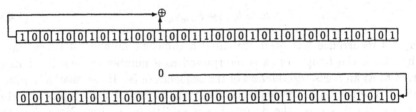

Figure 1.11: One Shift of a Feedback Shift Register

The two sources of bits to shift into the right-hand side of the register are called "taps". Thus, Figure 1.11 represents a two-tap generator. The basic idea of the feedback shift register can of course easily be extended to more than two taps.

## 1.6.1   Generalized Feedback Shift Registers and Variations

Feedback shift registers have an extensive theory (see, e.g., Golomb, 1982) because they have been used for some time in communications and cryptography.

A variation on the Tausworthe generator is suggested by Lewis and Payne (1973), who call their modified generator a *generalized feedback shift register* (GFSR) generator. In the GFSR method, the bits of the sequence from recurrence (1.39) form the bits in one binary position of the numbers being generated. The next binary position of the numbers being generated is filled with the same bit sequence but with a delay. By using the bit stream from the trinomial $x^4 + x + 1$ and the starting sequence that we considered before, and again forming 4-bit words by putting the bits into a fixed binary position with a delay of 3 between binary positions, we have

$$
\begin{aligned}
1010 &= 10 \\
1110 &= 14 \\
0011 &= 3 \\
0101 &= 5 \\
1111 &= 15 \\
0001 &= 1 \\
0010 &= 2 \\
0111 &= 7 \\
1000 &= 8 \\
1001 &= 9 \\
1011 &= 11 \\
1100 &= 12 \\
0100 &= 4 \\
1101 &= 13 \\
0110 &= 6,
\end{aligned}
$$

at which point the sequence repeats. Notice that the recurrence (1.41),

$$ x_i = x_{i-p} \oplus x_{i-p+q}, $$

still holds, where the $x$s are the numbers represented by interpreting the $l$-tuples as binary notation, and the exclusive-or operation is performed bitwise.

Lewis and Payne (1973) discussed methods of initializing the generator and gave programs both for the initialization and for the generation. Kirkpatrick and Stoll (1981) presented a faster way of initializing the generator and developed a program, R250, implementing a generator with $p = 250$ and $q = 103$ (that is, an R(103, 250)). This program is widely used among physicists. Empirical results for this generator are mixed. The generator performed well in tests reported by L'Ecuyer (1997) and by Kankaala, Ala-Nissila, and Vattulainen (1993), but other tests based on physical simulations have indicated some problems (see, for example, Ferrenberg, Landau, and Wong, 1992; Selke, Talapov, and Shchur, 1993; and Ziff, 1998).

Fushimi (1990) has studied GFSR methods in gener and has described some particular GFSRs that he showed to have good properties both theoretically and empirically. One particular case studied by Fushimi had $p = 521$, with $a_{521} = a_{32} = 1$ (with a slight modification in the definition of the recursion as $x_i = x_{i-3p} \oplus x_{i-3q}$). The generator, which is available in the IMSL Libraries, has a period of $2^{521} - 1$. This generator also performed well in tests reported by L'Ecuyer (1997) but not so well in one test by Matsumoto and Kurita (1994).

Ziff (1998) suggests that the problems arising in two-tap shift registers can be ameliorated by use of four-tap shift registers; that is, generators based on the recurrence

$$ x_i = x_{i-a} \oplus x_{i-b} \oplus x_{i-c} \oplus x_{i-d}. $$

He shows that four-tap sequences can also be constructed by decimation of two-tap sequences. See also Compagner (1991, 1995) for further discussion of multitap rules.

### Twisting the Bits

Another way of addressing the problems of two-tap shift registers has been proposed by Matsumoto and Kurita (1992, 1994). They modify the GFSR in the recurrence (1.41) by "twisting" the bit pattern in $x_{i-p+q}$. This is done by viewing the $x$s as $l$-vectors of zeros and ones and multiplying $x_{i-p+q}$ by an $l \times l$ matrix $A$. The recurrence then becomes

$$x_i = x_{i-p} \oplus A x_{i-p+q}. \tag{1.43}$$

They call this a *twisted GSFR generator*. Matsumoto and Kurita (1994) show how to choose $A$ and modify the output to achieve good uniformity properties. They also gave a relatively short C program implementing the method. Further empirical studies for this generator are needed. The idea of a twisted GFSR generator is related to the matrix congruential generators discussed in Section 1.4.2.

### The Mersenne Twister

Matsumoto and Nishimura (1998) describe a twisted generator based on the recurrence (1.43) that has approximately 100 terms in the characteristic polynomial of the matrix $A$. They give a C program for a version of this generator, called MT19937, that has a period of $2^{19937} - 1$ and 623-variate uniformity (see page 63). The Mersenne twister requires a rather complicated initialization procedure and the one suggested in the 1998 article has some flaws. The authors maintain a webpage

    http://www.math.keio.ac.jp/~matumoto/emt.html

that describes current work on the Mersenne twister and provides links for code and additional references. This generator is becoming widely used, but additional empirical studies of it are needed.

### Restriction that the Output Be in the Open Interval $(0, 1)$

In our discussion of computer implementation of congruential generators on page 29, we mentioned a technique to ensure that the result of the generator is less than 1. The same technique, of course, works for GFSR generators, but in these generators we have the additional concern of a 0. By using the obvious bit operations, which are inexpensive, it is usually fairly easy to prevent 0 from occurring, however.

## 1.6.2   Skipping Ahead in GFSR Generators

Golomb (1982) noted that the basic recurrence (1.41),

$$x_i = x_{i-p} \oplus x_{i-p+q},$$

implies

$$x_i = x_{i-2^e p} \oplus x_{i-2^e p+2^e q}$$

for any integer $e$. This provides a method of skipping ahead in a generalized feedback shift register generator by a fixed distance that is a power of 2. Aluru, Prabhu, and Gustafson (1992) described a leapfrog method using this relationship and applied it to parallel random number generators. Thus, for $k$ a power of 2, we have

$$x_i = x_{i-kp} \oplus x_{i-k(p-q)},$$

so we can generate a leapfrog sequence for GFSR generators analogous to equation (1.18) on page 24 for multiplicative congruential generators:

$$x_s, x_{s+k}, x_{s+2k}, \dots .$$

Although there is a dearth of published studies on the correlations among substreams generated by leapfrogging with a GFSR generator, serious correlations can exist. The same issues discussed on page 26 and illustrated in Figure 1.9 for leapfrogged substreams from a simple linear congruential generator are relevant in the case of GFSR generators also.

# 1.7   Other Sources of Uniform Random Numbers

Many different mechanisms for generating random numbers have been introduced. Some are simple variations of the linear congruential generator or the feedback shift register, perhaps designed for microcomputers or some other special environment. Various companies distribute proprietary random number generators, perhaps based on some physical process. Marsaglia (1995) uses some of them, and his assessment of those he used was that they were not of high quality. (Nevertheless, he combined them with other sources and produced pseudorandom streams of high quality.)

Digits in the decimal representations of approximations to transcendental numbers, such as $\pi$ or e, or simpler irrational numbers, such as $\sqrt{2}$, have often been suggested as streams of independent discrete uniform numbers (see Dodge, 1996, for example). A number is said to be *normal in base b* if, in its representation in base $b$, every nonnegative integer less than $b$ occurs with frequency $1/b$ and every subsequence of length $l$ occurs with frequency $1/b^l$, for any $l$. It is clear that only irrational numbers can be normal. A number that is normal in a given base may not be normal in another base. See Bailey and

Crandall (2001) for further discussion of normal numbers. Most statistical tests of streams from $\pi$ and e (usually in base 10) have not detected nonzero correlations or departures from uniformity. See Jaditz (2000) and NIST (2000) for summaries of some tests of the randomness in the expansions of these irrational numbers. Also see Exercise 2.7 on page 90.

## 1.7.1   Generators Based on Cellular Automata

John von Neumann introduced *cellular automata* as a way of modeling biological systems. A cellular automaton consists of a countable set of fixed sites, each with a given value chosen from a countable, usually finite, set. In discrete time steps, the values at all sites are updated simultaneously based on simple rules that use the values at neighboring sites. If the values at a step $i-1$ are $\ldots, b_{-2}^{(i-1)}, b_{-1}^{(i-1)}, b_0^{(i-1)}, b_1^{(i-1)}, b_2^{(i-1)}, \ldots$, the update rule is

$$b_j^{(i)} = \phi\left(b_{j-r}^{(i-1)},\ b_{j-r+1}^{(i-1)},\ \ldots,\ b_j^{(i-1)},\ \ldots,\ b_{j+r}^{(i-1)}\right)$$

for some function $\phi$.

We can think of this as a grid of elements $b_j^{(i)}$ that is filled starting with the given row $b_1^{(1)}, \ldots, b_k^{(1)}$ by following the update rule. Wolfram (1994, 2002) provides extensive descriptions of ways of constructing cellular automata and discussions of their properties.

Wolfram (1984) suggested a random number generator using the cellular automaton in which $b_j$ takes on the values 0 and 1, and

$$b_j^{(i)} \equiv \left(b_{j-1}^{(i-1)} + b_j^{(i-1)} + b_{j+1}^{(i-1)} + b_j^{(i-1)} b_{j+1}^{(i-1)}\right) \bmod 2. \qquad (1.44)$$

This is equivalent to a rule on a two-dimensional grid that generates the $j^{\text{th}}$ element in line $i$ from the previous line by taking the element from the previous line just to the left of the $j^{\text{th}}$ element if both the $j^{\text{th}}$ element and the element just to the right of the $j^{\text{th}}$ element are both 0; otherwise taking the $j^{\text{th}}$ element in line $i$ to be the opposite of that element just to the left. Wolfram calls this "rule 30".

The first line in the grid can be initiated in a random but reproducible (therefore pseudorandom) manner. Edge effects in equation (1.44) are handled by assigning 0s to $b$s having indices that are out of range.

Wolfram describes some ways of mapping a vector of bits to an integer and then, from a given integer characterized by a finite vector of bits, using the iteration (1.44) to generate a new number. The $b$s on any line, for example, could be the bits in the binary representation of an integer. A random number generator based on the rule 30 cellular automaton is available in the Mathematica software system.

## 1.7.2 Generators Based on Chaotic Systems

Because chaotic systems exhibit irregular and seemingly unpredictable behavior, there are many connections between random number generators and chaotic systems. Chaotic systems are generally defined by recursions, just like random number generators. It is difficult to define useful random processes based on chaotic systems, but some of the results of chaos theory may have some relevance to the understanding of pseudorandom generators.

Lüscher (1994) relates the Marsaglia–Zaman subtract-with-borrow generator to a chaotic dynamical system. This relationship allows an interpretation of the correlational defects of the basic generator as short-term effects and provides a way for skipping ahead in the generator to avoid those problems. The Lüscher generator has a long period and performed well in the statistical tests reported by Lüscher. James (1994) provides a portable Fortran program, RANLUX, implementing this generator.

## 1.7.3 Other Recursive Generators

Another family of combination generators with long periods are the so-called ACORN (additive congruential random number) generators described by Wikramaratna (1989). The $k^{\text{th}}$-order ACORN generator yields the sequence

$$u_{k,1},\ u_{k,2},\ u_{k,3}, \ldots$$

from the recursions

$$u_{0,i} = u_{0,i-1}, \qquad\qquad\quad \text{for } i \geq 1,$$
$$u_{j,i} \equiv \left(u_{j-1,i} + u_{j,i-1}\right) \bmod 1, \quad \text{for } i \geq 1, \text{ and } j = 1,2,\ldots k.$$

The ACORN generators depend heavily on the starting values $u_{j,0}$. Wikramaratna (1989) shows that initial values satisfying certain conditions can yield very long periods for this generator. Although the generator is easy to implement, the user must be concerned with the proper choice of seeds.

It is easy to modify existing random number generators or to form combinations of existing generators, and every year several new generators are suggested in the scientific literature. The *Journal of Computational Physics* and *Computer Physics Communications* provide new generators in abundance. For example, one proposed in the former journal is the Chebyshev generator, defined by Erber, Everett, and Johnson (1979) as

$$x_i = x_{i-1}^2/2,$$
$$u_i = \frac{1}{\pi} \arccos(x_i/2).$$

Although this generator has some desirable qualities, it does not hold up well in statistical tests. (See other articles in the same journal; for example, Hosack, 1986, and Wikramaratna, 1989.)

Some newly proposed generators merit further study, but the scientist generally must be very careful in choosing a generator to use. It is usually better to avoid new generators or ones for which there is not a large body of empirical evidence regarding their quality.

### 1.7.4   Tables of Random Numbers

Before the development of reliable methods for generating pseudorandom numbers, there were available a number of tables of random numbers that had been generated by some physical process or that had been constructed by systematically processing statistical tables to access midorder digits of the data that they contained. The most widely used of these were the "RAND tables" (RAND Corporation, 1955), which contain a million random digits generated by a physical process. The random digits in these tables have been subjected to many statistical tests, with no evidence that they did not arise from a discrete uniform distribution over the ten mass points.

Marsaglia (1995) has produced a CD-ROM that contains 4.8 billion random bits stored in sixty separate files. The random bits were produced by a combination of two or more deterministic random number (pseudorandom number) generators and three devices that use physical processes. Some of the files also have added bit streams from digitized music or pictures.

## 1.8   Combining Generators

Both the period and the apparent randomness of random number generators can often be improved by combining more than one generator. The shuffling methods of MacLaren and Marsaglia (1965) and others, described on page 21, may use two or more generators in filling a table and shuffling it.

Another way of combining generators is suggested by Collings (1987). His method requires a pool of generators, each maintaining its own stream. For each number in the delivered stream, an additional generator is used to select which of the pool of generators is to be used to produce the number.

The output streams of two or more generators can also be combined directly in various ways, such as by a linear combination followed by a modulo reduction. Each generator combined in this way is called a "source" generator. This is one of the simplest and most widely used ways of combining generators. The most common ones are two-source and three-source generators.

The generators that are combined can be of any type. One of the early combined generators, developed by George Marsaglia and called "Super-Duper", used a linear congruential generator and a Tausworthe generator. See Learmonth and Lewis (1973) for a description of Super-Duper. Another similar one that uses a linear congruential generator and two shift register generators is called "KISS" (see Marsaglia, 1995; it is included on the CD-ROM).

### Wichmann/Hill Generator

Wichmann and Hill (1982 and corrigendum, 1984) describe a three-source generator that is a combination of linear congruential generators. It is easy to program and has good randomness properties. The generator is

$$x_i \equiv 171x_{i-1} \bmod 30269,$$
$$y_i \equiv 172y_{i-1} \bmod 30307,$$
$$z_i \equiv 170z_{i-1} \bmod 30323,$$

and

$$u_i = \left(\frac{x_i}{30269} + \frac{y_i}{30307} + \frac{z_i}{30323}\right) \bmod 1.$$

The seed for this generator is the 3-vector $(x_0, y_0, z_0)$. The generator directly yields numbers $u_i$ in the interval $(0,1)$. The period is of the order of $10^{12}$. Notice that the final modular reduction is equivalent to retaining only the fractional part of the sum of three uniform numbers (see Exercise 1.4a, page 57).

Zeisel (1986) showed that the Wichmann/Hill generator is the same as a single multiplicative congruential generator with a modulus of $27\,817\,185\,604\,309$. (The reduction makes use of the Chinese Remainder Theorem; see Fang and Wang, 1994. See also L'Ecuyer and Tezuka, 1991.)

De Matteis and Pagnutti (1993) studied the higher-order autocorrelations in sequences from the Wichmann/Hill generator and found them to compare favorably with those from other good generators. This, together with the ease of implementation, makes the Wichmann/Hill generator a useful one for common applications. A straightforward implementation of this generator can yield a 0 (see Exercise 1.17, page 59).

### L'Ecuyer Combined Generators

L'Ecuyer (1988) suggests combining $k$ multiplicative congruential generators that have prime moduli $m_j$, such that $(m_j - 1)/2$ are relatively prime, and with multipliers that yield full periods. Let the sequence from the $j^{\text{th}}$ generator be $x_{j,1}, x_{j,2}, x_{j,3}, \ldots$. Assuming that the first generator is a relatively "good" one and that $m_1$ is fairly large, we form the $i^{\text{th}}$ integer in the sequence as

$$x_i \equiv \sum_{j=1}^{k} (-1)^{j-1} x_{j,i} \bmod (m_1 - 1).$$

The other moduli $m_j$ do not need to be large.

The normalization takes care of the possibility of a 0 occurring in this sequence:

$$u_i = \begin{cases} x_i/m_1 & \text{if } x_i > 0, \\ (m_1 - 1)/m_1 & \text{if } x_i = 0. \end{cases}$$

A specific generator suggested by L'Ecuyer (1988) is the following:

$$
\begin{aligned}
x_i &\equiv 40014x_{i-1} \bmod 2147483563, \\
y_i &\equiv 40692y_{i-1} \bmod 2147483399, \\
z_i &\equiv (x_i - y_i) \bmod 2147483563,
\end{aligned}
\tag{1.45}
$$

and

$$
u_i = 4.656613z_i \times 10^{-10}.
$$

The period is of the order of $10^{18}$. L'Ecuyer (1988) presented results of both theoretical and empirical tests that indicate that the generator performs well.

Notice that the difference between two discrete uniform random variables modulo the larger of the ranges of the two random variables is also a discrete uniform random variable (see Exercise 1.4c, page 57). Notice also that the normalization always yields numbers less than 1 because the normalizing constant is larger than 2147483563.

L'Ecuyer (1988) gives a portable program for the generator using the techniques that we have discussed for keeping the results of intermediate computations small (see page 27 and following).

L'Ecuyer and Tezuka (1991) analyze generators of this form and present one with similar size moduli and multipliers that seems slightly better than the generator (1.45).

L'Ecuyer (1996, 1999) studied combined multiple recursive random number generators and concluded that combinations of generators of order greater than 1 would perform better than the combinations of the simple order 1 generators. One that he recommends is a two-source combination of third-order linear congruential generators:

$$
\begin{aligned}
x_i &\equiv (63308x_{i-2} - 183326x_{i-3}) \bmod (2^{31} - 1), \\
y_i &\equiv (86098y_{i-1} - 539608y_{i-3}) \bmod 2145483479, \\
z_i &\equiv (x_i - y_i) \bmod (2^{31} - 1).
\end{aligned}
\tag{1.46}
$$

## 1.9  Properties of Combined Generators

A judicious combination of two generators may improve the apparent randomness properties. Because of the finite number of states in any generator, however, it is possible that the generators will magnify their separate bad properties. Consider, for example, combining the two streams (1.21) and (1.22) on page 26. The combined generator in that case would have a period of 1. That, of course, would be a very extreme case, and it would require extremely bad luck or extreme carelessness to combine those generators.

Use of combined generators is likely to make the process more difficult to predict; hence, intuitively, it is likely to make the process more "random". Of course, we also are concerned about uniformity. Does the use of combined generators increase the fidelity of the process to a sample from a uniform distribution?

We present below two heuristic arguments that sequences from combined generators are more "random". See also Marshall and Olkin (1979) for other discussions of combinations of random variables that may support the use of combined generators.

## A Heuristic Argument Based on the Density of the Uniform Distribution

Deng and George (1990) and Deng et al. (1997) provide arguments in favor of using combined generators based on probability densities that are almost constant. They consider sums of random variables whose distributions are "nearly" uniform and use a variation of the uniformity property of the modular reduction (1.8) on page 10: the fractional part of the sum of independent random variables one of which is $U(0,1)$ is also a $U(0,1)$ random variable. If $X_i$ are independently distributed with probability density functions $p_i$, and $|p_i(x) - 1| \leq \epsilon_i$ over $[0,1]$, then for the density $p_X$ of the reduced sum, $X = (\sum X_i) \bmod 1$, we have

$$|p(x) - 1| \leq \prod \epsilon_i$$

over $[0,1]$. This implies that the random variable $X$ is more nearly uniform in some sense than any of the individual $X_i$.

Deng and George (1992) also describe several other properties of uniform variates, some of which characterize the uniform distribution, that might be useful in improving uniform random number generators.

## A Heuristic Argument Based on $n$-dimensional Lattices

Consider two sequences produced by random number generators over a finite set $S$, which without loss we can consider to be the integers $1, 2, 3, \ldots, n$. A result due to Marsaglia (1985) suggests that the distribution of a sequence produced by a one-to-one function from $S \times S$ onto $S$ is at least as close to uniform as the distribution of either of the original sequences.

Let $p_1, p_2, p_3, \ldots, p_n$ be the probabilities of obtaining $1, 2, 3, \ldots, n$ associated with the first sequence corresponding to the random variable $X$, and let $q_1, q_2, q_3, \ldots, q_n$ be the corresponding probabilities associated with the second sequence that are realizations of the random variable $Y$. The distance of either sequence from the uniform distribution is taken to be the Euclidean norm of the vector of differences of the probabilities from the uniform $1/n$. Let $s$ be the vector with all $n$ elements equal to $1/n$, and let $p$ be the vector with elements $p_i$. Then, for example,

$$d(X) = \|p - s\| = \sqrt{\sum \left( p_i - \frac{1}{n} \right)^2}.$$

This is similar to the discrepancy measure discussed in Section 2.2, page 64. It can be thought of as a measure of uniformity. The one-to-one function from $S \times S$ onto $S$ can be represented as the operator $\circ$:

|     |     | $X$ |     |     |     |
|-----|-----|-----|-----|-----|-----|
| $\circ$ | 1 | 2 | 3 | $\cdots$ | $n$ |
| 1 | $z_{(1,1)}$ | $z_{(1,2)}$ | $z_{(1,3)}$ | $\cdots$ | $z_{(1,n)}$ |
| 2 | $z_{(2,1)}$ | $z_{(2,2)}$ | $z_{(2,3)}$ | $\cdots$ | $z_{(2,n)}$ |
| $Y$        3 | $z_{(3,1)}$ | $z_{(3,2)}$ | $z_{(3,3)}$ | $\cdots$ | $z_{(3,n)}$ |
| $\cdots$ | $\cdots$ | $\cdots$ | $\cdots$ | $\cdots$ | $\cdots$ |
| $n$ | $z_{(n,1)}$ | $z_{(n,2)}$ | $z_{(n,3)}$ | $\cdots$ | $z_{(n,n)}$ |

That is, $i \circ j = z_{(i,j)}$. In each row and each column, every integer from 1 to $n$ occurs as the value of some $z_{(i,j)}$ exactly once. The probability vector for the random variable $Z = X \circ Y$ is then given by

$$\Pr(Z = k) = \sum_i \Pr(X = i)\Pr(Y = j \circ k),$$

where $j$ is such that $i \circ j = k$. This expression can be written as $Mq$, where $M$ is the matrix whose rows $m_i^{\mathrm{T}}$ correspond to permutations of the elements $p_1, p_2, \ldots, p_n$, such that each $p_j$ occurs in only one column, and $q$ is the vector containing the elements $q_j$. Now

$$\sum_i \Pr(Y = j \circ k) = 1,$$

so each row (and also each column) of $M$ sums to 1, and $Ms = s$. For such a matrix $M$, we have for any vector $v = (v_1, v_2, \ldots, v_n)$,

$$
\begin{aligned}
\|Mv\|^2 &= \left(\sum m_{1i}v_i\right)^2 + \left(\sum m_{2i}v_i\right)^2 + \cdots \left(\sum m_{ni}v_i\right)^2 \\
&\leq \sum m_{1i}v_i^2 + \sum m_{2i}v_i^2 + \cdots \sum m_{ni}v_i^2 \\
&= \sum v_i^2 \\
&= \|v\|^2
\end{aligned}
$$

(1.47)

(see Exercise 1.18, page 59). Then, we have

$$
\begin{aligned}
d(X \circ Y) &= \|Mq - s\| \\
&= \|M(q - s)\| \\
&\leq \|q - s\| \\
&= d(Y).
\end{aligned}
$$

Heuristically, this increased uniformity implies greater "uniformity" of the subsequences of $X \circ Y$ than of subsequences of either $X$ or $Y$.

### Additional Considerations for Combined Generators

If the streams in a combination generator suffer similar irregularities in their patterns, the combination may not be able to overcome the problems. Combining some generators can actually degrade the quality of the output of both of them.

We have mentioned the important difference between the ranges [0, 1] and (0, 1) for the simulated uniform distribution. Combining generators exposes the naive implementer to this problem anew. Whenever two generators are combined, there is a chance of obtaining a 0 or a 1 even if the output of each is over (0, 1); see Exercise 1.17.

## 1.10 Independent Streams and Parallel Random Number Generation

We often want to have independent substreams of random numbers. This may be because we want to use blocking in a simulation experiment or because we want to combine the results of multiple experiments, possibly performed on different processors. Blocks are used in simulation experiments in much the same way as in ordinary experiments: to obtain a better measure of the underlying variation. In each block or each simulation run, an independent stream of random numbers is required.

Many of the Monte Carlo methods are "embarrassingly" parallel; they consist of independent computations, at least at some level, that are then averaged. The main issue for Monte Carlo methods performed in parallel is that the individual computations (that is, all computations except the outer averaging loop) be performed independently. For the random number generators providing data for parallel Monte Carlo computations, the only requirement over and above those of any good random number generator is the requirement of independence of the random number streams.

The methods for generating independent streams of random numbers are of three types. One type is based on skipping ahead in the stream either by using different starting points or by a leapfrog method, as discussed in Section 1.2 on page 24 for linear congruential generators and in Section 1.6 on page 43 for GFSR generators. Both kinds of skipping ahead can be done randomly or at a fixed distance. Neither the congruential generator nor the generalized feedback register generator has entirely satisfactory methods of skipping ahead for use in parallel random number generation. As we have already mentioned, the congruential generator also suffers from a very short period.

Various combination generators overcome the limitations of either the simple linear congruential generator or the GFSR generator, and combination generators also can be used effectively in generating independent streams. In Section 1.10.1 below, we discuss this kind of method, which involves skipping ahead or leapfrogging in a combination generator.

The third type of method uses different generators for different substreams, as discussed in Section 1.10.2 below.

## 1.10.1 Skipping Ahead with Combination Generators

Both independent streams and long periods can be achieved by a combination generator in which a simple skip-ahead method is implemented in one of the generators, either a congruential or a GFSR generator, and the other generator is a generator with a very long period. Wollan (1992) describes such a method for $k$ processors that combines a skipping multiplicative congruential generator with a lagged Fibonacci generator by subtraction modulo 1 of the normalized output. For the skipping congruential generator based on $x_i \equiv a x_{i-1} \bmod m$, the seed $a^j x_0$ is used for the $j^{\text{th}}$ processor, and the recurrence $x_i \equiv a^k x_{i-1} \bmod m$ is used in each processor. The other generator is the same in all processors. Using a generic long-period generator $f$, as in equation (1.1), which we assume returns values in the interval $(0, 1)$, we start the $j^{\text{th}}$ stream at $w_{j(0)} = a^j z_0 \bmod$ and $y_{j(0)}, \ldots, y_{j(1-k)}$ and continue the $j^{\text{th}}$ stream as

$$
\begin{aligned}
w_{j(i)} &\equiv a^k w_{j(i-1)} \bmod m, \\
v_{j(i)} &= w_{j(i)}/m, \\
y_{j(i)} &= f(y_{j(i-1)}, \cdots, y_{j(i-k)}), \\
x_{j(i)} &\equiv (v_{j(i)} - y_{j(i)}) \bmod 1.
\end{aligned} \tag{1.48}
$$

Wollan also allowed a single process to spawn a child process. Both the parent and the child process begin using $a^2$ in place of $a$ (i.e., the multiplier for both is $a^{2k}$ when one of the original $k$ processes spawns a new process), and the child process skips from the current value $x_i$ to $a x_i$.

The properties of this combination generator depend heavily on the properties of the generator $y_i = f(y_{i-1}, \cdots, y_{i-k})$. Although it is assumed that the streams of this generator are not identical, they may overlap. The congruential generator ensures that the final streams do not overlap. If some streams of $f$ were identical, the final streams would not overlap, but they might display some of the structural regularities of the congruential generator.

## 1.10.2 Different Generators for Different Streams

The problem with use of a single generator is that the period of each of the subsequences is shorter than the period of the underlying generator. Furthermore, we must be concerned about correlations among the subsequences. Use of different generators for the different streams may be a better approach if we can devise a method for selecting different generators and if we can ensure that all of the generators are good ones.

Deng, Chan, and Yuan (1994) propose the use of multiplicative congruential generators with different multipliers but with the same modulus. The $j^{\text{th}}$ stream is generated by the recursion

$$
x_{j(i)} \equiv a_j x_{j(i-1)} \bmod m. \tag{1.49}
$$

All of the multipliers, $a_j$, are primitive elements modulo $m$, and they are selected randomly. As we saw in equation (1.20), any primitive root can be obtained as a power of another, and the power is equivalent to a skip distance in a leapfrog method. We also saw that the skip distance must be relatively prime to the period in order for the sequence formed by leapfrogging to have the same period. Deng, Chan, and Yuan (1994) recommend only primes be used as the modulo in (1.49), so the period is $m - 1$. They form the multipliers in the generator (1.49) by use of another generator that yields random skip distances that are relatively prime to $m - 1$. Such distances are generated by use of a second multiplicative congruential generator,

$$k_j \equiv bk_{j-1} \bmod (m - 1), \tag{1.50}$$

and the corresponding multipliers are generated as

$$a_j \equiv a_0^{k_j} \bmod m.$$

There are several choices to be made in this scheme. The modulus $m$ should be as large as is practical. To have a large number of different streams, $m$ should have a large number of primitive roots. This is determined by the Euler totient function. Furthermore, in order to generate a large number of exponents (skip distances) to form the primitive roots, the maximum period of a multiplicative congruential generator with modulus $m - 1$ (which is composite) should be as large as possible. This is determined by the so-called Carmichael function (see Knuth, 1998). Deng, Chan, and Yuan (1994) evaluated these properties for $m$ of the form $2^{31} - r$ for $r$ a small positive integer. They found that a value of $r = 69$ yielded a prime and that for that choice there were the largest number of primitive roots and the largest maximum period for a modulus $2^{31} - r - 1$ of any other relatively small value of $r$. For a given value of $m$, a primitive root modulo $m$ is chosen for the multiplier in the generator (1.49), the multiplier in the generator (1.50) is chosen as a primitive element modulo $m - 1$, and $k_0$ is chosen relatively prime to $m - 1$.

There are other systematic approaches for forming generators with different multipliers or for completely different generators in the different streams. Mascagni and Srinivasan (2000) describe and give C code for a library of generators that operate in parallel by choosing different generators on different processors. The library, called SPRNG, includes linear congruential generators, feedback shift register generators, and lagged Fibonacci generators. More information, including the code, is available at

    http://sprng.cs.fsu.edu

## 1.10.3  Quality of Parallel Random Number Streams

Assessing the quality of software operating in parallel can be quite difficult. This is because, for optimal use of parallel processing resources, we do not control the allocation of tasks to processors. If random number generator substreams are allowed to be utilized in computations based on the progress of the job (that is, if the process is allowed to spawn new substreams), we cannot expect strict reproducibility. It is best to control which generators are used in which process. This can be done by giving to one specific processor the task of assigning substreams or generators to processes or deciding in advance the maximum number of processors.

# 1.11   Portability of Random Number Generators

A major problem with random number generation on computers is that programs for the same generators on different computers often yield different sequences.  For programs that make use of random number generators to be portable, the generator must yield the same sequence in all computer/compiler environments. It is desirable that the generators be portable to facilitate transfer of research and development efforts. Portability reduces the number of times that the wheel is reinvented as well as the amount of computer-knowledge overhead that burdens a researcher. The user can devote attention to the research problem rather than to the extraneous details of the computer tools used to address the problem.

The heterogeneous computing environment in which most scientists work has brought an increased importance to portability of software. Formerly, portability was a concern primarily for distributors of software, for users who may be switching jobs, or for computer installations changing or contemplating changing their hardware. With the widespread availability of personal computers, all computer users now are much more likely to use (or to attempt to use) the same program on more than one machine. There are both technical and tactical reasons for using a micro and a mainframe while working on the same problem. The technical reasons include the differences in resources (memory, CPU speed, software) available on micros and mainframes. These differences likely will continue. As new and better micros are introduced and more software is developed for them, new and better supercomputers will also be developed.

The availability of the different computers in different working environments such as home, lab, and office means that using multiple computers on a single problem can make more efficient use of one's time. These tactical reasons for using multiple computers will persist, and it will become increasingly commonplace for a researcher to use more than one computer.

Use of parallel processing can introduce additional concerns of portability unless we are very careful to control which generators are used in which process. If we control the assignment of generators to processors, then the issues of portability for parallel processing are no different from those for single-thread processing.

Many programming languages and systems come with built-in random number generators. The quality of the built-in generators varies widely (see Lewis and Orav, 1989, for analyses of some of these generators). Generators in the same software system, such as `rand()` in `stdlib.h` of the C programming language, may not generate the same sequence on different machines or even in different C compilers on the same machine. It is generally better to use a random number generator from a system such as the IMSL Library, which provides portability across different platforms.

The algorithms for random number generation are not always straightforward to implement, as we have discussed on page 27. Algorithms with relatively small operands, such as in Wichmann and Hill (1982) and the alternate generator of L'Ecuyer (1988), are likely to be portable and, in fact, can even be implemented on computers with 16-bit words.

We have mentioned on page 27 several subtle problems for implementing congruential generators. Other generators have similar problems, such as the 0 and 1 problem, of which most people who have never built generators used in large-scale simulations are not aware. See Gentle (1981) for further discussion of the issue of portability.

## 1.12 Summary

In this chapter, we have discussed several methods for generating a sequence $x_1, x_2, \ldots$ that *appears* to be a realization of a simple random sample from a $U(0, 1)$ distribution (or, alternatively and equivalently for practical purposes, either a realization of a simple random sample discrete uniform distribution over the integers $1, 2, \ldots, j$, where $j$ is some large integer, or else a realization of a simple random sample from a Bernoulli distribution, which as random bits would be used as the binary representation of the integers $1, 2, \ldots, j$). By discussing so many methods, I have implied that there is a place for various random number generators. I have made some comments about some methods having various flaws, but, in general, have not provided much guidance in identifying superior methods. In Chapter 2, we will discuss various ways of evaluating random number generators, but even with a number of considerations, we will not be able to identify a single "best" method.

In most areas of numerical computing, we seek a single algorithm (or perhaps a "polyalgorithm", which with little or no user interaction selects a specific method). In the case of random number generation, however, I believe that having a variety of "good" methods is desirable.

Each of the methods for pseudorandom number generation is some variation on the recursion in equation (1.1), coupled with a transformation to the set $\mathbb{F}$:

$$x_i = f(x_{i-1}, \cdots, x_{i-k}),$$
$$u_i = g(x_i, \cdots, x_{i+1-k}).$$

The generator requires the previous $k$ numbers to get each successive one and so requires a seed that consists of $k$ values. The purpose of $g(\cdot)$ is to produce a result in $(0,1) \cap \mathbb{F}$.

In many generators, such as the class of linear congruential generators and related generators, the $x_i$ are positive integers less than some number $m$, that is, they are elements of $\mathbb{I}$, so the function $g(\cdot)$ is simply

$$g(x_i, \cdots, x_{i+1-k}) = x_i/m.$$

(Recall that this operation in the computer must be done with some care.)

In other generators, such as those using feedback shift register methods in which $x_i, \cdots, x_{i+1-k}$ are 0s and 1s, $g(\cdot)$ is a rule that forms a number in $(0,1)$ by a binary representation using some subset of the 0s and 1s.

The most widely used generators for many years have been those in which $f(\cdot)$ is a linear congruential function,

$$x_i \equiv (a^{\mathsf{T}} x_{i-1} + c) \bmod m, \quad \text{with} \quad 0 \le x_i < m,$$

where $a$ is a $k$-vector of constants, and $x_{i-1}$ is the $k$-vector $(x_{i-1}, \cdots, x_{i-k})$. In the simple linear congruential generator, $k = 1$ (and often $c = 0$, in which case it is sometimes called the "multiple congruential generator"). The simple linear congruential generator has several desirable properties, but, because of its short period and poor uniformity in dimensions greater than 1, it should be used only in combinations with other generators.

The linear congruential generators with $k > 1$ (often called "multiple recursive generators") may have longer periods and generally good uniformity properties. The fifth-order multiple recursive generator (1.29) has performed well in many empirical tests.

In many of the better generators, the sequence $x_0, x_1, \ldots$ are $j$-tuples, and the function $f(\cdot)$ is a composition of $j-1$ sources. We have discussed specific generators of this type, such as simple shuffled generators and more complicated combinations of multiple sources. The combination generators may have very long periods and very good uniformity properties. The two-source combination of third-order multiple recursive generators (1.46) has performed well in many empirical tests.

The feedback shift register methods are based on a recurrence $f(\cdot)$ of 0s and 1s of the form of equation (1.37), or equivalently of equation (1.41), in which an operation of the function $y(\cdot)$ has already been applied to convert tuples of bits to integers. Prior to converting the bits to integers, a tuple of bits may be "twisted" by a linear transformation as in equation (1.43). The Mersenne twister has a very long period and appears to have very good distributional properties.

## Exercises

1.1. (a) In equation (1.3) on page 6, show that $S = [b^{e_{\min}-p}, 1-b^{-p}] \subset (0,1)$.

(b) How many computer numbers are there in $[b^j,\ b^{j+1}]$? Does it matter what $j$ is?

(c) How many computer numbers are there in $S$?

(d) How many computer numbers are there in $(0, .5)$?

(e) How many computer numbers are there in $(.5, 1)$?

1.2. Given the random variable $X$ in equations (1.4) and (1.5), determine its exact mean and variance in terms of $b$ and $p$. Now, add the restriction $X \neq 0$, and determine its exact mean and variance.

1.3. Use equations (1.6) and (1.7) to evaluate $\phi(160)$.

1.4. Modular reduction and uniform distributions.

(a) Let $Y$ be a random variable with a U(0, 1) distribution, let $k$ be a nonzero integer constant, and let $c$ be a real constant. Let

$$X \equiv (kY + c) \bmod 1, \text{ with } 0 \leq X \leq 1.$$

Show that $X$ has a U(0, 1) distribution. *Hint:* First, let $c = 0$ and consider $kY$; then, consider $T + c$, where $T$ is from U(0, 1).

(b) Prove a generalization of Exercise 1.4a in which the constant $c$ is replaced by a random variable with any distribution.

(c) Let $T$ be a random variable with a discrete uniform distribution with mass points $0, 1, \ldots, d - 1$. Let $W_1, W_2, \ldots, W_n$ be independently distributed as discrete uniform random variables with integers as mass points. Show that

$$T + \sum_{i=1}^{n} W_i \bmod d$$

has the same distribution as $T$. (The reduced modulus is used in this expression, of course.) *Hint:* First, consider $T + W_1$, and write $(T+W_1) \bmod d$ as $j+kd$, where $j$ and $k$ are integers with $0 \leq j \leq d-1$ (see also L'Ecuyer, 1988).

1.5. Use Fortran, C, or some other programming system to write a program to implement a generator using a multiplicative congruential method with $m = 2^{13} - 1$ and $a = 17$. Generate 500 numbers $x_i$. Compute the correlation of the pairs of successive numbers $x_{i+1}$ and $x_i$. Plot the pairs. On how many lines do the points lie? Now, let $a = 85$. Generate 500 numbers, compute the correlation of the pairs, and plot them. Now, look at the pairs $x_{i+2}$ and $x_i$. Compute their correlation.

1.6. Now, modify your program from Exercise 1.5 to implement a matrix congruential method

$$
\begin{bmatrix} x_{1i} \\ x_{2i} \end{bmatrix} \equiv \begin{bmatrix} a_{11} & a_{12} \\ a_{21} & a_{22} \end{bmatrix} \begin{bmatrix} x_{1,i-1} \\ x_{2,i-1} \end{bmatrix} \bmod m
$$

with $m = 2^{13} - 1$, $a_{11} = 17$, $a_{22} = 85$, and $a_{12}$ and $a_{21}$ variable. Letting $a_{12}$ and $a_{21}$ vary between 0 and 17, generate 500 vectors and compute their sample variance-covariances. Are the variances of the two elements in your vectors constant? Explain. What about the covariances? Can you see any relationship between the covariances and $a_{12}$ and $a_{21}$? Is there any reason to vary $a_{12}$ and $a_{21}$ separately? Can a lower triangular matrix (that is, one with $a_{21} = 0$) provide all of the flexibility of matrices with varying values of $a_{21}$?

1.7. Write a Fortran or C function to implement the multiplicative congruential generator (1.12) (RANDU) on page 19.

(a) Generate a sequence $x_i$ of length 20,002. For all triplets in your sequence, $(x_i, x_{i+1}, x_{i+2})$, in which $0.5 \le x_{i+1} \le 0.51$, plot $x_i$ versus $x_{i+2}$. Comment on the pattern of your scatterplot. (This is similar to the graphical analysis performed by Lewis and Orav, 1989.)

(b) Generate a sequence of length 1002. Use a program that plots points in three dimensions and rotates the axes to rotate the points until the 15 planes are clearly visible. (A program that could be used for this is the S-Plus function spin, for example.)

1.8. Using an analysis similar to that leading to equation (1.13) on page 19, determine the maximum number of different planes on which triplets from the generator (1.12) would lie if instead of 65 539, the multiplier were 65 541. Determine the number of different planes if the multiplier were 65 533. (Notice that both of these multipliers are congruent to 5 mod 8, as James, 1990, suggested.)

1.9. Write a Fortran or C function to use a multiplicative congruential method with $m = 2^{31} - 1$ and $a = 16\,807$ (the "minimal standard").

1.10. Write a Fortran or C function to use a multiplicative congruential method with $m = 2^{31} - 1$ and $a = 950\,706\,376$. Test your program for correctness (not for statistical quality) by using a seed of 1 and generating ten numbers. (This is a multiplier found by Fishman and Moore (1982, 1986) to be very good for use with the given modulus.)

1.11. Suppose that a sequence is generated using a linear congruential generator with modulus $m$ beginning with the seed $x_0$. Show that this sequence and the sequence generated with the seed $m - x_0$ are antithetic. *Hint:* Use induction.

1.12. Suppose that a sequence is generated by $x_{i+1} \equiv ax_i \bmod m$ and that a second sequence is generated by $y_{i+1} \equiv by_i \bmod m$, where

$$b \equiv a^{c-1} \bmod m$$

and $c$ is the period. Prove that the sequences are in reverse order.

1.13. For the generator $x_i \equiv 16\,807x_{i-1}\bmod(2^{31} - 1)$, determine the value $x_0$ that will yield the largest possible value for $x_1$. (This seed can be used as a test that the largest value yielded by the generator is less than 1. It is desirable to scale all numbers into the *open* interval $(0, 1)$ because the numbers from the uniform generator may be used in an inverse CDF method for a distribution with an infinite range. To ensure that this is the case, the value used for scaling must be greater than $2^{31} - 1$; see Gentle, 1990.)

1.14. In this exercise, use a package that supports computations for number theory, such as Maple.

   (a) Check that the multipliers in Exercises 1.5, 1.8, and 1.13 are all primitive roots of the respective moduli.

   (b) Determine all of the primitive roots of the modulus $2^{13} - 1$ of Exercise 1.5.

1.15. Suppose that for one stream from a given linear congruential generator the seed is 500 and that for another stream from the same generator the seed is 1000. What is the approximate correlation between the two streams?

1.16. Write a program to implement the cellular automaton (1.44) as a random number generator producing 32-bit strings. Seed the generator with a linear congruential generator, and generate 1000 32-bit numbers. Compute the average Hamming distance between successive numbers. Now, form integers from the strings in the stream, and plot overlapping pairs as in Figure 1.3. Do you see any evidence of structure?

1.17. Consider the Wichmann/Hill random number generator (page 47). Because the moduli are relatively prime, the generator cannot yield an exact zero. Could a computer implementation of this generator yield a zero? First, consider a computer that has only two digits of precision. The answer is obvious. Now, consider a computer with a more realistic number system (such as whatever computer you use most often). How likely is the generator to yield a 0 on this computer? Perform some computations to explore the possibilities. Can you make a simple adjustment to the generator to prevent a 0 from occurring?

1.18. Prove the inequality occurring in the array of equations (1.47) on page 50. *Hint:* Define an appropriate random variable $W$, and then use the fact that $(\mathrm{E}(W))^2 \leq \mathrm{E}(W^2)$.

1.19. Write pseudocode for the following random number generators for parallel processing computers. Be sure that your generators preserve independence of the separate streams. You can assume that the maximum number of processors, say $k$, is provided by the user. (What else does the user provide?)

(a) Write a program implementing the method of Wollan (1.48) using the "minimal standard" generator to produce independent streams and the cellular automaton (1.44) as the long-period generator.

(b) Write a program implementing the method of Deng, Chan, and Yuan (1994) using a modulus of $2^{31} - 1$.

(c) Test and compare your generators. Write a brief description comparing and contrasting the two methods (not just based on empirical results of using your generators but considering other issues such as user interface, quality concerns, etc.).

# Chapter 2

# Quality of Random Number Generators

Many Monte Carlo applications routinely require orders of $10^{15}$ or more random numbers and may use months of computing time on several workstations running simultaneously. This kind of research work that depends so heavily on random numbers emphasizes the need for high-quality random number generators.

An initial consideration for any generator is its period. As the number and complexity of Monte Carlo studies increase, our requirement for the period of the generator increases. The minimal standard generator of Park and Miller (1988) (see page 20) has a period of $2^{31} - 1$. As Deng and Lin (2000) argue, this is no longer sufficient for most serious work. Aside from issues of very large simulations, we must be concerned about the inability of a random number generator with a small period to visit all points in a sample space for a very simple problem such as generating a random sample from a relatively small population (see Exercise 6.2).

Only random number generators that have solid theoretical properties should even be considered. In addition, statistical tests should be performed on samples generated, and only generators whose output has successfully passed a battery of statistical tests should be used. Even so, often in Monte Carlo applications it is appropriate to construct ad hoc tests that are sensitive to departures from distributional properties that are important in the given application. For example, in using Monte Carlo methods to evaluate a one-dimensional integral, autocorrelations of order one may not be as harmful as they would be in evaluating a two-dimensional integral.

In many applications, of course, random numbers from other distributions, such as the normal or exponential, are needed. (We discuss methods for transforming the basic sequence of uniform random numbers into sequences from other distributions in Chapters 4 and 5.) If the routines for generating random deviates from nonuniform distributions use exact transformations from a uniform distribution, their quality depends almost solely on the quality of the underlying uniform generator. Because we can never fully assess that quality, however, it is often advisable to employ an ad hoc goodness-of-fit test for the transformations that are to be applied.

It is unlikely that a single generator or even a single class of generators should be chosen as the generator to be used in all applications. Even in a given application, it is a good idea to use more than one generator and, for each, to use more than one seed. The different generators and different seeds provide classification models for a Monte Carlo experiment. Standard analysis-of-variance methods can be used to ascertain agreement among the blocks of the experiment (see page 238). The blocks of the Monte Carlo experiment should not exhibit statistically significant differences. If they do, then what? This is the basic dilemma in testing random number generators. It may be wise to replicate the experiment with different random number generators. Spuriously significant results occur even when the distributional assumptions for a data-generating process hold, so the practice of just starting over introduces bias in statistical inference. In Monte Carlo studies, however, we must always be alert to the possibility that the distributional assumptions do not hold. In some cases, all that should be done is just to extend the Monte Carlo study, keeping new data separate from the old in order to continue the assessment of the quality of the pseudorandom numbers by comparisons of results.

In the paradigm of statistical hypothesis testing, but in a manner different from other applications of hypothesis testing, it is the null hypothesis that we hope to accept.

## 2.1   Properties of Random Numbers

The objective of a uniform random number generator is to produce samples of any given size that are indistinguishable from samples of the same size from a $U(0, 1)$ distribution. In evaluating a random number generator, therefore, we compare various properties of samples from the generator with the corresponding properties of samples from a $U(0, 1)$ distribution.

### One-Dimensional Uniform Frequencies

The most obvious properties for comparison are the relative frequencies in corresponding sets. What proportion of the output from a random number generator is in each of the intervals $(0.0, 0.1], (0.1, 0.2], \ldots, (0.9, 1.0)$, for example? Are these close enough to the expected proportion of 0.10 to be acceptable? In general, if we have a total of $n$ generated points, for any set of intervals with $n_i$ points occurring in the $i^{\text{th}}$ interval, the question is whether the observed relative frequencies, $f_i = n_i/n$, are close enough to the expected proportions, $\pi_i$, say.

**Independence and Its Consequences**

Another kind of property to study is the apparent mutual independence of the variates in the generated stream. There are various ways of approaching independence. One is by computing correlations between variates at fixed distances apart in the stream (that is, autocorrelations of fixed lags). Variates should exhibit zero autocorrelations at any length of lag. Zero autocorrelations do not imply independence; rather, they are a consequence of it. Nonzero autocorrelations imply lack of independence.

Independence of the variates results in other distributional relationships among variates in the sequence (for example, the numbers and lengths of substreams that are monotonically increasing or decreasing). Such substreams are called "runs". The generator should produce streams with runs that correspond with what is expected from a $U(0, 1)$ data-generating process.

Order statistics should behave similarly to their expectations. In any given sample of size $n$ from the generator, the $i^{\text{th}}$ order statistic should stochastically appear to be a beta random variable with parameters $i$ and $n - i + 1$ (see page 221). The differences in successive order statistics should stochastically appear to be ratios of exponential random variables.

### *d*-Variate Uniformity

Any of the independence properties of samples from a $U(0, 1)$ distribution are equivalent to frequency properties of some multivariate random variable that has uniform marginals over $(0, 1)$ and a diagonal variance-covariance matrix. The uniformity properties over various dimensions imply the other properties.

A simple, but useful, approach to form putative $d$-variate uniform random variables is to use successive substreams of length $d$. This is what is done in Chapter 1 for two dimensions and shown in Figures 1.3 and 1.5 (in those cases, for a discrete uniform distribution rather than for $U(0, 1)$). These multivariate numbers are then compared with properties expected from the true multivariate uniform distribution. In particular, the observed frequencies of the $d$-vectors within equal-sized regions are compared with the constant frequency expected from a $d$-variate uniform distribution. The sizes of the regions can be arbitrarily small, within the limits imposed by the finite computer arithmetic, of course. If the frequency is as expected, we say that the random number generator, or a stream that it produces, has the "$d$-variate uniformity" property. A stream with this property is also said to be "equidistributed in $d$ dimensions".

Any of the specific properties mentioned above could be assessed by testing the $d$-variate uniformity for $d = 1, 2, \ldots$. It is useful, nevertheless, to test for some of these properties specifically. A specific test is likely to be more powerful than the simple tests of $d$-variate uniformity. For example, as we saw in the discussion beginning on page 14, the distances between points in a stream generated by a linear congruential generator are not likely to be as we would expect from a $d$-variate uniform distribution, so we might devise a specific test for this property. L'Ecuyer (1998) reported that none of the otherwise "good" linear congruential generators that he tested performed satisfactorily with regard to the distribution of the nearest pairs of points.

## 2.2  Measures of Lack of Fit

The quality of a random number generator depends on how closely the properties of the output of the generator match the properties of an independent and stationary uniform data-generating process. To assess the quality, we need quantitative measures of differences of the properties of the output stream and an ideal stream. We call these measures of lack of fit, or, equivalently, of goodness of fit.

Some measures of lack of fit, or at least bounds for the measure, can be computed based on analysis of the algorithm used by the generator. For other measures, it is necessary to produce a stream of numbers from the generator and compute the measure based on that particular sample. To decide whether the measure is acceptable in that case requires statistical testing under the assumption that the pseudorandom stream is a realization of a random data-generating process.

### 2.2.1  Measures Based on the Lattice Structure

For a generator that yields $k$ different values, $d$-tuples of successive numbers in the output would ideally fall on $k^d$ points uniformly spaced through a $d$-dimensional hypercube. As we saw in Section 1.2.1, however, $d$-tuples of the output from pseudorandom number generators lie on somewhat restricted lattices in a $d$-dimensional hypercube.

As we mentioned in Section 1.2.2, it is best to avoid using a significant proportion of the full period of a generator, so both the lattice of the full sequence of numbers from a generator and how subsequences are distributed over the lattice are of interest.

It is instructive again to consider Figures 1.3 and 1.5 (pages 16 and 17), which show the lattices of the output of the two small generators. In Figure 1.3, not only is the lattice structure bad (all of the points are on just three lines that are far apart), but subsequences of points are not distributed well over the lattice. Notice that in the subsequence of the first five points (formed by overlapping numbers in the sequence), two pairs of points (1 and 3, and 4 and 5) are very close to each other. In Figure 1.5, on the other hand, both the full lattice and the first five points are more evenly distributed.

In assessing the lattice structure of the output from a random number generator, we may consider the full lattice, in which case the distance between adjacent hyperplanes is of interest, or we may consider a subsequence of the full period of the generator, in which case the distance between points that are close together is of interest.

## Distances Between Adjacent Hyperplanes

As we saw heuristically in Section 1.2, the smaller the maximum distance between the lines (or, in general, hyperplanes) determined by lattices such as those shown in Figures 1.3 and 1.5, the better the output of the generator covers the hypercube. Although it is not always easy to identify the lines or hyperplanes as we did in Section 1.2, in some cases, the maximum distance can be computed directly from the specification of the algorithm. The *spectral test* determines the maximum distance between adjacent parallel hyperplanes defined by the lattice.

The spectral test for a linear congruential generator does not use the output of the generator but rather computes values analogous to a discrete Fourier transform of the output using the recurrence formula of the congruential relationship itself.

For a linear congruential generator with multiplier $a$ and modulus $m$, let $\delta_d(m, a)$ represent the maximum distance between adjacent parallel hyperplanes in the $d$-dimensional lattice. Now, for any $d$-dimensional lattice with $m$ points, there is a lower bound, $\delta_d^*(m)$, on $\delta_d(m, a)$ (see Knuth, 1998); hence, a useful measure for a given multiplier for use with the modulus $m$ is the ratio

$$S_d(m, a) = \frac{\delta_d^*(m)}{\delta_d(m, a)}. \tag{2.1}$$

The closer this ratio is to 1, the better the generator is with respect to this measure.

Coveyou and MacPherson (1967) describe a spectral test method that computes the denominator in equation (2.1). Their method was improved by Dieter (1975) and Knuth (1998) and is described in detail on pages 98 through 103 of the latter reference. Hopkins (1983) gives a computer program to implement the Coveyou/MacPherson test for a full-period linear congruential generator. (Note that updated versions of the 1983 program are available.)

L'Ecuyer (1988) computes values of $S_d(2^{31} - 1, a)$ for some specific values of $d$ and $a$. For example, for the multiplier 16 807 (see Section 1.2), he gives

| $d$ | 2 | 3 | 4 | 5 | 6 |
|---|---|---|---|---|---|
| $S_d(2^{31} - 1,\ 16\,807)$ | .34 | .44 | .58 | .74 | .65 |

and for the multiplier 950 706 376 (see Exercise 1.10, page 58), he gives

| $d$ | 2 | 3 | 4 | 5 | 6 |
|---|---|---|---|---|---|
| $S_d(2^{31} - 1,\ 950\,706\,376)$ | .86 | .90 | .87 | .83 | .83 |

Based on $S_d$, the latter value of the multiplier appears to be better. The empirical studies of Fishman and Moore (1982, 1986) also found this value to be better for use with the given modulus.

**Lengths of Minkowski Reduced Basis Vectors**

Beyer, Roof, and Williamson (1971) and Beyer (1972) describe a measure of the uniformity of the lattice of the output of a random number generator in terms of the ratio of the shortest vector to the longest vector in the Minkowski reduced basis of the lattice. A basis for a lattice in $\mathbb{R}^d$ is a linearly independent generating set of the lattice. Given a set of linearly independent vectors $\{v_1, v_2, \ldots, v_d\}$ in $\mathbb{R}^d$, a lattice is the set of vectors $w$ of the form $\sum_{i=1}^d z_i v_i$, where $z_i$ are integers (see Figure 1.1, page 15). The set of vectors $\{v_i\}$ is a basis for the lattice. The basis is a Minkowski reduced basis if

$$\|v_k\| \leq \left\| \sum_{i=1}^d z_i v_i \right\|, \quad \text{for } 1 \leq k \leq d,$$

for all sets of $d$ integers $z_i$ such that the greatest common divisor of the set $\{z_k, z_{k+1}, \ldots, z_d\}$ is 1. (The symbol $\|v\|$ denotes the Euclidean norm of the vector $v$.)

The ratio of the length of the shortest vector to the longest length,

$$\frac{\|v_d\|}{\|v_1\|}, \tag{2.2}$$

is called the *Beyer ratio*. A large Beyer ratio (close to 1) indicates good uniformity.

It is not easy to compute the Minkowski reduced basis. (In fact, the problem has been shown to be NP-complete.) Afflerbach and Grothe (1985) give an algorithm for computing the basis and the Beyer ratio for linear congruential generators. They used the algorithm up to dimension 20. Eichenauer-Herrmann and Grothe (1990) give upper bounds for the ratio for linear congruential generators for higher dimensions.

**Other Measures on the Full Lattice**

There are various ways that we can assess the uniformity of coverage of the lattice. Basically, the closer the lattice is to a uniform grid of rotated hypercubes, the more uniform is its coverage.

Another assessment of the badness of the lattice related to the Beyer ratio is the *lattice test* of Marsaglia (1972a), which is the ratio of the longest edge of the smallest parallelepiped in $d$ dimensions to that of the shortest edge. Atkinson (1980) implemented Marsaglia's lattice test and illustrated its use on some generators. Eichenauer and Niederreiter (1988), however, give examples that show that this lattice test fails to detect some generators with poor lattice structure.

**Distances Between Individual Points**

Because in any application we do not use the full sequence of the generator, we may be interested in properties of the subset of lattice points in the subsequence used. The nearest point test described by L'Ecuyer, Cordeau, and Simard (2000) is very sensitive to certain departures from randomness. See their paper for more details.

## 2.2.2 Differences in Frequencies and Probabilities

One of the most important properties of a random number generator is the relative frequencies $f_i$ of the output stream over various regions. As we mentioned earlier, if the frequencies match those of a $U(0,1)$ distribution in all dimensions (that is, if the generator has $d$-variate uniformity for all $d$), then the other properties of the uniform distribution follow. An important measure of the quality of a random number generator, therefore, is how closely the frequencies over various regions match the probabilities $\pi_i$ of those same regions under a $U(0,1)$ distribution. Rather than comparing $f_i$ and $\pi_i$ over discrete regions, we may compare their densities, $p_n(x)$ and $p(x)$, over continuous domains. In the familiar chi-squared tests, we use discrete regions; in the Kolmogorov–Smirnov and Anderson–Darling tests, we use continuous domains.

There are two major issues here. One is whether to use discrete regions and, if so, how to form the regions, and the other is how to compare a set of observed frequencies with probabilities that correspond to the $U(0,1)$ distribution. The most obvious way of forming discrete regions is, for $d = 1, 2, \ldots$, to form a rectangular grid over the $d$-dimensional unit hypercube. If each axis is divided into $k$ intervals, the number of cells in the grid is $k^d$. This approach for forming regions is quickly swamped by the curse of dimensionality. There are many ways of transforming the output of the random number generator, forming regions on the transformed values, and determining the probabilities of those regions for random variables formed by the same transformations applied to a random sample from the $U(0,1)$ distribution.

The goodness-of-fit tests discussed in Section 2.3.1 use regions formed in a variety of ways.

The next major issue is how to compare a set of relative frequencies $f_i$ with a set of probabilities $\pi_i$. There are two general approaches. One is to form a summary number from all of the relative frequencies and compare that with the summary number from all of the probabilities:

$$s(f_1, f_2, \ldots) - s(\pi_1, \pi_2, \ldots).$$

One such summary number is called "entropy", which we will discuss below.

Another approach is to measure differences in the individual relative frequencies and probabilities by some function $\delta(f_i, \pi_i)$ and then to compute a summary number from that:

$$g(\delta(f_1, \pi_1), \delta(f_2, \pi_2), \ldots).$$

Many familiar statistical goodness-of-fit tests, such as the chi-squared test and the Kolmogorov–Smirvov test for discrete distributions, are of this form. For the chi-squared statistic, the regions are nonoverlapping, $\delta(f_i, \pi_i) = n(f_i - \pi_i)^2 / \pi_i$, and $g(\cdot)$ is just a simple sum. For the Kolmogorov–Smirnov statistic, the regions form a nested sequence, $\delta(f_i, \pi_i) = |f_i - \pi_i|$, and $g(\cdot)$ is the maximum. (The more familiar Kolmogorov–Smirnov test over continuous domains uses $|P_n(x) - P(x)|$, and $g(\cdot)$ is a supremum.)

### Cumulative Relative Frequencies and Probabilities

For comparing relative frequencies and probabilities over continuous domains, we compare functions. One of the most fundamental functions characterizing a probability distribution is the cumulative distribution function, or CDF, which for the random variable $X$ we denote as $P_X$, or often just as $P$, and define as

$$P(x) = \Pr(X \le x),$$

where "Pr" represents "probability". At all points where $P$ is differentiable, its derivative is the density $p(x)$ mentioned above.

The corresponding function for a given sample of size $n$ is called the empirical cumulative distribution function, or ECDF, and is denoted as $P_n(x)$. For a sample of points $x_1, \ldots, x_n$, the ECDF is

$$P_n(x) = \frac{1}{n} \sum_{i=1}^{n} I_{(-\infty, x]}(x_i) \tag{2.3}$$

for the indicator function $I_{(-\infty, x]}(\cdot)$. We also use "$P_n(\cdot)$" to denote a similar function whose argument is a set:

$$P_n(S) = \frac{1}{n} \sum_{i=1}^{n} I_S(x_i).$$

In the context of our previous discussion, the regions of interest for the CDF and ECDF are the orthants defined by a given $d$-vector $x_0$ as $\{x : x \le x_0\}$.

The CDF and ECDF are much easier to visualize and understand in the one-dimensional case. As we see below, one-dimensional CDFs and ECDFs have applications even when the regions are multidimensional.

### Differences in Summary Measures: Entropy

An important property of a data-generating process yielding $k$ discrete values with discrete probabilities $\pi_1, \ldots, \pi_k$ is its *entropy*, defined as

$$h = - \sum_{i=1}^{k} \pi_i \log \pi_i. \tag{2.4}$$

Letting the $\pi_i$ correspond to the probabilities of various sets in the outcome space of a $U(0, 1)$ distribution, we can compare the entropy of the stream produced by the random number generator with the entropy of variables from the

U(0, 1). The simplest sets to consider are equal-length subintervals of (0, 1), in which case all of the $\pi_i$s are just equal to the length of the subintervals. (The maximum entropy for any given $k$ is attained when the probabilities are equal.)

We can also define entropy for a set of points in terms of relative frequencies in a disjoint covering class of subsets. Consider a set of points $x_1, \ldots, x_n$ and a collection of sets $S_1, \ldots, S_k$. Let $n_i = \#\{x_j : x_j \in S_i\}$. If all $n_i > 0$, the entropy of the set of points $x_j$ with respect to the collection of sets $S_i$ can be defined analogously to equation (2.4) by using the relative frequencies $f_i = n_i/n$ in place of $\pi_i$. A different collection of sets would yield a different value of the entropy.

For any set of points or for any nondegenerate probability distribution, various collections of sets can be defined, and the entropy of the points and the entropy of the probability distribution can be defined with respect to each collection of sets. The differences in the entropy or just the differences in the relative frequencies and the probabilities provide a measure of the lack of fit of the point set to the expectations of the probability distribution with respect to that collection of sets. An overall measure of the lack of fit would have to be based on various collections of sets.

### Sum of Normed Squared Individual Differences: Chi-Squared

The chi-squared statistic is one of the most commonly used measures of differences between observed relative frequencies $f_i$ and hypothesized probabilities $\pi_i$. For $k$ regions, the statistic is

$$\chi_c^2 = n \sum_{i=1}^{k} \frac{(f_i - \pi_i)^2}{\pi_i}. \tag{2.5}$$

This measure is heuristically appealing. Under certain restrictive assumptions about the random sample yielding the $f_i$, the statistic has a chi-squared distribution. Under much less restrictive assumptions, which we can usually reasonably assume are satisfied, the statistic has an asymptotic chi-squared distribution.

### Maximum Individual Difference: Discrepancy

This measure is called *discrepancy*, and it is defined in terms of a measure of the extreme deviation of the relative frequencies of a given set of points from the entropy for a uniform distribution for a collection of intervals. The discrepancy of a set of points $S = \{x_1, x_2, \ldots, x_n\}$ in the unit hypercube is

$$D_n(S) = \sup_{J} |F_n(J) - V(J)|, \tag{2.6}$$

where the supremum is taken over all subintervals $J$ of the unit hypercube, $F_n(J)$ is the number of points in $S \cap J$ divided by $n$, and $V(J)$ is the volume of $J$. The discrepancy is a measure of the uniformity of spread of a given set of points within another set of points.

For a given generator of points, we may consider the discrepancy of arbitrary sets of points within a $d$-dimensional unit hypercube. In that case, we use the notation $D^{(d)}$.

## 2.2.3   Independence

Just as important as the correspondence between observed frequencies and theoretical probabilities is the lack of systematic patterns in the sequence of variates. The pseudorandom numbers should appear to be a realization of *independent and identically distributed* (i.i.d.) random variables. The simplest measures that may indicate lack of independence are correlations or covariances between the variates at fixed lags in the generated sequence:

$$a_k = \text{Cov}(x_i, x_{i-k}).$$

Zero correlations or covariances do not imply independence, of course, but they are necessary conditions for independence.

The i.i.d. condition is equivalent to the equality of probabilities of *all* transformations in all dimensions, so the methods of Section 2.2.2 can also be used to assess independence by forming various types of regions for comparisons.

### Long-Range Autocorrelations

In general, for most generators, the autocorrelations die out very early (that is, they become close to zero at very small lags), and then the autocorrelations grow again at large lags. At a lag equal to the period, the correlation becomes 1, of course. For some generators, the correlation may become relatively large at some intermediate lags.

In most cases, the autocorrelations are not large enough to be statistically significant for samples of reasonable size if the generated sequence were assumed to arise from a random (instead of pseudorandom) process. The problem in statistical assessment of long-range autocorrelations is the sample size required for any meaningful tests. Absent assumptions about short-range autocorrelations, sequences long enough to compute long-lag autocorrelations from a given starting point are necessary. The fact that the autocorrelations are not statistically significant does not mean that these systematic autocorrelations would not be so bad as to invalidate a Monte Carlo study. More discussion of the general problem of long-range autocorrelations in sequences from multiplicative congruential generators is provided by Eichenauer-Herrmann and Grothe (1989) and by De Matteis and Pagnutti (1990).

## 2.3 Empirical Assessments

In addition to analyzing the algorithm itself, we may evaluate a specific generator by analyzing the output stream that it produces. This approach has the additional advantage of assessing the program that is actually used, so the question of whether the program correctly implements the algorithm does not arise. There are basically two general kinds of properties of a generator to assess: the elements in the *set* of deviates without regard to the order in which they were generated (that is, the "static" properties) and the patterns in the *sequence* of the deviates (that is, the "dynamic" properties).

An important source of information about the quality of a generator is specific simulations. Whatever is known about the physics of a problem that is simulated can be compared with the results of a simulation. It was this kind of study that led to the discovery of problems with RANDU (see Coldwell, 1974).

Empirical tests of random number generators include both statistical tests of the output stream and the anecdotal evidence of specific simulations.

### 2.3.1 Statistical Goodness-of-Fit Tests

In statistical tests of random number generators, the null hypothesis is either that the elements of a given sequence of real numbers are independently and identically distributed as $U(0, 1)$ or else that the elements of a given sequence of bits are independently and identically distributed as Bernoulli with parameter $1/2$. The alternative hypothesis is that the sequence does not have that distribution.

This alternative is uncountably composite. This breadth of possibilities means that there cannot be a uniformly most powerful test. We need a suite of statistical tests. Even so, of course, not all alternatives can be addressed.

#### Statistical Tests for Making Decisions

Another problem with the use of statistical goodness-of-fit tests is that the paradigm of statistical testing is not appropriate for this application. In most applications of statistical hypothesis tests, we set up a strawman null hypothesis. A "positive result" is to reject this null hypothesis. We can describe our decision in terms of a fixed $\alpha$ level that controls the probability of a type I error or else similarly determine a p-value for the test procedure. In testing random number generators, we hope for a "negative result"; that is, we do not want to reject the null hypothesis.

When many tests are performed, however, too many negative results is evidence that the generated variates are not independent realizations from the hypothesized distribution. This observation gives rise to a second-order test: perform a selected goodness-of-fit test many times and then perform a goodness-of-fit test on either the sample of computed test statistics or on the sample p-values. We could also perform third-order tests, and so on, ad infinitum.

Due to the nature of the alternative hypothesis (that is, the ways in which the null hypothesis may be false), second-order tests may be more powerful for some alternatives.

**Test Statistics and Their Distributions**

Any of the measures discussed in Section 2.2, as well as a number of other statistics, could be used to develop a statistical goodness-of-fit test. In any goodness-of-fit tests, we must decide what is an extreme value; that is, what is the stronger evidence that the null hypothesis is not true. Because there are many ways in which the null hypothesis may not be true, some thought should be given to the type of test statistic that may be sensitive to various alternative hypotheses.

The first requirement is that we know the distribution of the test statistic under the null hypothesis. When a test is applied multiple times in the case where the *null hypothesis is true*, the ECDF of the test statistic should correspond to the cumulative CDF of the test statistic in the null case. Figure 2.1 shows a plot of an ECDF and the corresponding ECDF for a test statistic. (This plot is for a chi-squared test applied to the "minimal standard" generator.)

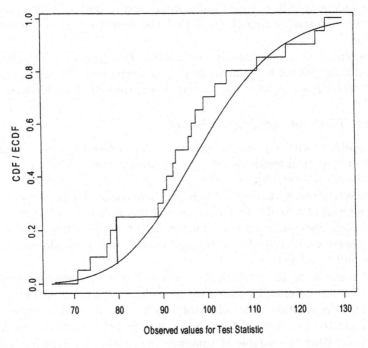

Figure 2.1: ECDF and Null CDF of a Test Statistic

Another way of comparing a set of observed values of a test statistic with what would be expected if the null hypothesis is true is to use a q–q plot of the observed values with the null distribution of the test statistic, as in Figure 2.2. In a q–q plot, the vertical axis is scaled by the CDF of the null distribution. The plot in Figure 2.2 is for the same chi-squared test statistics as in Figure 2.1. Note that the large gap in the quantiles occurs at the point at which the CDF and ECDF are farthest apart.

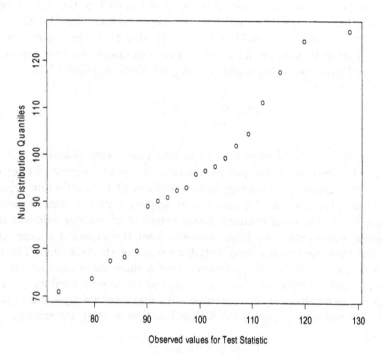

Figure 2.2: q–q Plot of Chi-Squared Values

As a second-order test, we could test for whether the observed values of the test statistic conform to the expected values under the null hypothesis. (Using a Kolmogorov–Smirnov test, for example, we get a test statistic of 0.1746, corresponding to the maximum separation in the graphs in Figure 2.1. The p-value for such a Kolmogorov–Smirnov test statistic is 0.5197.) Under the null hypothesis, the p-value has a $U(0, 1)$ distribution; hence, we could also (and equivalently) do a goodness-of-fit test on the observed p-values.

The statistical tests for random number generators are usually goodness-of-fit tests for various combinations and transformations of the numbers in the output stream. These tests are of infinite variety. We discuss some of these tests below. Some tests, such as chi-squared tests and Kolmogorov–Smirnov tests, apply directly to the distribution of a random variable and the sample. Other tests can be based on chi-squared and Kolmogorov–Smirnov tests for various transformations of the sample.

## Chi-Squared Goodness-of-Fit Tests

In a chi-squared goodness-of-fit test, observed counts of events are compared with counts of occurrences of those events that would be expected under the null hypothesis to be tested. If the null hypothesis, for example, is that a random variable has a uniform $(0,1)$ distribution, we may take 100 observations on the random variable, count how many are in each of the ten intervals $(0, 0.1]$, $(0.1, 0.2]$, $\ldots$, $(0.9, 1.0)$, and compare those counts with the expected numbers, which in this case would be 10 in each interval. If the observed numbers are significantly different from the expected numbers, we have reason to reject the null hypothesis. Formally, we compute, from equation (2.5),

$$\chi_c^2 = \sum_{i=1}^{k} \frac{(o_i - e_i)^2}{e_i},$$

where $k$ is the number of intervals (10 in this case), and $o_i$ and $e_i$ are, respectively, the observed and expected numbers in the $i^{\text{th}}$ interval. Under the hypothesis of independent sampling from a uniform $(0,1)$ distribution, $\chi_c^2$ is a realization of a random variable that has an approximate chi-squared distribution with $k - 1$ degrees of freedom. Large values of $\chi_c^2$ provide evidence that the observed counts differ by large amounts from the expected counts; that is, evidence that the hypothesized distribution is not the true one. The test is performed by computing the probability that a chi-squared random variable with $k - 1$ degrees of freedom would be as large as the observed value, $\chi_c^2$. This probability, the p-value, can be computed with many software packages. If we let x take the value of $\chi_c^2$, using the S-Plus function pchisq, for example, the probability is

$$1 - \text{pchisq}(\text{x}, \text{k} - 1);$$

or, using the IMSL Fortran function chidf, the probability is

$$1 - \text{chidf}(\text{x}, \text{k} - 1).$$

If the p-value is smaller than some preset value, say 0.05, the null hypothesis is rejected.

## Kolmogorov–Smirnov Tests

A Kolmogorov–Smirnov test (or just "Kolmogorov test") compares the empirical cumulative distribution function, or ECDF, with the hypothesized cumulative distribution function, or CDF, $P(x)$.

The comparison of the two functions used by the Kolmogorov–Smirnov test is based on

$$\sup_x |P_n(x) - P(x)|, \qquad (2.7)$$

which is the *Kolmogorov distance* between two probability distributions. (In this case, one of the probability distributions is the discrete uniform distribution with probability $1/n$ at each of the points in the sample.) The discrepancy in equation (2.6) on page 69 is similar to the Kolmogorov distance except that differences are not accumulated. In discrepancy, differences in individual cells are considered. All possible cells are formed, however.

An easy statistical test for a basic random number generator is just a goodness-of-fit test for uniformity in various dimensions. A test for uniformity in one dimension is just based on a one-dimensional histogram. For higher dimensions, the bins are usually made to correspond to groups formed by subsequences of the output of the generator; for example, in two dimensions, each successive pair of numbers constitutes a single bivariate observation. The discrepancy in equation (2.6) is a multivariate Kolmogorov distance. There are obviously other ways of forming multivariate observations by grouping variates in a univariate sequence.

Goodness-of-fit tests can also be performed over subregions of the support of the distribution. This may be particularly useful when the hypothesized distribution has infinite support. The tail behavior is often not fit well.

S-Plus and R provide a function `ks.gof` and the IMSL Fortran Library provides a subroutine `ksone` that compute the Kolmogorov–Smirnov statistic and determine its p-value.

### Other Tests Based on $|P_n(x) - P(x)|$

Other relatively powerful goodness-of-fit tests can be developed based on the differences in $P_n(x)$ and $P(x)$. A class of such tests is based on the weighted, integrated quadratic difference

$$\int_{-\infty}^{\infty} (P_n(x) - P(x))^2 w(x) dP(x). \qquad (2.8)$$

When $w(x) = 1$, this is the Cramér–von Mises statistic, often denoted as $W^2$. For

$$w(x) = \frac{1}{P(x)(1 - P(x))},$$

this is the Anderson–Darling statistic, usually denoted by $A^2$. The distance measures in the Anderson–Darling form are different for different distributions. Because of the denominator in $w(x)$, the Anderson–Darling statistic puts more weight in the tails of the distribution. Also, because it is specific for the hypothesized distribution, the Anderson–Darling test based on $A^2$ is likely to be more powerful that the standard Kolmogorov–Smirnov test (see D'Agostino, 1986, and Stephens, 1986). For these reasons, it is recommended that the Anderson–Darling test be used instead of the Kolmogorov–Smirnov test in testing random number generators.

For a sample of size $n$ in which the $i^{\text{th}}$ order statistic is denoted as $x_{(i)}$, the Anderson–Darling statistic can be computed as

$$A^2 = -n - \frac{1}{n} \sum_{j=1}^{n} (2j-1)\Big( \log\big(F(x_{(j)})\big) + \log\big(1 - F(x_{(n-j+1)})\big)\Big). \qquad (2.9)$$

The distribution is much harder to work out, and of course it is different for different hypothesized distributions and for different sample sizes. Critical values and p-values for a variety of distributions have been computed. These computations have been performed in various ways, sometimes by Monte Carlo estimation (see Chapter 7 and Exercise 7.4 on page 272). Stephens (1986) gives tables for some common distributions, including the normal and the exponential. Sinclair and Spurr (1988) have suggested approximations based on cumulants. More recently, saddlepoint approximations have been used. Unfortunately, the Anderson–Darling test is not as widely available in standard software packages as the Kolmogorov–Smirnov test.

## Goodness-of-Fit Tests in Different Regions

After developing basic tests to compare a sample to a hypothesized distribution, the simplest modification that we can make is to divide the distribution into different regions and then apply the same tests in the separate regions. If $X \sim U(0,1)$, then, for example,

$$X|X \in (0,.1) \sim U(0,.1).$$

The same idea applies to any distribution. Sometimes, it is useful to perform a test based on $|P_n(x) - P(x)|$, such as a Kolmogorov–Smirnov test, in various regions of the distribution because the fit may be better in some regions than in others. The idea is shown in Figure 2.3.

We may be able to relate regions of the test statistics to specific types of departure from the null hypothesis. This would be much more meaningful than just a simple "reject" or "not reject". As we mentioned before, sometimes a distribution may not be simulated well in the tails. In using separate tests in different regions, however, as in any approach in which multiple tests are being performed, we must be careful in attaching any meaning to a p-value.

## Statistical Tests of Stochastic Processes

Dynamic statistical tests are for the independence of the output. These tests address patterns in the output stream; an example is a test for serial correlation of various lags. As we indicated above, these patterns would show up in tests for $d$-variate uniformity, but specific tests such as we discuss below can be more powerful.

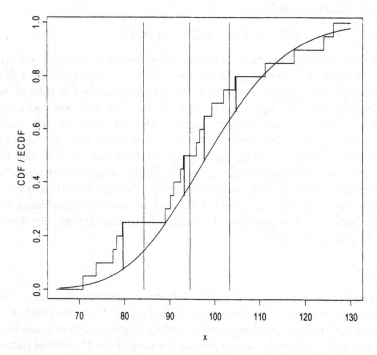

Figure 2.3: Kolmogorov–Smirnov Tests in Different Regions

## Runs Tests

Some of the most effective of the dynamic tests are runs tests. A runs test uses the number and/or length of successive runs in which the values of the sequence are either increasing or decreasing. There are different ways to count the runs. To illustrate one way of counting runs, consider the sequence

$$1, 5, 4, 1, 3, 1, 3, 4, 7.$$

There are five different runs in this sequence:

up : 1, 5;   down : 5, 4, 1;   up : 1, 3;   down : 3, 1;   up : 1, 3, 4, 7.

The number of runs could be used as a test statistic, but a more powerful test can be constructed using the lengths of the runs up and/or the runs down. The lengths of the runs up as shown above are 2, 2, and 4. With small integers such as in this example, an issue may be how to handle ties (that is, two adjacent numbers that are equal). This issue is not relevant when the test is applied to the output of a generator that simulates a continuous uniform distribution.

A more common way of counting the lengths of the runs up (and analogously the runs down) is to identify breakpoints where the sequence no longer continues increasing (or decreasing). The lengths of runs are then the lengths of the subsequences between breakpoints. For the sequence above, we have breakpoints as indicated below:

$$1, 5 \mid 4 \mid 1, 3 \mid 1, 3, 4, 7$$

This gives one run up of length 1; two runs up of length 2; and one run up of length 4. The runs test (more properly, the runs up test and the runs down test) is based on the covariance matrix of the random number of runs of each length. Usually, only runs up to some fixed length, such as 7, are considered. (Runs greater than that are lumped together; the covariance matrix can be adjusted to fix the test.) The test statistic is the quadratic form in the counts with the covariance matrix, which has an asymptotic chi-squared distribution with degrees of freedom equal to the number of different lengths of runs counted. Grafton (1981) gives a Fortran program to compute the chi-squared test statistic for the runs test (with runs of length 1 through 6) and the significance level of the test statistic. The program is available from `statlib` as the *Applied Statistics* algorithm AS 157.

### Serial Test

The serial test is performed on sequences of $d$ bits in a sample of $n$ bits. (There is more than one version of this test, which differ slightly from one another. We will describe one that is commonly used in testing random number generators.) For the $n$ bits $b_1, \ldots, b_n$ in the output stream, for each of the $2^d$ possible patterns in a substream of $d$ bits, let $\nu_{i_1 \cdots i_d}$ be the number of times that the pattern $i_1, \ldots, i_d$ appears in the augmented stream $b_1, \ldots, b_n, b_1, \ldots, b_{d-1}$. We then form

$$\Psi_d^2 = \frac{2^d}{n} \sum_{i_1 \cdots i_d} \left( \nu_{i_1 \cdots i_d} - \frac{n}{2^d} \right)^2$$

$$= \frac{2^d}{n} \sum_{i_1 \cdots i_d} \nu_{i_1 \cdots i_d}^2 - n. \tag{2.10}$$

Now, form the first backward difference in $d$,

$$\Delta \Psi_d^2 = \Psi_d^2 - \Psi_{d-1}^2, \tag{2.11}$$

and the second backward difference,

$$\Delta^2 \Psi_d^2 = \Psi_d^2 - 2\Psi_{d-1}^2 + \Psi_{d-2}^2. \tag{2.12}$$

Then, under the null hypothesis that the bit stream is a simple random sample from a Bernoulli distribution with parameter 0.5, $\Delta \Psi_d^2$ has an asymptotic (in $n$) chi-squared distribution with $2^{d-1}$ degrees of freedom, and $\Delta^2 \Psi_d^2$ has an asymptotic chi-squared distribution with $2^{d-2}$ degrees of freedom. This null distribution forms the basis for a test called the generalized serial test (see Kato, Wu, and Yanagihara, 1996b).

## Tests of Matrix Ranks

An interesting type of test is based on the ranks of binary matrices (that is, matrices whose elements are zeros and ones) formed from the output of the generator. The procedure is to form matrices from bit streams, to determine the ranks of the matrices, and to compare the observed ranks to what would be expected based on the distribution of ranks of matrices whose entries are independent Bernoulli random variables with parameter $1/2$. For vectors and matrices whose elements are members of a Galois field, the ordinary rules of vector/matrix operations hold, with all operations being defined in terms of the operations of field. The matrices used in this test are formed from bit streams and so are based in $\mathbb{G}(2)$.

For an $n \times m$ random binary matrix, Kovalenko (1972) gives the probability function for the rank of the matrix as

$$p(r) = 2^{r(n+m-r)-nm} \prod_{j=0}^{r-1} \frac{(1 - 2^{j-n})(1 - 2^{i-m})}{1 - 2^{j-r}}, \quad \text{for } r = 0, \dots, \min(n, m)$$

(2.13)

(see also Marsaglia and Tsay, 1985). It is easy to determine the rank of a binary matrix using elementary row or column operations. There is no concern for rounding, as is the case in computing the rank of a matrix over the field of the reals. Notice that the probability that the rank of binary matrices is less than full is nonzero. (That probability is 0 for matrices whose elements are randomly chosen from dense subsets of $\mathbb{R}$ and is very close to 0 for matrices whose elements are randomly chosen from $\mathbb{F}$.)

Most reasonable linear congruential generators pass binary matrix rank tests. Generalized feedback shift register generators tend to fail the tests because they tend to generate matrices with full rank; that is, they are "too independent".

## Test Suites

Different tests are sensitive to different types of departures from the null hypothesis of independent uniformity. We need a set of tests that spans the space of alternative hypotheses. All that it takes to devise an empirical test is a little imagination. Tests can be based on familiar everyday random processes such as occur in gaming. A "poker test", for example, is a chi-squared goodness-of-fit test for simulated poker hands. The idea is to devise a scheme to use a random number generator for simulating deals from a poker deck and then to compare the sampled frequency of various poker hands with the expected frequency of such hands. Specific tests for dynamic properties are often more difficult to construct.

Some important suites of statistical tests for random number generators are the tests of Fishman and Moore (1982, 1986), the tests of Vattulainen, Ala-Nissila, and Kankaala (1994, 1995), the DIEHARD tests (Marsaglia, 1985, 1995), the NIST Test Suite (NIST, 2000), and TestU01 (some descriptions in

L'Ecuyer, 1998). I briefly describe the tests in DIEHARD and the NIST Test Suite below. The TestU01 suite includes the tests from DIEHARD and NIST, as well as others, so I recommend use of TestU01. It currently includes over 60 tests. Fishman and Moore's tests are goodness-of-fit tests on various transformations of the sample. They applied the tests to linear congruential generators with the same modulus but different multipliers, $2^{31} - 1$, in order to identify good multipliers for that widely used modulus. The tests of Vattulainen, Ala-Nissila, and Kankaala are based on physical models (see Section 2.3.2).

Given a set of tests, it is interesting to know to what extent the tests address the same properties; that is, how independent are the tests from each other. Banks (1998) used factor analysis to address this question. He applied a number of tests to a variety of generators and sequences from those generators and then performed a factor analysis on the results to study the extent of independence of the tests.

In addition to having a suite of statistical tests for generators, it is useful to have a suite of generators or data streams. The suite would include a selection of commonly used generators, a selection of generators or data streams that seem to satisfy the null hypothesis to the extent that they have been tested, and some generators or data streams that violate the null hypothesis in prescribed ways. Marsaglia's DIEHARD distribution CD includes several generators and bit streams, and the NIST package also includes "reference" generators. Again, the TestU01 package is more extensive than either DIEHARD or NIST in this regard. TestU01 also includes facilities for combining basic generators.

### DIEHARD Tests

Marsaglia (1985) describes a battery of eighteen goodness-of-fit tests called "DIEHARD". Most of these tests are performed on integers in the interval $(0, 2^{31} - 1)$ that are hypothesized to be realizations of a discrete uniform distribution with mass points being the integers in that range.

The "p-values" reported by the DIEHARD tests are not the same as the usual p-values in goodness-of-fit tests; rather, they are the values of the CDF of the test statistic. A large "p-value" for a chi-squared test therefore corresponds to what is usually a small (that is, a significant) p-value. A small "p-value" reported by a chi-squared test in the DIEHARD suite indicates close agreement to the expected frequencies. In either case, however, the DIEHARD test would indicate that the stream generated by the random number generator is suspect. In one case it would be because the frequencies do not correspond to the theoretical ones, and in the other case it would be because the frequencies match the theoretical ones too well.

Some of the tests report a single "p-value", while other tests actually perform the test many times and compare the distribution of the test statistic with the theoretical distribution of the test statistic under the null hypothesis; that is, some of the tests are second-order tests.

These tests cover a variety of possibilities for assessing randomness of particular features exhibited by a sequence of numbers. The tests are:

- birthday spacings test
  For this test, choose $m$ birthdays in a year of $n$ days. List the spacings between the birthdays. If $j$ is the number of values that occur more than once in that list, then $j$ has an asymptotic Poisson distribution with mean $m^3/(4n)$. Various groups of bits in the binary representation of the hypothesized uniform integers are used to simulate birthday spacings; goodness-of-fit tests are performed, yielding CDF values; other groups of bits are used and tests performed to yield more CDF values; and so on. Then, a goodness-of-fit test is performed on the CDF values.

- overlapping 5-permutation test
  Each set of five consecutive integers can be in one of 120 states for the 5! possible orderings of five numbers. The test is on the observed numbers of states and transition frequencies.

- binary rank test for 31 × 31 matrices
  The leftmost 31 bits of 31 random integers from the test sequence are used to form a 31 × 31 binary matrix over the field $\mathbb{G}(2)$. The rank of the matrix is determined. Each time that this is done, counts are accumulated for matrices with ranks of 31, 30, 29, and 28 or less. A chi-squared test is performed for these four outcomes.

- binary rank test for 32 × 32 matrices
  This is similar to the test above.

- binary rank test for 6 × 8 matrices
  This is similar to the tests above except that the rows of the matrix are specified bytes in the integer words.

- bit stream test
  Using the stream of bits from the random number generator, form 20-bit words, beginning with each successive bit (that is, the words overlap). The bit stream test compares the observed number of missing 20-bit words in a string of $2^{21}$ overlapping 20-bit words with the approximate distribution of that number.

- overlapping-pairs-sparse-occupancy test
  For this test, two-letter "words" are formed from an alphabet of 1024 letters. The letters in a word are determined by a specified ten bits from a 32-bit integer in the sequence to be tested, and the bits defining the letters overlap. The test counts the number of missing words (that is, combinations that do not appear in the entire sequence being tested). The count has an asymptotic normal distribution, and that is the basis for the goodness-of-fit test, when many sequences are used.

- overlapping-quadruples-sparse-occupancy test
  This is similar to the test above. The null distribution of the test statistic is very complicated; interestingly, a parameter of the null distribution of the test statistic was estimated by Monte Carlo methods.

- DNA test
  This is similar to the tests above except that it uses ten-letter words built on a four-letter alphabet (the DNA alphabet).

- count-the-1s test on a stream of bytes
  This test is based on the binomial (8, 0.5) distribution of 1s in uniform random bytes. Rather than testing this directly, however, counts of 1s are grouped into five sets: $\{0, 1, 2\}$, $\{3\}$, $\{4\}$, $\{5\}$, and $\{6, 7, 8\}$; that is, if a byte has no 1s, exactly one 1, or exactly two 1s, it is counted in the first group. Next, overlapping sequences of length 5 are formed, and the counts of each of the $5^5$ combinations are obtained. A chi-squared test is performed on the counts. The test takes into account the covariances of the counts and so is asymptotically correct. If the groups of counts of 1s are thought of as five letters and the groups of bytes thought of as five-letter words, the output of the sequence under test is similar to the typing of a "monkey at a typewriter hitting five keys" (with rather strange probabilities), so the test is sometimes called the "monkey at a typewriter test".

- count-the-1s test for specific bytes
  This is similar to the test above.

- parking lot test
  This test is based on the results of randomly packing circles of radius 1 about a center located randomly within a square of side 100. The test is based on the number of "cars parked" (i.e., nonoverlapping circles packed) after a large number of attempts. The distribution of the test statistic, and hence its critical values, is determined by simulation.

- minimum distance test
  This test is based on the minimum distance between a large number of random points in a square. If the points are independent uniform, then the square of the minimum distance should be approximately exponentially distributed with mean dependent on the length of the side of the square and the number of points. A chi-squared goodness-of-fit test is performed on the CDF values of a large number of tests for exponentiality.

- 3-D spheres test
  In this test, 4000 points are chosen randomly in a cube of edge 1000. At each point, a sphere is centered that is large enough to reach the next closest point. The volume of the smallest such sphere is approximately exponentially distributed with mean $\frac{120\pi}{3}$. Thus, the radius cubed is exponential with mean 30. (The mean was obtained by extensive simulation.)

A chi-squared goodness-of-fit test is performed on the CDF values of a large number of tests for exponentiality.

- squeeze test
  This test is performed on floating-point numbers hypothesized to be from a U(0, 1) distribution. Starting with $k = 2^{31}$, the test finds $j$, the number of iterations necessary to reduce $k$ to 1, using the reduction $k = \lceil k * U \rceil$, with $U$ from the stream being tested. Such $j$s are found 100,000 times, and then counts for the number of times $j$ was $\leq 6, 7, \ldots, 47, \geq 48$ are used to provide a chi-squared test.

- overlapping sums test
  This test is performed on floating-point numbers hypothesized to be from a U(0, 1) distribution. Sums of overlapping subsequences of 100 uniforms are tested for multivariate normality by a chi-squared test several times and then a chi-squared goodness-of-fit test is performed on the CDF values of the chi-squared tests.

- craps test
  This test simulates games of craps and counts the number of wins and the number of throws necessary to end each game and then performs chi-squared goodness-of-fit tests on the observed counts.

- runs test
  This is the test of runs up and runs down described on page 77.

Source code for these tests is available in a CD-ROM (Marsaglia, 1995) and at

`http://stat.fsu.edu/~geo/diehard.html`

The input to the test program is a binary file consisting of the 32-bit (unsigned) integers produced by the random number generator. Because this is not the sequence that a random number generation program usually yields, use of the program often requires complicated conversions. If the output has been represented as floating-point numbers, it may be impossible to use the DIEHARD test programs. Also, if the output is signed integers (of 31 bits), straightforward use of the DIEHARD programs does not give valid statistics.

McCullough (1999) reports results of the DIEHARD tests on the random number generators in SAS, SPSS, and S-Plus. All three generators performed poorly on the count-the-1s test on a stream of bytes. S-Plus also did not do well on the overlapping quadruples test, the DNA test, and the count-the-1s test for specific bytes.

## The NIST Test Suite

The NIST Test Suite, described in NIST (2000), includes sixteen tests. Most of these tests are performed on sequences of bits that are hypothesized to be independent realizations of a Bernoulli process.

The list below indicates the types of tests. Some of these tests are very similar to tests in the DIEHARD suite. In many cases, the reader should be able to devise one or more tests of the indicated type. In other cases, a reference is given in the list below. The tests are:

- frequency test

- frequency tests within blocks

- runs test
  (This is based on the lengths of runs of bits.)

- test for longest run of ones in a block

- random binary matrix rank test

- discrete Fourier transform test

- nonoverlapping template matching test

- overlapping template matching test

- Maurer's universal statistical test
  (This is described by Maurer, 1992.)

- Lempel–Ziv complexity test
  (This is described by Ziv and Lempel, 1977.)

- linear complexity test

- serial test

- approximate entropy test
  (This is based on a sample analog of equation (2.4). The sample $f_i$s are the observed counts in various intervals.)

- cumulative sum test

- random excursions test

- random excursions variant test

Descriptions of these tests, examples of their use, and source code are available at

http://csrc.nist.gov/rng/

## TestU01

The TestU01 suite is very extensive. It includes the tests from DIEHARD and NIST and several other tests that uncover problems in some generators that pass DIEHARD and NIST with flying colors. Some of the interesting tests in TestU01 are based on the distribution of the nearest pairs of points in various dimensions (see L'Ecuyer, 1998).

TestU01, written in C, is much easier to use than DIEHARD or the NIST suite. A minimal amount of knowledge of C is required to use TestU01. The package comes with a number of additional functions, including some standard generators and utility programs for processing bit streams.

The tests of TestU01 are grouped into three batteries, "Small Crush", "Crush", and "Big Crush". The time required to test a single generator with Big Crush can be over ten hours on an upper-end PC, but Small Crush is two orders of magnitude faster. (DIEHARD is an order of magnitude faster than Small Crush. The speed of the test batteries depends on the number of tests and their complexity.) A reasonable procedure for testing a generator is to use Small Crush first; if the generator fails, stop; otherwise, use Crush; if the generator fails, stop; otherwise, use Big Crush.

The code and user's manual for TestU01 are available from

    http://www.iro.umontreal.ca/~lecuyer/

## Interpretation of Statistical Tests

Because the empirical tests are statistical tests, the ordinary principles of hypothesis testing apply. The results of a battery of tests must be interpreted carefully. As Marsaglia (1995) states (sic):

> By all means, do not, as a Statistician might, think that a $p < .025$ or $p > .975$ means that the RNG has "failed the test at the .05 level". Such $p$'s happen among the hundreds that DIEHARD produces, even with good RNG's. So keep in mind that "p happens".

Fortunately, a "Statistician" understands statistical hypothesis testing better than some other testers of random number generators. If the null hypothesis is true, and if the statistical test is exact, we expect $|p| < 0.05$ 5% of the time.

In some cases, the kind of departure from randomness that a particular test addresses is obvious; in other cases, however, it is not as clear what kind of nonrandomness a given test may detect. In addition, the relationships of various tests to each other are not well-understood. Factor analysis of the tests in the DIEHARD and NIST suites has shown them to be surprisingly independent, except on very poor generators, which fail all tests (see Banks, 1998).

## 2.3.2   Comparisons of Simulated Results with Statistical Models in Physics

Another way that random number generators are tested is in their usage in simulations in which some of the output can be compared with what established theory would suggest. There are several physical processes in statistical mechanics that could be used for testing random number generators. One of the most widely used models in computational physics is the Ising model (see Section 7.9). This model can be solved analytically in two dimensions. Vattulainen, Ala-Nissila, and Kankaala (1995) describe a suite consisting of four tests based on physical models. Two tests use clusters and autocorrelations from a two-dimensional Ising model. The two others are based on random walks on lattices.

Ferrenberg, Landau, and Wong (1992) used some of the generators that meet the Park and Miller (1988) minimal standard (see page 20) to perform several simulation studies in which the correct answer was known. Their simulation results suggest that even some of the "good" generators could not be relied on in some simulations. In complex studies, it is difficult to trace unexpected results to errors in the computer code or to some previously unknown phenomenon. Because of sampling error in simulations, there is always a question of whether results are due to a sample that is unusual. Vattulainen, Ala-Nissila, and Kankaala (1994) studied the methods of Ferrenberg, Landau, and Wong and determined that the anomalous results were indeed due to defects in the random number generator.

Vattulainen, Ala-Nissila, and Kankaala (1994, 1995) conducted further studies on these generators as well as generators of other types and found that their simulations often did not correspond to reality. Selke, Talapov, and Shchur (1993), however, found that the Park–Miller minimum standard generator performed much better than the R250 generator in a specific simulation of particle behavior. Vattulainen et al. (1995) study eight widely available generators. For the tests in their study, the Park–Miller minimum standard generator (which is the default generator in the IMSL Libraries), R250, and G05FAF from the Nag Library were found to be acceptable. The generally inconclusive results lead us to repeat the advice given above to employ an ad hoc goodness-of-fit test for the specific application.

## 2.3.3   Anecdotal Evidence

The null hypothesis in the statistical tests of random number generators is that "the generator is performing properly"; therefore, failure to reject is not confirmation that the generator is a good one. There have been many statistical studies of the properties of given sets of random number generators. The tests reported in the literature have often been inconclusive, however.

Every Monte Carlo study in which there is additional information available about expected results provides evidence about the quality of the random number generators used. We call this anecdotal evidence. The first indication that the RANDU generator had problems came from anecdotal evidence in a simulation with results that did not correspond to theory (see Coldwell, 1974). Although statistical measures such as standard deviations are sometimes used in comparing simulation results with theoretical results, there is often no attempt to quantify the probability of a type I error.

### 2.3.4  Tests of Random Number Generators Used in Parallel

For parallel random number generators, in addition to the usual concerns about randomness and correlations within a single stream, we must ensure that the correlations between streams are small.

A major problem in random number generation in parallel is the inability to synchronize the computations. This is easy to appreciate even in a simple case of acceptance/rejection; if one processor must do more rejections than another, that processor will get behind the other one. Cuccaro, Mascagni, and Pryor (1994), Vattulainen (1999), and Srinivasan, Mascagni, and Ceperley (2003) describe some approaches for testing parallel random number generators.

## 2.4  Programming Issues

In addition to the quality of the algorithm for generation of random numbers, there are also many issues relating to the quality of the computer implementation of the random number generator. Because of the limitations of representing numbers exactly, a computer program rarely corresponds exactly to a mathematical expression. Sometimes, a poor program completely negates the positive qualities of the algorithm that it was intended to implement. The programming considerations relevant to implementing a random number generator are often subtle and somewhat different from those arising in other mathematical software (see Gentle, 1990).

To the extent that widely used and well-tested software for random number generation is available, it should be used instead of software developed ad hoc. Software for random number generation is discussed in Chapter 8.

## 2.5  Summary

There are many issues to consider in assessing the quality of a random number generator. It is easy to state what is desired for the basic uniform generator. The first desideratum is obvious:

- the output should be essentially indistinguishable from a sample from a uniform distribution.

This has nothing to do with how the generator is implemented, whether it is in a computer program or in some other form. It does, however, have implications for the period of a deterministic, cyclic pseudorandom number generator implemented in a computer program. The period should be very long. The specific meaning of "long" in this context depends on the size of problems in which random number generators find application. The size of the problems grows from year to year, thus our criterion for "long" changes.

Another desideratum for a random number generator is not so obvious:

- a sequence should be reproducible on (almost) any computer.

This quality, of course, means that a pseudorandom number generator is to be preferred to a random one.

There are specific desirable qualities that arise from the use of the random number generator in the computer:

- neither 0 nor 1 should occur in the simulation of samples from $U(0,1)$;

- the generator should be efficient in its use of computing resources;

- the generator should be easy to use.

The last two qualities listed above have additional specific implications:

- the generator should not require user input with qualities that are difficult to assess;

- it should be possible to generate long substreams that do not overlap.

Although, as we have suggested, before using a random number generator, it is wise to apply ad hoc goodness-of-fit tests that may uncover problems in that particular application, it is desirable that the quality of the output of the generator not be dependent on specific transformations or on a specific seed.

The last quality listed above makes the generator more useful in parallel implementations.

Finally, an additional concern is the quality of the computer program:

- the computer program must faithfully implement the generator.

Bugs in computer programs are all too common, and no matter how good the underlying algorithm, if the program is incorrect, the generator may not be good.

# Exercises

2.1. Devise a "dice test" (see Marsaglia's "craps test" in Section 2.3). Your test should accept the output of a uniform random number generator (you can assume either discrete or continuous) and simulate the roll of a pair of dice to obtain a total (i.e., the pairs (1,3), (2,2), (2,2), and (3,1) are not distinguished). For a specified number of rolls, your test should be a chi-squared test with 10 degrees of freedom (meaning that the number of rolls should be greater than some fixed number).

Implement your test in the language of your choice. Then, apply it to some random number generators (for example, the generator you wrote in Exercise 1.5, and whatever system random number is available to you, such as in S-Plus or the IMSL Libraries).

Now, put your dice test in the inner loop of a test program to perform the test many times, each time retaining the $p$-value. At the end of the inner loop, do a chi-squared test on the $p$-values. Does this testing of the output of tests make sense? What about testing the output of tests of the output of tests?

2.2. Write a program for a test for binary ranks. The input to your program is $n$, the number of rows, $m$, the number of columns in the matrices to be formed and tested, and a vector of 0s and 1s from whose first $nm$ elements the matrices are to be formed. (It is irrelevant whether you form the matrices in a row major or column major pattern, but let us say that the first $n$ elements of the vector become the first column, the next $n$ elements become the second column, and so on. Another consideration is the storage format of the input vector. For efficiency, it should be binary, but for ease of programming, you can use regular fixed-point storage.) Use elementary row and column operations to diagonalize the matrix (see Gentle, 1998, page 88 and following, for example). The number of 1s on the diagonal of the transformed matrix is the rank. For a given size of matrices and input vector, compute the frequencies of ranks from 1 to $\min(n, m)$. Do a chi-squared test that these frequencies match the expected frequencies from equation (2.13). (You should write a separate program module to compute the probability mass function.)

(a) Write a brief but clear user document for your test program.

(b) Perform the binary matrix rank test on $31 \times 31$ matrices produced by the generator that you wrote in Exercise 1.5. (You may have to modify the program for the generator slightly to get a bit stream.) Write a report of your results.

(c) Binary matrices are defined over $\mathbb{G}(2)$. If the elements have independent identical discrete uniform distributions (they are Bernoulli random variables with parameter $1/2$), the probability mass function for the rank of an $n \times m$ matrix is given in equation (2.13). Suppose, instead, that the elements are in $\mathbb{R}$ and have independent identical $U(0, 1)$ distributions. What is the probability mass function for the rank of an $n \times m$ matrix in that case?

2.3. Sullivan (1993) suggested the following test of a random number generator. For each of the seeds $1, 2, \ldots, N$, generate a sequence of length $n$.

Let $I$ represent the index of the maximum element in each sequence. If $n$ is considerably less than the period of the generator, we would expect $I$ to be uniformly distributed over the integers $1, 2, \ldots, n$.

Using $N = 100$ and $n = 10000$, perform this test on the "minimal standard" generator that you wrote in Exercise 1.9. What are your results? What is wrong with this test?

2.4. Use standard test suites on the "minimal standard" generator that you wrote in Exercise 1.9. You must prepare the output appropriately for these test suites.

   (a) Obtain Marsaglia's DIEHARD test suite, and apply all tests to your implementation of the "minimal standard" generator. Write a clear and complete report of your tests. (You should describe the generator, how you prepared its output for input to the DIEHARD tests, the results of each test, and how you interpret the results.)

   (b) Obtain the NIST test package, and apply all tests to your implementation of the "minimal standard" generator. Write a clear and complete report of your tests.

   (c) Obtain the TestU01 test package, and apply Small Crush to your implementation of the "minimal standard" generator. Should you proceed to use Crush on this generator? Write a clear and complete report of your tests.

2.5. Marsaglia's distribution of the DIEHARD test suite includes a number of standard random number generators (some good, some not so good). Select ten of these generators to test. Some of the DIEHARD tests produce multiple "p-values". (Recall the reported "p-value" is the CDF of the test statistic evaluated at the realized value.) For each test, decide on just one "p-value" to use, and for each generator that you have chosen, run each of the DIEHARD tests 20 times and save the "p-values". This yields 200 "p-values" for each test.

   (a) Analyze relationships among the eighteen tests. Do the different tests appear to be testing different things?

   (b) Perform a multivariate analysis of variance to decide whether the variation between random number generators is significant.

2.6. Test the Fortran or C RANDU generator that you wrote in Exercise 1.7 of Chapter 1. Devise any statistical tests you want. (See Section 2.3, and use your imagination.) Does your generator pass all of your tests? (Do not answer "no" until you have found a realistic statistical test that RANDU actually fails.)

2.7. As mentioned in Chapter 1, $\pi$ seems to be a normal number in any base (that is, the digits in an approximation of $\pi$ in any base seem to be independently and uniformaly distributed). Bailey and Crandall (2001) give a recursion for the digits in a hexadecimal representation of $\pi$. Let $x_0 = 0$, and, for $i = 1, 2, \ldots$,

$$x_i = \left( 16x_{i-1} + \frac{120i^2 - 89i + 16}{512i^4 - 1024i^3 + 712i^2 - 206i + 21} \right) \bmod 1.$$

Then, take $d_i = \lfloor 16x_i \rfloor$ as the $i^{\text{th}}$ hexadecimal digit. Apply the Crush suite of TestU01 to these digits (alternatively, of course, apply the test to the $x$s.) Does $\pi$ appear to be normal in base 16?

2.8. Briefly, for each of the points listed in Section 2.5, beginning on page 87, how would you rate the following classes or types of generators?

   (a) linear congruential generators;

   (b) generalized feedback generators;

   (c) generators based on cellular automata;

   (d) generators based on chaotic systems.

# Chapter 3

# Quasirandom Numbers

In statistical applications, there is a fundamental tension in random sampling between the need for randomness as a basis for making inferences from the sample and the requirement that the sample be "representative" of the sample space. In survey sampling, stratification is used to make the sample more representative and simultaneously reduce sampling variance. This idea can be extended, but as the number of strata increases, the sample size in each stratum decreases. In the limit, there is no randomness remaining in the sampling. (See Section 7.5.2 for discussion of stratification in Monte Carlo applications with pseudorandom numbers.)

In pseudorandom number generation, we emphasize the simulation of a random process. Sometimes, we may be concerned about whether the sample is representative or whether it is an "outlier", but our primary concern is that the process simulate a random process. Because we are simulating a random process, the results are estimates and are subject to sampling variation just as a random sampling process would be.

In an attempt to ensure that a random sample is representative, we may constrain the pseudorandom numbers to have the same mean as the population or so that other properties match, as we discuss in Section 7.5.4. In this chapter, we discuss ways of drawing representative samples by systematic traversal of the sample space.

Some of the properties of expectation and variance of results can carry over reasonably well to pseudorandom numbers, but any attempt to make a sample more representative must change our use of the sample variance in making inferences.

## 3.1  Low Discrepancy

For a deterministic sequence, instead of the expectation of the estimator, we might consider its limit, so we might take into account the discrepancy (equation (2.6), page 69).

For pseudorandom numbers generated by any of the methods that we have discussed, the limit is taken over a cyclic finite set, and the supremum in the discrepancy is a maximum. Suppose that, instead of the pseudorandom numbers resulting from the generators that we have discussed, we explicitly address the problem of discrepancy. Then, instead of being concerned with $E(\widehat{\theta})$, we might consider

$$\lim_{n\to\infty} \widehat{\theta} = \lim_{n\to\infty} (b-a) \frac{\sum f(y_i)}{n}.$$

A moment's pause, of course, tells us that this kind of limit does not apply to the real world of computing, with its finite set of numbers. Nevertheless, it is useful to consider a deterministic sequence with low discrepancy. The objective is that any finite subsequence fill the space uniformly.

These sequences are called *quasirandom* sequences.

Quasirandom sequences correspond to samples from a $U(0,1)$ distribution. (Contrast this statement with the statement that "pseudorandom sequences *simulate random samples* from a $U(0,1)$ distribution".) The techniques of Chapters 4 and 5 can therefore be used to generate quasirandom sequences that correspond to samples from nonuniform distributions. For the methods that yield one nonuniform deviate from each uniform deviate, such as the inverse CDF method, everything is straightforward. For other methods that use multiple independent uniform deviates for each nonuniform deviate, the quasirandom sequence may be inappropriate. The quasirandom method does not simulate independence.

## 3.2   Types of Sequences

Whereas pseudorandom sequences or pseudorandom generators attempt to simulate randomness, quasirandom sequences are decidedly *not* random. The objective for a (finite) pseudorandom sequence is for it to "look like" a sequence of realizations of i.i.d. uniform random variables, but for a (finite) quasirandom sequence the objective is that it fill a unit hypercube as uniformly as possible. Several such sequences have been proposed, such as van der Corput sequences, Halton sequences (Halton, 1960), Faure sequences, Sobol' sequences (Sobol', 1967, 1976), and Niederreiter sequences (Niederreiter, 1988).

### 3.2.1   Halton Sequences

A Halton sequence is formed by reversing the digits in the representation of some sequence of integers in a given base. (This is a generalization of a van der Corput sequence.) Although this can be done somewhat arbitrarily, a straightforward way of forming a $d$-dimensional Halton sequence $x_1, x_2, \ldots,$ where $x_i = (x_{i1}, x_{i2}, \ldots, x_{id})$, is first to choose $d$ bases, $b_1, b_2, \ldots, b_d$, perhaps the first $d$ primes. The $j^{\text{th}}$ base will be used to form the $j^{\text{th}}$ component of each vector in the sequence. Then, begin with some integer $m$ and

1. choosing $t_{mj}$ suitably large, represent $m$ in each base:

$$m = \sum_{k=0}^{t_{mj}} a_{jk}(m)b_j^k, \quad j = 1, \ldots d,$$

2. form

$$x_{ij} = \sum_{k=0}^{t_{mj}} a_{jk}(m)b_j^{k-t_{mj}-1}, \quad j = 1, \ldots d,$$

3. set $m = m + 1$ and repeat.

Suppose that, for example, $d = 3$, $m = 15$, and we use the bases 2, 3, and 5. We form $15 = 1111_2$, $15 = 120_3$, and $15 = 30_5$ and deliver the first $x$ as $(0.1111_2, 0.021_3, 0.03_5)$ or $(0.937500, 0.259259, 0.120000)$. Continuing in this way, the first five 3-vectors are shown in Table 3.1. Even in this small example, we can see how each element in the vectors in the sequence is achieving a uniform coverage and how the elements are not synchronized. We also note, however, that larger values of the base yield undesirable monotone subsequences. The length of the monotone subsequences corresponding to the base $b_j$ is obviously $b_j$ because that is the length of a sequence in which the leading digit does not change.

The Halton sequences are acceptably uniform for lower dimensions, up to about 10. For higher dimensions, however, the quality of the Halton sequences degrades rapidly because the two-dimensional planes occur in cycles with decreasing periods.

Various generalized Halton sequences have been proposed and studied by Braaten and Weller (1979), Hellekalek (1984), Faure (1986), and Krommer and Ueberhuber (1994). The basic idea of the Faure sequences is to permute the $a_{jk}(m)$s in step 2.

Kocis and Whiten (1997) suggest a "leaped Halton sequence". In this method. the cycles of the Halton sequence are destroyed by using only every $l^{th}$ Halton number, where $l$ is a prime different from all of the bases $b_1, b_2, \ldots, b_d$.

Table 3.1: Three-Dimensional Halton Numbers, Starting with $m = 15$

| $m$ | vector in bases 2, 3, and 5 | x |
|---|---|---|
| 15 | $(0.1111_2, \ 0.021_3, 0.03_5)$ | $(0.937500, 0.259259, 0.120000)$ |
| 16 | $(0.00001_2, 0.121_3, 0.13_5)$ | $(0.031250, 0.592593, 0.320000)$ |
| 17 | $(0.10001_2, 0.221_3, 0.23_5)$ | $(0.531250, 0.925926, 0.520000)$ |
| 18 | $(0.01001_2, 0.002_3, 0.33_5)$ | $(0.281250, 0.074074, 0.720000)$ |
| 19 | $(0.11001_2, 0.102_3, 0.43_5)$ | $(0.781250, 0.407407, 0.920000)$ |

## 3.2.2   Sobol' Sequences

A Sobol' sequence is based on a set of "direction numbers", $\{v_i\}$. The $v_i$ are

$$v_i = \frac{m_i}{2^i},$$

where the $m_i$ are odd positive integers less than $2^i$, and the $v_i$ are chosen so that they satisfy a recurrence relation using the coefficients of a primitive polynomial in the Galois field $\mathbb{G}(2)$,

$$f(z) = z^p + c_1 z^{p-1} + \cdots + c_{p-1} z + c_p$$

(compare equation (1.38), page 38). For $i > p$, the recurrence relation is

$$v_i = c_1 v_{i-1} \oplus c_2 v_{i-2} \oplus \cdots \oplus c_p v_{i-p} \oplus \lfloor v_{i-p}/2^p \rfloor,$$

where $\oplus$ denotes bitwise binary exclusive-or. An equivalent recurrence for the $m_i$ is

$$m_i = 2 c_1 m_{i-1} \oplus 2^2 c_2 m_{i-2} \oplus \cdots \oplus 2^p c_p m_{i-p} \oplus m_{i-p}.$$

As an example, consider the primitive polynomial (1.42) from page 39,

$$x^4 + x + 1.$$

The corresponding recurrence is

$$m_i = 8 m_{i-3} \oplus 16 m_{i-4} \oplus m_{i-4}.$$

If we start with $m_1 = 1$, $m_2 = 1$, $m_3 = 3$, and $m_4 = 13$, for example, we get

$$
\begin{aligned}
m_5 &= 8 \oplus 16 \oplus 1 \\
&= 01000(\text{binary}) \oplus 10000(\text{binary}) \oplus 00001(\text{binary}) \\
&= 11001(\text{binary}) \\
&= 25.
\end{aligned}
$$

The $i^{\text{th}}$ number in the Sobol' sequence is now formed as

$$x_i = b_1 v_1 \oplus b_2 v_2 \oplus b_3 v_3 \oplus \cdots,$$

where $\cdots b_3 b_2 b_1$ is the binary representation of $i$.

Antonov and Saleev (1979) show that equivalently the Sobol' sequence can be formed as

$$x_i = g_1 v_1 \oplus g_2 v_2 \oplus g_3 v_3 \oplus \cdots, \tag{3.1}$$

where $\cdots g_3 g_2 g_1$ is the binary representation of a particular *Gray code* evaluated at $i$. (A Gray code is a function, $G(i)$, on the nonnegative integers such that the binary representations of $G(i)$ and $G(i+1)$ differ in exactly one bit; that is, with a Hamming distance of 1.) The binary representation of the Gray code used by Antonov and Saleev is

$$\cdots g_3 g_2 g_1 = \cdots b_3 b_2 b_1 \oplus \cdots b_4 b_3 b_2.$$

(This is the most commonly used Gray code, which yields function values $0, 1, 3, 2, 6, 7, 5, 4, \ldots$ .) The Sobol' sequence from equation (3.1) can be generated recursively by

$$x_i = x_{i-1} \oplus v_r,$$

where $r$ is determined so that $b_r$ is the rightmost zero bit in the binary representation of $i - 1$.

The uniformity of a Sobol' sequence can be very sensitive to the starting values, especially in higher dimensions. Bratley and Fox (1988) discuss criteria for starting values, $m_1, m_2, \ldots$ . (The starting values used in the example with the primitive polynomial above satisfy those criteria.)

### 3.2.3 Comparisons

Empirical comparisons of various quasirandom sequences that have been reported in the literature are somewhat inconclusive. Sarkar and Prasad (1987) compare the performance of pseudorandom and quasirandom sequences in the solution of integral equations by Monte Carlo methods and find no difference in the performance of the two quasirandom sequences that they studied: the Halton and Faure sequences. Fox (1986), on the other hand, finds the performance of Faure sequences to be better than that of Halton sequences. This is supported by the study by Bratley and Fox (1988), who also find that the performance of the Sobol' sequence is roughly the same as that of the Faure sequence. The empirical results reported in Bratley, Fox, and Niederreiter (1992) show inconclusive differences between Sobol' sequences and Niederreiter sequences. In an application in financial derivatives, Papageorgiou and Traub (1996) compared the performance of a generalized Faure sequence with a Sobol' sequence. They concluded that the Faure sequence was superior in that problem. Some of the inferior performace of Sobol' sequences has been attributed to poor starting values (see Jäckel, 2002).

### 3.2.4 Variations

Tezuka (1993) describes an analog of Halton sequences and a generalization of Niederreiter sequences and then shows that these extensions are related.

Quasirandom sequences that cover domains other than hypercubes have also been studied. Beck and Chen (1987) review work on low-discrepancy sequences over various types of regions.

Fang and Wang (1994) defined a type of quasirandom sequence of higher dimensions that they called NT-nets ("number theory" nets). Fang and Li (1997) gave a method for generating random orthogonal matrices based on an NT-net.

Combination generators (see Section 1.8, page 46) for quasirandom sequences can also be constructed. Braaten and Weller (1979) describe a method in which a pseudorandom generator is used to scramble a quasirandom sequence. It generally would not be useful to combine another sequence with a quasirandom sequence except to shuffle the quasirandom sequence. The shuffling generator could be either a quasirandom generator or a pseudorandom generator, but a pseudorandom generator would be more likely to be effective. The use of a pseudorandom generator may provide a method of estimating variances in the results of Monte Carlo studies. When a combination generator is composed of one or more quasirandom generators and one or more pseudorandom generators, it is called a hybrid generator.

### 3.2.5  Computations

Halton sequences are easy to generate (Exercise 3.1 asks you to write a program to do so). Fox (1986) gives a program for Faure sequences, Bratley and Fox (1988) give a program for Sobol' sequences (using Gray codes, as mentioned above), and Bratley, Fox, and Niederreiter (1994) give a program for Niederreiter sequences.

## 3.3   Further Comments

There are several applications of Monte Carlo methods reported in the literature that use quasirandom numbers. For example, Shaw (1988) uses quasirandom sequences instead of the usual pseudorandom sequences for evaluating integrals arising in Bayesian inference. Do (1991) uses quasirandom sequences in a Monte Carlo bootstrap. Quasirandom sequences seem to be widely used in applications in finance. Joy, Boyle, and Tan (1996), empirically comparing the use of Faure sequences with pseudorandom sequences in valuing financial options of various types, found that the quasirandom sequences had better convergence rates in that application.

### Variances and Error Bounds

An important property of numerical computations is the rate of convergence. In a deterministic algorithm, an upper bound on the error can often be given in terms of some simple expression (usually problem specific, however, so of questionable utility). These bounds often decrease initially as a function of some aspect of the number of computations (perhaps the number of iterations or the number of terms in a series), and the rate of convergence is expressed as a function of the number of computations. In algorithms that ostensibly use random sampling (that is, ones using pseudorandom numbers), it may not be possible to determine an upper bound on the error. The nature of the methods are fundamentally different. In deterministic algorithms, we *approximate* a result using finite truncations of expressions with an infinite number of components.

(The simplest of these finite truncations is due to the finite precision arithmetic used in the computer.) In random algorithms, *estimate* a result using a random sample, and the error depends on the realizations of the random sampling. In this case, we quantify the "error" in terms of the variance of the estimator. If the error is taken to be proportional to the standard deviation, it usually decreases proportionally to the square root of the size of the sample.

Use of quasirandom sequences is more similar to use of a deterministic algorithm than it is to use of pseudorandom sequences. Although in some cases it is simple to make comparisons between the performance of pseudorandom and quasirandom sequences, the fundamental difference in the nature of the error bounds appropriate for Monte Carlo methods and for other numerical algorithms must be recognized. We comment further on these distinctions in Section 7.3.

Hickernell (1995) compares a certain type of error bound for quadrature using pseudorandom and quasirandom sequences and shows that the quasirandom sequences resulted in smaller bounds for errors. Bouleau and Lépingle (1994), quoting Pagès and Xiao, give comparative sample sizes required by pseudorandom and Halton and Faure sequences to achieve the same precision for quadrature in various dimensions from 2 to 20. Precision, in this case, is approximated using the asymptotic convergence of the quadrature formulas.

Variance is not a meaningful concept in quasirandom sequences because of the systematic way in which they are formed. In applications, however, variances may be needed to form confidence regions for results. Ökten (1998) suggests use of random samples from quasirandom sequences in order to estimate a confidence interval. It is not clear how useful these methods would be in practice. In most cases, it is better to rely on asymptotic deterministic error bounds for results computed using quasirandom sequences.

### Nonuniform Random Deviates

In many applications of pseudorandom or quasirandom sequences, the simulated or approximated $U(0,1)$ numbers are transformed into numbers similar to what might be expected when sampling from a distribution that is not uniform (a normal distribution, for example). In Chapter 4, we discuss methods for these transformations. We see that one of these methods is a direct transformation involving a single uniform deviate to produce a single nonuniform deviate. For this method, we can use a quasirandom sequence and we get a quasirandom sequence. In some sense, that sequence also has desirable distributional properties in the probability space of the transformation. For the other methods of transforming uniform deviates into nonuniform deviates (that is, the methods that use more than one uniform deviate to produce one nonuniform deviate—the methods most widely used), the patterns in the quasirandom sequence render the methods invalid.

**Further Reading**

The texts by Niederreiter (1992) and Tezuka (1995) have extensive discussions
of quasirandom sequences. Jäckel (2002) dissusses some important issues in the
use of quasirandom sequences, particularly the Sobol' sequences.

# Exercises

3.1. Write a subprogram to generate $d$-dimensional Halton sequences using the
integers $m$, $m+1$, ..., $m+n-1$ and using the first $d$ primes as the basis
for each succeeding component of the $d$-tuples. Your subprogram should
accept as input $m$, $d$, $n$, and possibly the declared first dimension of the
output array. Your subprogram should output an $n \times d$ array.

3.2. Use the subprogram from Exercise 3.1 to evaluate the integral

$$\int_0^1 \int_0^2 y \sin(\pi x)\, dy\, dx$$

for increasing values of $n$. How do the errors in your answers depend on
$n$? (What is the order of the error?)

3.3. Apply the dice test you constructed in Exercise 2.1 to 1000 pairs generated
by the Halton sequence program in Exercise 3.1.

3.4. (a) Write a program to perform a chi-squared goodness-of-fit test of $d$-
uniformity in the unit hypercube. For an input of $n$ $d$-vectors, form
$n^{d/(d+1)}$ equal-sized hypercubes by subdividing the unit interval of
each axis into $n^{1/(d+1)}$ equal-sized lengths. Use the counts in the
hypercubes in your chi-squared goodness-of-fit test.

   (b) Run the test program you wrote in Exercise 3.4a 100 times on $10^6$
five-dimensional vectors generated by the Halton sequence program
in Exercise 3.1. Make a q–q plot of the 100 computed chi-squared
statistics versus a true chi-squared random variable (of how many degrees of freedom?). Does your Halton generator produce acceptable
pseudorandom numbers? Why or why not?

3.5. Design and develop a hybrid combination generator that uses a pseudorandom number generator, perhaps such as you wrote in Exercise 1.10
of Chapter 1 (page 58), to shuffle the output of a quasirandom number
generator (perhaps the Halton sequence generator in Exercise 3.1). Be
sure to provide appropriate user control to set the initial state of the
combination generator.

3.6. Briefly, for each of the points listed in Section 2.5, beginning on page 87,
how would you rate quasirandom generators? (Compare Exercise 2.8.)

# Chapter 4

# Transformations of Uniform Deviates: General Methods

Sampling of random variates from a nonuniform distribution is usually done by applying a transformation to uniform variates. Each realization of the nonuniform random variable might be obtained from a single uniform variate or from a sequence of uniforms. Some methods that use a sequence of uniforms require that the sequence be independent; other methods use a random walk sequence, a Markov chain.

For some distributions, there may be many choices of algorithms for generation of random numbers. The algorithms differ in speed, accuracy, storage requirements, and complexity of coding. Some of the faster methods are approximate, but given the current cost of computing, the speedup resulting from an additional approximation beyond the approximation resulting from the ordinary finite-precision representation is not worth the accuracy loss. All of the methods that we discuss in this chapter are exact, so any approximation is a result of the ordinary rounding and truncation necessary in using a computer to simulate real numbers or infinite sets. Occasionally, a research paper will contend that the quality of the random numbers generated by some particular method, such as Box–Muller or the ratio-of-uniforms, is bad, but the quality ultimately depends only on the quality of the underlying uniform generator. A particular method, however, may exacerbate some fault in the uniform generator, and it is always a good idea to conduct a goodness-of-fit test for the specific distribution of interest. This is an ad hoc test, as we advocate in Chapter 2.

After accuracy, the next most important criterion is speed. The speed of a random number generation algorithm has two aspects: the setup time and the generation time. In most cases, the generation time is the more important component to optimize. Whenever the setup time is significant, the computer program can preserve the variables initialized, so that if the function is called again with the same parameters, the setup step can be bypassed. In a case of relatively expensive setup overhead, a software system may provide a second function for the same distribution with less setup time.

The two other criteria mentioned above, storage requirements and complexity of coding, are generally of very little concern in selecting algorithms for production random number generators.

Another important issue is whether the algorithm can be implemented in a portable manner, as discussed in Section 1.11.

The methods discussed in this chapter are "universal" in the sense that they apply to many different distributions. (Some authors call these "black box" methods.) Some of these methods are better than others for a particular distribution or for a particular range of the distribution. These techniques are used, either singly or in combination, for particular distributions. We discuss these methods in Chapter 5.

Some of these methods, especially those that involve inverting a function, apply directly only to univariate random variables, whereas other methods apply immediately to multivariate random variables.

The descriptions of the algorithms in this chapter are written with an emphasis on clarity, so they should not be incorporated directly into program code without considerations of the efficiency. These considerations generally involve avoiding unnecessary computations. This may mean defining a variable not mentioned in the algorithm description or reordering the steps slightly.

## 4.1  Inverse CDF Method

For the random variable $X$, the *cumulative distribution function*, or *CDF*, is the function $P_X$ defined by

$$P_X(x) = \Pr(X \le x),$$

where $\Pr(A)$ represents the probability of the event $A$. Two important properties are immediately obvious: the CDF is nondecreasing, and it is continuous from the right.

### Continuous Distributions

If $X$ is a scalar random variable with a continuous CDF $P_X$, then the random variable

$$U = P_X(X)$$

has a U(0, 1) distribution. (This is easy to show; you are asked to do that in Exercise 4.1, page 159.) This fact provides a very simple relationship with a uniform random variable $U$ and a random variable $X$ with distribution function $P$:

$$X = P_X^{-1}(U). \tag{4.1}$$

Use of this straightforward transformation is called the *inverse CDF* technique. The reason it works can be seen in Figure 4.1; over a range for which the

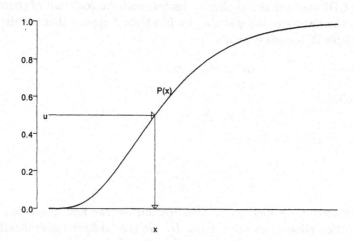

Figure 4.1: The Inverse CDF Method to Convert a Uniform Random Number to a Number from a Continuous Distribution

derivative of the CDF (the density) is large, there is more probability of realizing a uniform deviate.

The inverse CDF relationship exists between any two continuous (nonsingular) random variables. If $X$ is a continuous random variable with CDF $P_X$ and $Y$ is a continuous random variable with CDF $P_Y$, then

$$X = P_X^{-1}(P_Y(Y))$$

over the ranges of positive support. Use of this kind of relationship is a matching of "scores" (that is, of percentile points) of one distribution with those of another distribution. In addition to the uniform distribution, as above, this kind of transformation is sometimes used with the normal distribution.

Whenever the inverse of the distribution function is easy to compute, the inverse CDF method is a good one. It also has the advantage that basic relationships among a set of uniform deviates (such as order relationships) may result in similar relationships among the set of deviates from the other distribution.

Because it is relatively difficult to compute the inverse of some distribution functions of interest, however, the inverse CDF method is not as commonly used as its simplicity might suggest. Even when the inverse $P^{-1}$ exists in closed form, evaluating it directly may be much slower than use of some alternative method for sampling random numbers. On the other hand, in some cases when $P^{-1}$ does not exist in closed form, use of the inverse CDF method by solving the equation

$$P(x) - u = 0$$

may be better than the use of any other method.

## Discrete Distributions

The inverse CDF method also applies to discrete distributions, but of course we cannot take the inverse of the distribution function. Suppose that the discrete random variable $X$ has mass points

$$m_1 < m_2 < m_3 < \ldots$$

with probabilities

$$p_1, \; p_2, \; p_3, \; \ldots$$

and with the distribution function

$$P(x) = \sum_{i \ni m_i \leq x} p_i.$$

To use the inverse CDF method for this distribution, we first generate a realization $u$ of the uniform random variable $U$. We then deliver the realization of the target distribution as $x$, where $x$ satisfies the relationship

$$P(x_{(-)}) < u \leq P(x), \tag{4.2}$$

where $x_{(-)}$ is a value arbitrarily close to, but less than, $x$. Alternatively, we have

$$x = \min\{t, \text{ s.t. } u \leq P(t)\}. \tag{4.3}$$

This is illustrated in Figure 4.2.

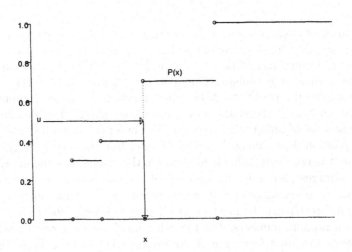

Figure 4.2: The Inverse CDF Method to Convert a Uniform Random Number to a Number from a Discrete Distribution

An example of a common and very simple application of the inverse CDF technique is for generating a random deviate from a Bernoulli distribution with parameter $\pi$, as in Algorithm 4.1. The probability function for the Bernoulli distribution with parameter $\pi$ is

$$p(x) = \pi^x(1-\pi)^{1-x}, \quad \text{for } x = 0, 1, \tag{4.4}$$

where $0 < \pi < 1$.

## Algorithm 4.1 Generating a Bernoulli Deviate by the Inverse CDF

1. Generate $u$ from a U(0, 1) distribution.

2. If $u < \pi$, then
    2.a. deliver 0;
   otherwise,
    2.b. deliver 1. ∎

Without loss of generality, we often assume that the mass points of a discrete distribution are the integers $1, 2, 3, \ldots$. The special case in which there are $k$ mass points and they all have equal probability is called the discrete uniform distribution, and the use of the inverse CDF method is particularly simple: the value is $\lceil uk \rceil$.

The evaluation of the inverse CDF involves a search. Rather than starting the search at some arbitrary point, it is usually more efficient to start it at some point with a higher probability of being near the solution. Either the mean or the mode is usually a good place to begin the search, which then proceeds up or down as necessary.

### Table Lookup

Using the inverse CDF method for a general discrete distribution is essentially a table lookup; it usually requires a search for the $x$ in equation (4.2). The search may be performed sequentially or by some tree traversal. Marsaglia (1963), Norman and Cannon (1972), and Chen and Asau (1974) describe various ways of speeding up the table lookup. Improving the efficiency of the table-lookup method is often done by incorporating some aspects of the *urn method*, in which the distribution is simulated by a table (an "urn") that contains the mass points in proportion to their population frequency. In the urn, each $p_i$ is represented as a rational fraction $n_i/N$, and a table of length $N$ is constructed with $n_i$ pointers to the mass point $i$. A discrete uniform deviate, $\lceil uN \rceil$, is then used as an index to the table to yield the target distribution.

The method of Marsaglia implemented by Norman and Cannon (1972) involves forming a table partitioned in such a way that the individual partitions can be sampled with equal probabilities. In this scheme, the probability $p_i$ associated with the $i^{\text{th}}$ mass point is expressed to a precision $t$ in the base $b$ as

$$p_i \approx \sum_{j=1}^{t} d_{ij}b^{-j}, \tag{4.5}$$

with $0 \leq d_{ij} < b$. We assume that there are $n$ mass points. (We frequently assume a finite set of mass points when working with discrete distributions. Because we can work with only a finite number of different values on the computer, it does not directly limit our methods, but we should be aware of any such exclusion of rare events and in some cases must modify our methods to be able to model rare events.) Let

$$
\begin{aligned}
P_0 &= 0, \\
P_j &= b^{-j} \sum_{i=1}^{n} d_{ij} \quad \text{for } j = 1, 2, \ldots, t, \\
N_0 &= 0, \\
N_k &= b^{-j} \sum_{j=1}^{k} \sum_{i=1}^{n} d_{ij} \quad \text{for } k = 1, 2, \ldots, t.
\end{aligned}
$$

Now, form a partitioned array and, in the $j^{\text{th}}$ partition, store $d_{ij}$ copies of $i$ (remember that the mass points are taken to be the integers; they could just as well be indexes). The $j^{\text{th}}$ partition is in locations $N_{j-1}+1$ to $N_j$. There are $N_t$ storage locations in all. Norman and Cannon (1972) show how to reduce the size of the table with a slight modification to the method. After the partitions are set up, Algorithm 4.2 generates a random number from the given distribution.

### Algorithm 4.2 Marsaglia/Norman/Cannon Table Lookup for Sampling a Discrete Random Variate

1. Generate $u$ from a $U(0,1)$ distribution, and represent it in base $b$ to $t$ places:

$$
u = \sum_{j=1}^{t} d_j b^{-j}.
$$

2. Find $m$ such that

$$
\sum_{j=0}^{m-1} P_j \leq u < \sum_{j=0}^{m} P_j.
$$

3. Take as the generated value the contents of location

$$
\sum_{j=1}^{m} d_j b^{m-j} - \left( b^m \sum_{j=0}^{m-1} P_j - N_{m-1} \right) + 1.
$$

In this algorithm, the $j^{th}$ partition is chosen with probability $P_j$. The probability of the $i^{th}$ mass point is

$$
\Pr(X = i) = \sum_{j=1}^{t} \Pr(j^{th} \text{ partition is chosen}) \times
$$

$$
\Pr(i \text{ is chosen from } j^{th} \text{ partition})
$$

$$
= \sum_{j=1}^{t} P_j \frac{d_{ij}}{\sum_{k=1}^{n} d_{kj}}
$$

$$
= \sum_{j=1}^{t} b^{-j} \sum_{i=1}^{n} d_{ij} \frac{d_{ij}}{\sum_{k=1}^{n} d_{kj}}
$$

$$
= \sum_{j=1}^{t} d_{ij} b^{-j}
$$

$$
\approx p_i.
$$

Norman and Cannon (1972) give a program to implement this algorithm that forms an equivalent but more compact partitioned array.

Chen and Asau (1974) give a hashing method using a "guide table". The guide table contains $n$ values $g_i$ that serve as indexes to the $n$ mass points in the CDF table. The $i^{th}$ guide value is the index of the largest mass point whose CDF value is less than $i/n$:

$$
g_i = \max_{\sum_{k=1}^{j} p_k < i/n} j.
$$

After the guide table is set up, Algorithm 4.3 generates a random number from the given distribution.

**Algorithm 4.3 Sampling a Discrete Random Variate Using the Chen and Asau Guide Table Method**

1. Generate $u$ from a U$(0,1)$ distribution, and set $i = \lceil un \rceil$.

2. Set $x = g_i + 1$.

3. While $\sum_{k=1}^{x} p_k > u$, set $x = x - 1$. ∎

**Efficiency of the Inverse CDF for Discrete Distributions**

Rather than using a stored table of the mass points of the distribution, we may seek other efficient methods of searching for the $x$ in equation (4.2). The search can often be improved by knowledge of the relative magnitude of the probabilities of the points. The basic idea is to begin at a point with a high probability of satisfying the relation (4.2). Obviously, the mode is a good place to begin the search, especially if the probability at the mode is quite high.

For many discrete distributions of interest, there may be a simple recursive relationship between the probabilities of adjacent mass points:

$$p(x) = f(p(x-1)) \quad \text{for } x > x_0,$$

where $f$ is some simple function (and we assume that the mass points differ by 1, and $x_0$ is the smallest value with positive mass). In the Poisson distribution (see page 188), for example,

$$p(x) = \theta p(x-1)/x \quad \text{for } x > 0.$$

For this case, Kemp (1981) describes two approaches. One is a "build-up search" method in which the CDF is built up by the recursive computation of the mass probabilities. This is Algorithm 4.4.

**Algorithm 4.4  Build-Up Search for Discrete Distributions**

0. Set $t = p(x_0)$.

1. Generate $u$ from a U(0,1) distribution, and set $x = x_0$, $p_x = t$, and $s = p_x$.

2. If $u \leq s$, then
    2.a. deliver $x$;
    otherwise,
    2.b. set $x = x+1$, $p_x = f(p_x)$, and $s = s + p_x$, and return to step 2. ∎

The second method that uses the recursive evaluation of probabilities to speed up the search is a "chop-down" method in which the generated uniform variate is decreased by an amount equal to the CDF. This method is given in Algorithm 4.5.

**Algorithm 4.5  Chop-Down Search for Discrete Distributions**

0. Set $t = p(x_0)$.

1. Generate $u$ from a U(0,1) distribution, and set $x = x_0$ and $p_x = t$.

2. If $u \leq p_x$, then
    2.a. deliver $x$;
    otherwise,
    2.b. set $u = u - p_x$, $x = x+1$, and $p_x = f(p_x)$, and return to step 2. ∎

Either of these methods could be modified to start at some other point, such as the mode.

**Interpolating in Tables**

Often, for a continuous random variable, we may have a table of values of the cumulative distribution function but not have a function representing the CDF over its full range. This situation may arise in applications in which a person familiar with the process can assign probabilities for the variable of interest yet may be unwilling to assume a particular distributional form. One approach to this problem is to fit a continuous function to the tabular values and then use the inverse CDF method on the interpolant. The simplest interpolating function, of course, is the piecewise linear function, but second- or third-degree polynomials may give a better fit. It is important, however, that the interpolant be monotone. Guerra, Tapia, and Thompson (1976) describe a scheme for approximating the CDF based on an interpolation method of Akima (1970). Their procedure is implemented in the IMSL routine rngct.

**Multivariate Distributions**

The inverse CDF method does not apply to a multivariate distribution, although marginal and conditional univariate distributions can be used in an inverse CDF method to generate multivariate random variates. If the CDF of the multivariate random variable $(X_1, X_2, \ldots, X_d)$ is decomposed as

$$P_{X_1 X_2 \ldots X_d}(x_1, x_2, \ldots, x_d) =$$
$$P_{X_1}(x_1) P_{X_2|X_1}(x_2|x_1) \cdots P_{X_d|X_1 X_2 \ldots X_{d-1}}(x_d|x_1, x_2, \ldots, x_{d-1})$$

and if the functions are invertible, the inverse CDF method is applied sequentially using independent realizations of a U(0,1) random variable, $u_1, u_2, \ldots, u_d$:

$$
\begin{aligned}
x_1 &= P_{X_1}^{-1}(u_1), \\
x_2 &= P_{X_2|X_1}^{-1}(u_2), \\
\ldots &\quad \ldots \\
x_d &= P_{X_d|X_1 X_2 \ldots X_{d-1}}^{-1}(u_d).
\end{aligned}
$$

The modifications of the inverse CDF for discrete random variables described above can be applied if necessary.

# 4.2  Decompositions of Distributions

It is often useful to break up the range of the distribution of interest using one density over one subrange and another density over another subrange. More generally, we may represent the distribution of interest as a mixture distribution that is composed of proportions of other distributions. Suppose that the probability density or probability function of the random variable of interest,

$p(\cdot)$, can be represented as

$$p(x) = \sum_{j=1}^{k} w_j p_j(x), \qquad (4.6)$$

where the $p_j(\cdot)$ are density functions or probability functions of random variables, the union of whose support is the support of the random variable of interest. We require

$$w_j \geq 0$$

and

$$\sum_{j=1}^{k} w_j = 1.$$

The random variable of interest has a *mixture distribution*.

If the $p_j$ are such that the pairwise intersections of the supports of the distributions are all null, the mixture is a *stratification*.

To generate a random deviate from a mixture distribution, first use a single uniform to select the component distribution, and then generate a deviate from it. The mixture can consist of any number of terms. To generate a sample of $n$ random deviates from a mixture distribution of $d$ distributions, consider the proportions to be the parameters of a $d$-variate multinomial distribution. The first step is to generate a single multinomial deviate, and then generate the required number of deviates from each of the component distributions.

Any decomposition of $p$ into the sum of nonnegative integrable functions yields the decomposition in equation (4.6). The nonnegative $w_i$ are chosen to sum to 1.

For example, suppose that a distribution has density $p(x)$, and for some constant $c$, $p(x) \geq c$ over $(a, b)$. Then, the distribution can be decomposed into a mixture of a uniform distribution over $(a, b)$ with proportion $c(b - a)$ and some leftover part, say $g(x)$. Now, $g(x)/(1 - c(b - a))$ is a probability density function. To generate a deviate from $p$:

> with probability $c(b - a)$,
>> generate a deviate from U$(a, b)$;
> otherwise,
>> generate a deviate from the density $\dfrac{1}{1 - c(b - a)} g(x)$.

If $c(b - a)$ is close to 1, we will generate from the uniform distribution most of the time, so even if it is difficult to generate from $g(x)/(1 - c(b - a))$, this decomposition of the original distribution may be useful.

Another way of forming a mixture distribution is to consider a density similar to equation (4.6) that is a conditional density,

$$p(x|y) = y p_1(x) + (1 - y) p_2(x),$$

where $y$ is the realization of a Bernoulli random variable, $Y$. If $Y$ takes a value of 0 with probability $w_1/(w_1 + w_2)$, then the density in equation (4.6) is the marginal density. This conditional distribution yields

$$
\begin{aligned}
p_X(x) &= \int p_{X,Y}(x,y)\,dy \\
&= \sum_y p_{X|Y=y} \Pr(Y = y) \\
&= w_1 p_1(x) + w_2 p_2(x),
\end{aligned}
$$

as in equation (4.6).

More generally, for any random variable $X$ with a distribution parameterized by $\theta$, we can think of the parameter as being the realization of a random variable $\Theta$. Some common distributions result from mixing other distributions; for example, if the gamma distribution is used to generate the parameter in a Poisson distribution, a negative binomial distribution is formed. Mixture distributions are often useful in their own right; for example, the beta-binomial distribution (see page 187) can be used to model overdispersion.

## 4.3 Transformations that Use More than One Uniform Deviate

Methods for generating random deviates by first decomposing the distribution of interest require the use of more than one uniform deviate for each deviate from the target distribution. Most other methods discussed in this chapter also require more than one uniform deviate for each deviate of interest. For such methods we must be careful to avoid any deleterious effects of correlations in the underlying uniform generator. An example of a short-range correlation occurs in the use of a congruential generator,

$$ x_i \equiv a x_{i-1} \bmod m, $$

when $x_{i-1}$ is extremely small. In this case, the value of $x_i$ is just $a x_{i-1}$ with no modular reduction. A small value of $x_{i-1}$ may correspond to some extreme intermediate value in one of the constituent distributions in the decomposition of the density in equation (4.6). Because $x_i = a x_{i-1}$, when $x_i$ is used to complete the transformation to the variate of interest, it may happen that the extreme values of that variate do not cover their appropriate range.

As a simple example, consider a method for generating a variate from a double exponential distribution. One way to do this is to use one uniform variate to generate an exponential variate (using one of the methods that we discuss below) and then use a second uniform variate to decide whether to change the sign of the exponential variate (with probability $1/2$). Suppose that the method for generating an exponential variate yields an extremely large value if the underlying uniform variate is extremely small. (The method given by equation (5.10) on page 176 does this.) If the next uniform deviate from the basic generator is used to determine whether to change the sign, it may happen that all of the extreme double exponentials generated have the same sign.

Many such problems arise because of a poor uniform generator; a particular culprit is a multiplicative congruential generator with a small multiplier. Use of a high-quality uniform generator generally solves the problem. A more conservative approach may be to use a different uniform generator for each uniform deviate used in the generation of a single nonuniform deviate. For this to be effective, each generator must be of high quality, of course.

Because successive numbers in a quasirandom sequence are constructed so as to span a space systematically, such sequences generally should not be used when more than one uniform deviate is transformed into a single deivate from another distribution. The autocorrelations in the quasirandom sequence may prevent certain ranges of values of the transformations from being realized.

A common way in which uniform deviates are transformed to deviates from nonuniform distributions is to use one uniform random number to make a decision about how to use another uniform random number. The decision is often based on a comparison of two floating-point numbers. In rare cases, because of slight differences in rounding to a finite precision, this comparison may result in different decisions in different computer environments. The different decisions can result in the generation of different output streams from that point on. Our goal of completely portable random number generators (Section 1.11) may not be achieved when comparisons are made between two floating-point numbers that might differ in the least significant bits on different systems.

## 4.4   Multivariate Uniform Distributions with Nonuniform Marginals

Suppose that $p_X$ is a continuous probability density function, and consider the set

$$S = \{(x, u), \text{ s.t. } 0 \leq u \leq p_X(x)\}. \tag{4.7}$$

Let $(X, U)$ be a bivariate random variable with uniform distribution over $S$. Its density function is

$$p_{XU}(x, u) = I_S(x, u). \tag{4.8}$$

The conditional distribution of $U$ given $X = x$ is $U(0, p_X(x))$, and the conditional distribution of $X$ given $U = u$ is also uniform with density

$$p_{X|U}(x|u) = I_{\{t, \text{ s.t. } p_X(t) \geq u\}}(x).$$

The important fact, which we see by integrating $u$ out of the density in equation (4.8), is that the marginal distribution of $X$ has density $p_X$.

This can be seen in Figure 4.3, where the points are uniformly distributed over $S$, but the marginal histogram of the $x$ values corresponds to the density $p_X$.

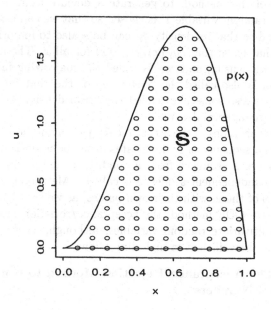

Figure 4.3: Support of a Bivariate Uniform Random Variable $(X, U)$ Having a Marginal with Density $p(x)$

These facts form the basis of methods of generating random deviates from various nonuniform distributions. The effort in these methods is expended in getting the bivariate uniform points over the region $S$. In most cases, this is done by generating bivariate points uniformly over some larger region and then rejecting those points that are not in the region $S$.

This same approach is valid if the random variable $X$ is a vector. In this case, we would identify a higher-dimensional region $S$ with a scalar $u$ and a vector $x$ corresponding respectively to a scalar uniform random variable and the vector random variable $X$.

## 4.5 Acceptance/Rejection Methods

To generate realizations of a random variable $X$, an *acceptance/rejection method* makes use of realizations of another random variable $Y$ having probability density $g_Y$ similar to the probability density of $X$, $p_X$. The basic idea is that selective subsamples from samples from one distribution are stochastically equivalent to samples from a different distribution. The acceptance/rejection technique is one of the most important methods in random number generation, and it occurs in many variations.

**Majorizing the Density**

In the basic form of the method, to generate a deviate from a distribution with density $p_X$, a random variable $Y$ is chosen so that we can easily generate realizations of it and so that its density $g_Y$ can be scaled to majorize $p_X$ using some constant $c$; that is, so that $cg_Y(x) \geq p_X(x)$ for all $x$. The density $g_Y$ is called the *majorizing* density, and $cg_Y$ is called the majorizing function. The majorizing function is also called the "envelope" or the "hat function". The majorizing density is also sometimes called the "trial density", the "proposal density", or the "instrumental density".

There are many variations of the acceptance/rejection method. The method described here uses a sequence of i.i.d. variates from the majorizing density. It is also possible to use a sequence from a conditional majorizing density. A method using a nonindependent sequence is called a Metropolis method (and there are variations of these, with their own names, as we see below).

Unlike the inverse CDF method, the acceptance/rejection method applies immediately to multivariate random variables, although, as we will see, the method may not be very efficient in high dimensions.

**Algorithm 4.6 The Acceptance/Rejection Method to Convert Uniform Random Numbers**

1. Generate $y$ from the distribution with density function $g_Y$.

2. Generate $u$ from a $U(0,1)$ distribution.

3. If $u \leq p_X(y)/cg_Y(y)$, then
     3.a. take $y$ as the desired realization;
   otherwise
     3.b. return to step 1.                                           ∎

It is easy to see that the random number delivered by Algorithm 4.6 has a density $p_X$. (In Exercise 4.2, page 160, you are asked to write the formal proof.) The pairs $(u, y)$ that are accepted follow a bivariate uniform distribution over the region $S$ in equation (4.7).

Figure 4.4 illustrates the functions used in the acceptance/rejection method. (Figure 4.4 shows the same density used in Figure 4.3 with a different scaling of the axes. The density is the beta distribution with parameters 3 and 2. In Exercise 4.3, page 160, you are asked to write a program implementing the acceptance/rejection method with the majorizing density shown.)

The acceptance/rejection method can be visualized as choosing a subsequence from a sequence of independently and identically distributed (i.i.d.) realizations from the distribution with density $g_Y$ in such a way that the subsequence has density $p_X$, as shown in Figure 4.5.

If we ignore the time required to generate $y$ from the dominating density $g_Y$, the closer $cg_Y(x)$ is to $p_X(x)$ (that is, the closer $c$ is to its lower bound of 1), the faster the acceptance/rejection algorithm will be. The proportion of

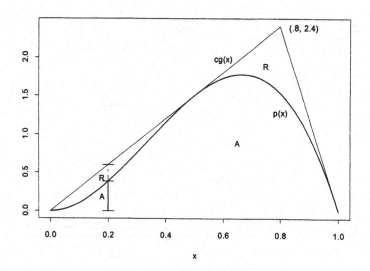

Figure 4.4: The Acceptance/Rejection Method to Convert Uniform Random Numbers

acceptances to the total number of trials is the ratio of the area marked "A" in Figure 4.4 to the total area of region "A" and region "R". Because $p_X$ is a density, the area of "A" is 1, so the relevant proportion is

$$1/(r+1), \qquad (4.9)$$

where $r$ is the area between the curves. This ratio only relates to the efficiency of the acceptance; other considerations in the efficiency, of course, involve the amount of computation necessary to generate from the majorizing density.

The random variable corresponding to the number of passes through the steps of Algorithm 4.6 until the desired variate is delivered has a geometric distribution (equation (5.21) on page 189, except beginning at 1 instead of 0) with parameter

$$\pi = 1/(r+1).$$

Selection of a majorizing function involves the principles of function approximation with the added constraint that the approximating function be a

| i.i.d. from $g_Y$ | $y_i$ | $y_{i+1}$ | $y_{i+2}$ | $y_{i+3}$ | $\cdots$ | $y_{i+k}$ | $\cdots$ |
|---|---|---|---|---|---|---|---|
| accept? | no | yes | no | yes | $\cdots$ | yes | $\cdots$ |
| i.i.d. from $p_X$ | | $x_j$ | | $x_{j+1}$ | $\cdots$ | $x_{j+l}$ | $\cdots$ |

Figure 4.5: Acceptance/Rejection

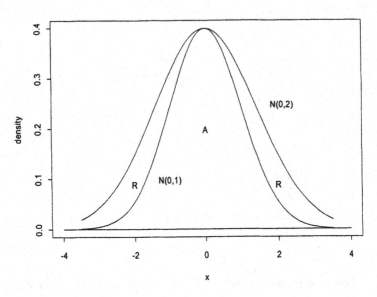

Figure 4.6: Normal (0, 1) Density with a Normal (0, 2) Majorizing Density

probability density from which it is easy to generate random variates. Often, $g_Y$ is chosen to be a very simple density, such as a uniform or a triangular density. When the dominating density is uniform, the acceptance/rejection method is similar to the "hit-or-miss" method (see Exercise 7.2, page 271).

The acceptance/rejection method can be used for multivariate random variables, in which case the majorizing distribution is also multivariate. For higher dimensions, however, the acceptance ratio (4.9) may be very small. Consider the use of a normal with mean 0 and variance 2 as a majorizing density for a normal with mean 0 and variance 1, as shown in Figure 4.6. A majorizing density like this with a shape more closely approximating that of the target density is more efficient. (This majorizing function is just chosen for illustration. An obvious problem in this case would be that if we could generate deviates from the N(0, 2) distribution, then we could generate ones from the N(0, 1) distribution, and we would not use this method.) In the one-dimensional case, as shown in Figure 4.6, the acceptance region is the area under the lower curve, and the rejection region is the thin shell between the two curves. The acceptance proportion (4.9) is $1/\sqrt{2}$. (Note that $c = \sqrt{2}$.) In higher dimensions, even a thin shell contains most of the volume, so the rejection proportion would be high. In $d$ dimensions, use of a multivariate normal with a diagonal variance-covariance matrix with all entries equal to 2 as a majorizing density to generate a multivariate normal with a diagonal variance-covariance matrix with all entries equal to 1 would have an acceptance proportion of only $1/\sqrt{d}$.

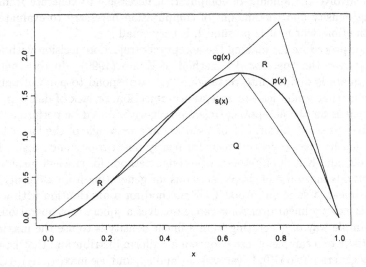

Figure 4.7: The Acceptance/Rejection Method with a Squeeze Function

## Reducing the Computations in Acceptance/Rejection: Squeeze Functions

A primary concern in reducing the number of computations in the acceptance/rejection method is to ensure that the proportion of acceptances is high; that is, that the ratio (4.9) is close to one. Two other issues are the difficulty in generating variates from the majorizing density and the speed of the computations to determine whether to accept or to reject.

If the target density, $p$, is difficult to evaluate, an easy way of speeding up the process is to use simple functions that bracket $p$ to avoid the evaluation of $p$ with a high probability. This method is called a "squeeze" (see Marsaglia, 1977). This allows quicker acceptance. The squeeze function is often a linear or piecewise linear function. The basic idea is to do pretests using simpler functions. Most algorithms that use a squeeze function only use one below the density of interest. Figure 4.7 shows a piecewise linear squeeze function for the acceptance/rejection setup of Figure 4.4. For a given trial value $y$, before evaluating $p_X(y)$ we may evaluate the simpler $s(y)$. If $u \leq s(y)/cg_Y(y)$, then $u \leq p_X(y)/cg_Y(y)$, so we can accept without computing $p_X(y)$. Pairs $(y, u)$ lying in the region marked "Q" allow for quick acceptance.

The efficiency of an acceptance/rejection method with a squeeze function depends not only on the area between the majorizing function and the target density, as in equation (4.9), but also on the difference in the total area of the acceptance region, which is 1, and the area under the squeeze function (that is, the area of the region marked "Q"). The closer this area is to 1, the more effective is the squeeze. These ratios of areas relate only to the efficiency of the acceptance and the quick acceptance. Other considerations in the efficiency, of course, involve the amount of computation necessary to generate from the majorizing density and the amount of computation necessary to evaluate the squeeze function, which, it is presumed, is very small.

Another procedure for making the acceptance/rejection decision with fewer computations is the "patchwork" method of Kemp (1990). In this method, the unit square is divided into rectangles that correspond to pairs of uniform distributions that would lead to acceptance, rejection, or lack of decision. The full evaluations for the acceptance/rejection algorithm need be performed only if the pair of uniform deviates to be used are in a rectangle of the latter type.

For a density that is nearly linear (or nearly linear over some range), Marsaglia (1962) and Knuth (1998) describe some methods for efficient generation. These methods make use of simple methods for generating from a density that is exactly linear. Use of an inverse CDF method for a distribution with a density that is exactly linear over some range involves a square root operation, but another simple way of generating from a linear density is to use the maximum order statistic of a sample of size two from a uniform distribution; that is, independently generate two $U(0,1)$ variates, $u_1$ and $u_2$, and use $\max(u_1, u_2)$. (Order statistics from a uniform distribution have a beta distribution; see Section 6.4.1, page 221.) Following Knuth's development, suppose that, as in Figure 4.8, the density over the interval $(s, s+h)$ is bounded by two parallel lines,

$$l_1(x) = a - b(x - s)/h$$

and

$$l_2(x) = b - b(x - s)/h.$$

Consider the density $p(x)$ shown in Figure 4.8. Algorithm 4.7, which is Knuth's method, yields deviates from the distribution with density $p$. Notice the use of the maximum of two uniform deviates to generate from an exactly linear density. By determining the probability that the resulting deviate falls in any given interval, it is easy to see that the algorithm yields deviates from the given density. You are asked to show this formally in Exercise 4.8, page 161. (The solution to the exercise is given in Appendix B.)

### Algorithm 4.7 Sampling from a Nearly Linear Density

1. Generate $u_1$ and $u_2$ independently from a $U(0,1)$ distribution. Set $u = \min(u_1, u_2)$, $v = \max(u_1, u_2)$, and $x = s + hu$.

2. If $v \leq a/b$, then
   2.a. go to step 3;
   otherwise,
   2.b. if $v > u + p(x)/b$, go to step 1.

3. Deliver $x$.

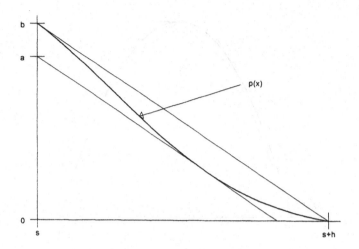

Figure 4.8: A Nearly Linear Density

Usually, when we take advantage of the fact that a density is nearly linear, it is not the complete density that is linear, but rather the nearly linear density is combined with other densities to form the density of interest. The density shown in Figure 4.8 may be the density over some interval $(s, s+h)$ so $\int_s^{s+h} p(x)\,\mathrm{d}x = p < 1$. (See the discussion of mixtures of densities in Section 4.2.)

For densities that are concave, we can also very easily form linear majorizing and linear squeeze functions. The majorizing function is a polygon of tangents and the squeeze function is a polygon of secants, as shown in Figure 4.9 (for the density $p(x) = \frac{4}{3}(1-x^2)$ over $[-1,1]$). Any number of polygonal sections could be used in this approach. The tradeoffs involve the amount of setup and house-keeping for the polygonal sections and the proportion of total rejections and the proportion of easy acceptances. The formation of the majorizing and squeeze functions can be done adaptively or sequentially, as we discuss on page 151.

## Acceptance/Rejection for Discrete Distributions

There are various ways that acceptance/rejection can be used for discrete distributions. One advantage of these methods is that they can be easily adapted to changes in the distribution. Rajasekaran and Ross (1993) consider the discrete random variable $X_s$ such that

$$
\begin{aligned}
\Pr(X_s = x_i) &= p_{si} \\
&= \frac{a_{si}}{a_{s1} + a_{s2} + \cdots a_{sk}}, \quad i = 1, \ldots, k.
\end{aligned}
$$

(If $\sum_{i=1}^{k} a_{si} = 1$, the numerator $a_{si}$ is the ordinary probability $p_{si}$ at the mass point $i$.) Suppose that there exists an $a_i^*$ such that $a_i^* \leq a_{si}$ for $s = 1, 2, \ldots$ and

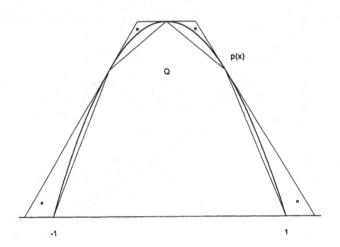

Figure 4.9: Linear Majorizing and Squeeze Functions for a Concave Density

$b > 0$ such that $\sum_{i=1}^{k} a_{si} \geq b$ for $s = 1, 2, \ldots$. Let

$$a^* = \max\{a_i^*\},$$

and let

$$P_{si} = a_{si}/a^* \quad \text{for } i = 1, \ldots, k.$$

The generation method for $X_s$ is shown in Algorithm 4.8.

### Algorithm 4.8 Acceptance/Rejection Method for Discrete Distributions

1. Generate $u$ from a U(0,1) distribution, and let $i = \lceil ku \rceil$.

2. Let $r = i - ku$.

3. If $r \leq P_{si}$, then
   3.a. take $i$ as the desired realization;
   otherwise,
   3.b. return to step 1.                                          ∎

Suppose that for the random variable $X_{s+1}$, $p_{s+1,i} \neq p_{si}$ for some $i$. (Of course, if this is the case for mass point $i$, it is also necessarily the case for some other mass point.) For each mass point for which the probability changes, reset $P_{s+1,i}$ to $a_{s+1,i}/a^*$ and continue with Algorithm 4.8.

Rajasekaran and Ross (1993) also gave two other acceptance/rejection type algorithms for discrete distributions that are particularly efficient for use with distributions that may be changing. The other algorithms require slightly more preprocessing time but yield faster generation times than Algorithm 4.8.

## Variations of Acceptance/Rejection

There are many variations of the basic acceptance/rejection method, and the idea of selection of variates from one distribution to form a sample from a different distribution forms the basis of several other methods discussed in this chapter, such as formation of ratios of uniform deviates, use of the characteristic function, and various uses of Markov chains.

Wallace (1976) introduced a modified acceptance/rejection method called *transformed rejection*. In the transformed acceptance/rejection method, the steps of Algorithm 4.6 are combined and rearranged slightly. Let $G$ be the CDF corresponding to the dominating density $g$. Let $H(x) = G^{-1}(x)$, and let $h(x) = \mathrm{d}\,H(x)/\mathrm{d}x$. If $v$ is a U$(0, 1)$ deviate, step 1 in Algorithm 4.6 is equivalent to $y = H(v)$, so we have Algorithm 4.9.

### Algorithm 4.9 The Transformed Acceptance/Rejection Method

1. Generate $u$ and $v$ independently from a U$(0, 1)$ distribution.

2. If $u \leq p(H(v))h(v)/c$, then
       2.a. take $H(v)$ as the desired realization;
   otherwise,
       2.b. return to step 1. ∎

Marsaglia (1984) describes a method very similar to the transformed acceptance/rejection method: use ordinary acceptance/rejection to generate a variate $x$ from the density proportional to $p(H(\cdot))h(\cdot)$ and then return $H(x)$. The choice of $H$ is critical to the efficiency of the method, of course. It should be close to the inverse of the CDF of the target distribution, $P^{-1}$. Marsaglia called this the *exact-approximation* method. Devroye (1986a) calls the method *almost exact inversion*.

## Other Applications of Acceptance/Rejection

The acceptance/rejection method can often be used to evaluate an elementary function at a random point. Suppose, for example, that we wish to evaluate $\tan(\pi U)$ for $U$ distributed as U$(-.5, .5)$. A realization of $\tan(\pi U)$ can be simulated by generating $u_1$ and $u_2$ independently from U$(-1, 1)$, checking if $u_1^2 + u_2^2 \leq 1$, and, if so, delivering $u_1/u_2$ as $\tan(\pi u)$. (To see this, think of $u_1$ and $u_2$ as sine and cosine values.) Von Neumann (1951) gives an acceptance/rejection method for generating sines and cosines of random angles. An example of evaluating a logarithm can be constructed by use of the equivalence of an inverse CDF method and an acceptance/rejection method for sampling an exponential random deviate. (The methods are equivalent in a stochastic sense; they are both valid, but they will not yield the same stream of deviates.)

These methods of evaluating deterministic functions are essentially the same as using the "hit-or-miss" Monte Carlo method described in Exercise 7.2 on page 271 to evaluate an integral.

Generally, if reasonable numerical software is available for evaluating special functions, it should be used rather than using Monte Carlo methods to estimate the function values.

## Quality and Portability of Acceptance/Rejection Methods

Acceptance/rejection methods, like any method for generating nonuniform random numbers, are dependent on a good source of uniform deviates. Hörmann and Derflinger (1993) illustrate that small values of the multiplier in a congruential generator for the uniform deviates can result in poor quality of the output from an acceptance/rejection method. Of course, we have seen that small multipliers are not good for generating uniform deviatess. (See the discussion about Figure 1.3, page 16.) Hörmann and Derflinger rediscover the method of expression (1.23) and recommend using it so that larger multipliers can be used in the linear congruential generator.

Acceptance/rejection methods generally use two uniform deviates to decide whether to deliver one variate of interest. In implementing an acceptance/rejection method, we must be aware of the cautionary note in Section 4.3, page 111. If the $y$ in Algorithm 4.6 is special (extreme, perhaps) and results from a special value from the uniform generator, the $u$ generated subsequently may also be special and may almost always result in the same decision to accept or to reject. Thus, we may get either an abundance or a deficiency of special values for the distribution of interest.

Because of the comparison of floating-point numbers (that occurs in step 3 of Algorithm 4.6), there is a chance that an acceptance/rejection method may yield different streams on different computer systems or in implementations in different precisions. Even if the computations are carried out correctly, the program is inherently nonportable and the results may not be strictly reproducible because if a comparision on one system at a given precision results in acceptance and the comparison on another system results in rejection, the two output streams will be different. At best, the streams will be the same except for a few differences; at worst, however, because of how the output is used, the results will be different beginning at the point at which the acceptance/rejection decision is different. If the decision results in the generation of another random number (as in Algorithm 4.10 on page 126), the two output streams can become completely different.

## Acceptance/Rejection for Multivariate Distributions

The acceptance/rejection method is one of the most widely applicable methods for random number generation. It is used in many different forms, often in combination with other methods. It is clear from the description of the algorithm that the acceptance/rejection method applies equally to multivariate distributions. (The uniform random number is still univariate, of course.)

As we have mentioned, however, for higher dimensions, the rejection proportion may be high, and thus the efficiency of the acceptance/rejection method may be low.

**Example of Acceptance/Rejection: A Bivariate Gamma Distribution**

Becker and Roux (1981) defined a bivariate extension of the gamma distribution that serves as a useful model for failure times for two related components in a system. (The model is also a generalization of a bivariate exponential distribution introduced by Freund, 1961; see Steel and Le Roux, 1987.) The probability density is given by

$$
p_{X_1 X_2}(x_1, x_2) = \begin{cases}
\begin{aligned}
&\lambda_2 \left(\Gamma(\alpha_1)\,\Gamma(\alpha_2)\,\beta_1^{\alpha_1}\beta_2^{\alpha_2}\right)^{-1} \times \\
&x_1^{\alpha_1-1} \left(\lambda_2(x_2 - x_1) + x_1\right)^{\alpha_2-1} \times \\
&\exp\left(-(\tfrac{1}{\beta_1} + \tfrac{1}{\beta_2} - \tfrac{\lambda_2}{\beta_2})x_1 - \tfrac{\lambda_2}{\beta_2}x_2\right) \quad \text{for} \ \ 0 \le x_1 \le x_2,
\end{aligned} \\[2em]
\begin{aligned}
&\lambda_1 \left(\Gamma(\alpha_1)\,\Gamma(\alpha_2)\,\beta_1^{\alpha_1}\beta_2^{\alpha_2}\right)^{-1} \times \\
&x_2^{\alpha_2-1} \left(\lambda_1(x_1 - x_2) + x_2\right)^{\alpha_1-1} \times \\
&\exp\left(-(\tfrac{1}{\beta_1} + \tfrac{1}{\beta_2} - \tfrac{\lambda_1}{\beta_1})x_2 - \tfrac{\lambda_1}{\beta_1}x_1\right) \quad \text{for} \ \ 0 \le x_2 < x_1,
\end{aligned} \\[2em]
0 \hspace{12em} \text{elsewhere.}
\end{cases}
$$

$$(4.10)$$

The density for $\alpha_1 = 4$, $\alpha_2 = 3$, $\beta_1 = 3$, $\beta_2 = 1$, $\lambda_1 = 3$, and $\lambda_2 = 2$ is shown in Figure 4.10.

It is a little more complicated to determine a majorizing density for this distribution. First of all, not many bivariate densities are familiar to us. The density must have support over the positive quadrant. A bivariate normal density might be tried, but the $\exp(-(u_1 x_1 + u_2 x_2)^2)$ term in the normal density dies out more rapidly than the $\exp(-v_1 x_1 - v_2 x_2)$ term in the gamma density. The normal cannot majorize the gamma in the limit.

We may be concerned about covariance of the variables in the bivariate gamma distribution, but the fact that the variables have nonzero covariance is of little concern in using the acceptance/rejection method. The main thing, of course, is that we determine a majorizing density so that the probability of acceptance is high. We can use a bivariate density of independent variables as the majorizing density. The density would be the product of two univariate densities. A bivariate distribution of independent exponentials might work. Such a density has a maximum at $(0,0)$, however, and there would be a large volume between the bivariate gamma density and the majorizing function formed from a bivariate exponential density. We can reduce this volume by choosing a bivariate uniform over the rectangle with corners $(0,0)$ and $(z_1, z_2)$. Our majorizing

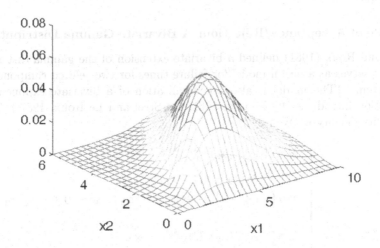

Figure 4.10: A Bivariate Gamma Density, Equation (4.10)

density then is composed of two densities, a bivariate exponential,

$$
g_1(y_1, y_2) = \begin{cases} \frac{1}{v} \exp\left(-\frac{y_1}{\theta_1} - \frac{y_2}{\theta_2}\right) & \text{for } y_1 > z_1 \text{ and } y_2 > 0 \\ & \text{or } y_1 > 0 \text{ and } y_2 > z_2, \\ 0 & \text{elsewhere,} \end{cases} \tag{4.11}
$$

where the constant $v$ is chosen to make $g_1$ a density, and a bivariate uniform,

$$
g_2(y_1, y_2) = \begin{cases} \frac{1}{z_1 z_2} & \text{for } 0 < y_1 \le z_1 \text{ and } 0 < y_2 \le z_2, \\ 0 & \text{elsewhere.} \end{cases} \tag{4.12}
$$

Next, we choose $\theta_1$ and $\theta_2$ so that the bivariate exponential density can majorize the bivariate gamma density. This requires that

$$
\frac{1}{\theta_1} \ge \max\left(\left(\frac{1}{\beta_1} + \frac{1}{\beta_2} - \frac{\lambda_2}{\beta_2}\right), \frac{\lambda_1}{\beta_1}\right),
$$

with a similar requirement for $\theta_2$. Let us choose $\theta_1 = 1$ and $\theta_2 = 2$. Next, we choose $z_1$ and $z_2$ as the mode of the bivariate gamma density. This point is $(4\frac{1}{3}, 2)$. We now choose $c$ so that $cg_1(z_1, z_2) \ge p(z_1, z_2)$.
    The method is:

1. Generate $u$ from a $U(0,1)$ distribution.

2. Generate $(y_1, y_2)$ from a bivariate exponential density such as (4.11) except over the full range; that is, with $v = \theta_1 \theta_2$.

3. If $(y_1, y_2)$ is outside of the rectangle with corners $(0,0)$ and $(z_1, z_2)$, then
    3.a. if $u \leq p(y_1, y_2)/cg_1(y_1, y_2)$, then
        3.a.i. deliver $(y_1, y_2)$;
    otherwise,
        3.a.ii. go to step 1;
otherwise,
    3.b. generate $(y_1, y_2)$ as bivariate uniform deviates in that rectangle
    and if $u \leq p(y_1, y_2)/(cy_1 y_2)$, then
        3.b.i. deliver $(y_1, y_2)$;
    otherwise,
        3.b.ii. go to step 1.

The majorizing density could be changed so that it is closer to the bivariate gamma density. In particular, instead of the uniform density over the rectangle with a corner on the origin, a pyramidal density that is closer to the bivariate gamma density could be used.

# 4.6 Mixtures and Acceptance Methods

In practice, in acceptance/rejection methods, the density of interest $p$ and/or the majorizing density are often decomposed into mixtures. If the mixture for the density is a stratification, it may be possible to have simple majorizing and squeeze functions within each stratum. Ahrens (1995) suggested using a stratification into equal-probability regions (that is, the $w_j$s in equation (4.6) are all constant) and then using constant majorizing and squeeze functions in each stratum. There is, of course, a tradeoff in gains in high probability of acceptance (because the majorizing function is close to the density) and/or in efficiency of the evaluation of the acceptance decision (because the squeeze function is close to the density) and the complexity introduced by the decomposition. Decomposition into regions where the density is nearly constant almost always will result in overall gains in efficiency. If the decomposition is into equal-probability regions, the random selection of the stratum is very fast.

There are many ways in which mixtures can be combined with acceptance/rejection methods.

Suppose that the density of interest, $p$, may be written as

$$p(x) = w_1 p_1(x) + w_2 p_2(x),$$

and suppose that there is a density $g$ that majorizes $w_1 p_1$; that is, $g(x) \geq w_1 p_1(x)$ for all $x$. Kronmal and Peterson (1981, 1984) consider this case and propose the following algorithm, which they call the acceptance/complement method.

## Algorithm 4.10 The Acceptance/Complement Method to Convert Uniform Random Numbers

1. Generate $y$ from the distribution with density function $g$.

2. Generate $u$ from a $U(0,1)$ distribution.

3. If $u > w_1 p_1(y)/g(y)$, then generate $y$ from the density $p_2$.

4. Take $y$ as the desired realization.     ∎

We discussed nearly linear densities and gave Knuth's algorithm for generating from such densities as Algorithm 4.7. Devroye (1986a) gives an algorithm for a special nearly linear density; namely, one that is almost flat. The method is based on a simple decomposition using the supremum of the density. (In practice, as we have indicated in discussing other techniques, this method would probably be used for a component of a density that has already been decomposed.) To keep the description simple, assume that the range of the random variable is $(-1, 1)$ and that the density $p$ satisfies

$$\sup_x p(x) - \inf_x p(x) \le \frac{1}{2}$$

over that interval. Now, because $p$ is a density, we have

$$0 \le \inf_x p(x) \le \frac{1}{2} \le \sup_x p(x)$$

and

$$\sup_x p(x) \le 1.$$

Let $p^* = \sup_x p(x)$, and decompose the target density into

$$p_1(x) = p(x) - \left(p^* - \frac{1}{2}\right)$$

and

$$p_2(x) = \left(p^* - \frac{1}{2}\right)$$

The method is shown in Algorithm 4.11.

## Algorithm 4.11 Sampling from a Nearly Flat Density

1. Generate $u$ from $U(0,1)$.

2. Generate $x$ from $U(-1,1)$.

3. If $u > 2(p(x) - (p^* - \frac{1}{2}))$, then generate $x$ from $U(-1,1)$.

4. Deliver $x$.     ∎

Another variation on the general theme of acceptance/rejection applied to mixtures was proposed by Deák (1981) in what he called the "economical method". To generate a deviate from the density $p$ using this method, an auxiliary density $g$ is used, and an "excess area" and a "shortage area" are defined. The excess area is where $g(x) > p(x)$, and the shortage area is where $g(x) \leq p(x)$. We define two functions $p_1$ and $p_2$:

$$\begin{aligned} p_1(x) &= g(x) - p(x) \quad \text{if} \quad g(x) - p(x) < 0, \\ &= 0 \quad \text{otherwise}, \end{aligned}$$

$$\begin{aligned} p_2(x) &= p(x) - g(x) \quad \text{if} \quad p(x) - g(x) \geq 0, \\ &= 0 \quad \text{otherwise}. \end{aligned}$$

Now, we define a transformation $T$ that will map the excess area into the shortage area in a way that will yield the density $p$. Such a $T$ is not unique, but one transformation that will work is

$$T(x) = \min \left\{ t, \text{ s.t. } \int_{-\infty}^{x} p_1(s)\, ds = \int_{-\infty}^{t} p_2(s)\, ds \right\}.$$

Algorithm 4.12 shows the method.

## Algorithm 4.12 The Economical Method to Convert Uniform Random Numbers

1. Generate $y$ from the distribution with density function $g$.

2. If $p(y)/g(y) < 1$, then
   2.a. generate $u$ from a U(0, 1) distribution;
   2.b. if $u \leq p(y)/g(y)$, then replace $y$ with $T(y)$.

3. Take $y$ as the desired realization.                    ∎

Using the representation of a discrete distribution that has $k$ mass points as an equally weighted mixture of $k$ two-point distributions, Deák (1986) develops a version of the economical method for discrete distributions. (See Section 4.8, page 133, on the alias method for additional discussion of two-point representations.)

Marsaglia and Tsang (1984) give a method that involves forming a decomposition of a density into horizontal slices with equal areas. For a unimodal distribution, they first form two regions, one on each side of the mode, prior to the slicing decomposition. They call a method that uses this kind of decomposition the "ziggurat method".

Marsaglia and Tsang (1998) also describe a decomposition and transformation that they called the "Monty Python method", in which the density (or a part of the density) is divided into three regions as shown in the left-hand plot in Figure 4.11. If the density of interest has already been decomposed into a mixture, the part to be decomposed further, $p(x)$, is assumed to have been scaled to integrate to 1. The density in Figure 4.11 may represent the right-hand side of a $t$ distribution, for example, but the function shown has been scaled to integrate to 1. The support of the density is transformed if necessary to begin at 0. One region is now rotated and stretched to fit into an area within a rectangle above another region, as shown in the right-hand plot in Figure 4.11.

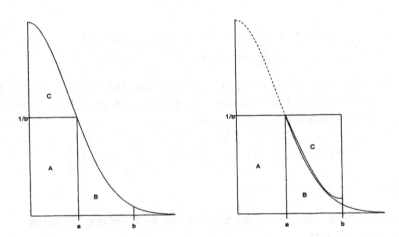

Figure 4.11: The Monty Python Decomposition Method

The key parameter in the Monty Python method is $b$, the length of the base of a rectangle that has an area of 1. The portion of the distribution represented by the density above $1/b$ between 0 and $p^{-1}(1/b)$ (denote this point by $a$) is transformed into a region of equal area between $a$ and $b$ bounded from below by the function

$$g(x) = \frac{1}{b} - cp(b - x) - d,$$

where $c$ and $d$ are chosen so the area is equal to the original and $g(x) \geq p(x)$ over $(a, b)$. This implies that the tail area beyond $b$ is equal to the area between $p(x)$ and $g(x)$. If a point $(x, y)$, chosen uniformly over the rectangle, falls in region A or B, then $x$ is delivered; if it falls in C, then $b - x$ is delivered; otherwise, $x$ is discarded, and a variate is generated from the tail of the distribution.

The efficiency of this method obviously depends on a choice of $b$ in which the decomposition minimizes the tail area. Marsaglia and Tsang (1998) suggest an improvement that may allow a better choice of $b$. Instead of the function $g(x)$, a polynomial, say a cubic, is determined to satisfy the requirements of majorizing $p(x)$ over $(a, b)$ and having an area equal to the original areal of C. This allows more flexibility in the choice of $b$. This is the same as the transformation in the exact-approximation method of Marsaglia referred to earlier.

# 4.7 Ratio-of-Uniforms Method

Kinderman and Monahan (1977) discuss a very useful relationship among random variables $U$, $V$, and $V/U$. If $(U,V)$ is uniformly distributed over the set

$$C = \left\{ (u,v),\ \text{s.t.}\ 0 \le u \le \sqrt{h\left(\frac{v}{u}\right)} \right\}, \qquad (4.13)$$

where $h$ is a nonnegative integrable function, then $V/U$ has probability density proportional to $h$. Use of this relationship is called a *ratio-of-uniforms* method.

It is easy to see that this relationship holds. For $U$ and $V$ as given, their joint density is $p_{UV}(u,v) = I_C(u,v)/c$, where $c$ is the area of $C$. Let $X = U$ and $Y = V/U$. The Jacobian of the transformation is $x$, so the joint density of $X$ and $Y$ is $p_{XY}(x,y) = x I_C(x,y)/c$. Hence, we have

$$p_{XY}(x,y) = \frac{x}{c} I_{\left[0,\ \sqrt{h(y)}\right]}(x),$$

and integrating out $x$, we get

$$
\begin{aligned}
p_Y(y) &= \int_0^{\sqrt{h(y)}} \frac{x}{c}\,\mathrm{d}x \\
&= \frac{1}{2c} h(y).
\end{aligned}
$$

In practice, we may choose a simple geometric region that encloses $C$, generate a uniform point in the rectangle, and reject a point that does not satisfy

$$u \le \sqrt{h\left(\frac{v}{u}\right)}.$$

The larger region enclosing $C$ is called the majorizing region because it is similar to the region under the majorizing function in acceptance/rejection methods.

The ratio-of-uniforms method is very simple to apply, and it can be quite fast.

If $h(x)$ and $x^2 h(x)$ are bounded in $C$, a simple form of the majorizing region is the rectangle

$$\{(u,v),\ \text{s.t.}\ 0 \le u \le b,\ c \le v \le d\},$$

where

$$
\begin{aligned}
b &= \sup_x \sqrt{h(x)}, \\
c &= \inf_x x\sqrt{h(x)}, \\
d &= \sup_x x\sqrt{h(x)}.
\end{aligned}
$$

This yields the method shown in Algorithm 4.13.

## Algorithm 4.13 Ratio-of-Uniforms Method (Using a Rectangular Majorizing Region for Continuous Variates)

1. Generate $u$ and $v$ independently from a $U(0,1)$ distribution.

2. Set $u_1 = bu$ and $v_1 = c + (d - c)v$.

3. Set $x = v_1/u_1$.

4. If $u_1^2 \leq h(x)$, then
   4.a. take $x$ as the desired realization;
   otherwise,
   4.b. return to step 1.                                            ∎

Figure 4.12 shows a rectangular region and the area of acceptance for the same density used to illustrate the acceptance/rejection method in Figure 4.4.

The full rectangular region as defined above has a very low proportion of acceptances in the example shown in Figure 4.12. There are many obvious ways of reducing the size of this region. A simple reduction would be to truncate the rectangle by the line $v = u$, as shown. Just as in other acceptance/rejection methods, there is a tradeoff in the effort to generate uniform deviates over a region with a high acceptance rate and the wasted effort of generating uniform deviates that will be rejected. The effort to generate *only* in the acceptance region is likely to be slightly greater than the effort to invert the CDF.

Wakefield, Gelfand, and Smith (1991) give a generalization of the ratio-of-uniforms method by introducing a strictly increasing, differentiable function $g$ that has the property $g(0) = 0$. Their method uses the fact that if $(U, V)$ is uniformly distributed over the set

$$C_{h,g} = \left\{ (u, v), \text{ s.t. } 0 \leq u \leq g\left( ch\left( \frac{v}{g'(u)} \right) \right) \right\},$$

where $c$ is a positive constant and $h$ is a nonnegative integrable function as before, then $V/g'(U)$ has a probability density proportional to $h$.

## Ratio-of-Uniforms and Acceptance/Rejection

Stadlober (1990, 1991) considers the relationship of the ratio-of-uniforms method to the ordinary acceptance/rejection method and applied the ratio-of-uniforms method to discrete distributions. If $(U, V)$ is uniformly distributed over the rectangle

$$\{(u, v), \text{ s.t. } 0 \leq u \leq 1, -1 \leq v \leq 1\},$$

and $X = sV/U + a$, for any $s > 0$, then $X$ has the density

$$g_X(x) = \begin{cases} \dfrac{1}{4s}, & a - s \leq x \leq a + s, \\[2ex] \dfrac{s}{4(x - a)^2} & \text{elsewhere}, \end{cases}$$

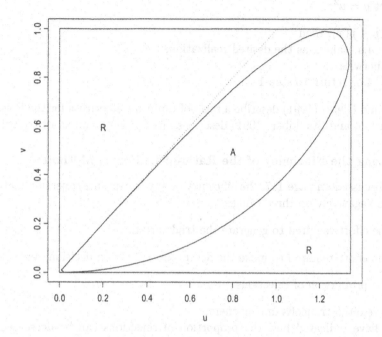

Figure 4.12: The Ratio-of-Uniform Method (Same Density as in Figure 4.4)

and the conditional density of $Y = U^2$, given $X$, is

$$
g_{Y|X}(y|x) = \begin{cases} 1 & \text{for} \quad a - s \le x \le a + s, \quad \text{and} \quad 0 \le y \le 1, \\[2mm] \dfrac{(x-a)^2}{s^2} & \text{for} \quad x > |a + s|, \qquad\quad \text{and} \quad 0 \le y \le \dfrac{s^2}{(x-a)^2}, \\[2mm] 0 & \text{elsewhere.} \end{cases}
$$

The conditional distribution of $Y$ given $X = x$ is uniform on $(0, 4sg(x))$, and the ratio-of-uniforms method is an acceptance/rejection method with a table mountain majorizing function.

### Ratio-of-Uniforms for Discrete Distributions

Stadlober (1990) gives the modification of the ratio-of-uniforms method in Algorithm 4.14 for a general discrete random variable with probability function $p(\cdot)$.

### Algorithm 4.14 Ratio-of-Uniforms Method for Discrete Variates

1. Generate $u$ and $v$ independently from a U(0, 1) distribution.

2. Set $x = \lfloor a + s(2v - 1)/u \rfloor$.

3. Set $y = u^2$.

4. If $y \leq p(x)$, then
    4.a. take $x$ as the desired realization;
    otherwise,
    4.b. return to step 1. ∎

Ahrens and Dieter (1991) describe a ratio-of-uniforms algorithm for the Poisson distribution, and Stadlober (1991) describes one for the binomial distribution.

### Improving the Efficiency of the Ratio-of-Uniforms Method

As we discussed on page 117, the efficiency of any acceptance/rejection method depends negatively on three things:

- the effort required to generate the trial variates;

- the effort required to make the acceptance/rejection decision; and

- the proportion of rejections.

There are often tradeoffs among them.

We have indicated how the proportion of rejections can be decreased by forming the majorizing region so that it is closer in shape to the shape of the acceptance region. This generally comes at the cost of more effort to generate trial variates. The increase in effort is modest if the majorizing region is a polygon. Leydold (2000) described a systematic method of forming polygonal majorizing regions for a broad class of distributions ($T$-concave distributions, see page 152).

The effort required to make the acceptance/rejection decision can be reduced in the same manner as a squeeze in acceptance/rejection. If a convex polygonal set interior to the acceptance region can be defined, then acceptance decisions can be made quickly by comparisons with linear functions. For a class of distributions, Leydold (2000) described a systematic method for forming interior polygons from construction points defined by the sides of a polygonal majorizing region.

### Quality of Random Numbers Produced by the Ratio-of-Uniforms Method

The ratio-of-uniforms method, like any method for generating nonuniform random numbers, is dependent on a good source of uniforms. The special relationships that may exist between two successive uniforms when one of them is an extreme value can cause problems, as we indicated on pages 111 and 122. Given a high-quality uniform generator, the method is subject to the same issues of floating-point computations that we discussed on page 122.

Afflerbach and Hörmann (1992) and Hörmann (1994b) indicate that, in some cases, output of the ratio-of-uniforms method can be quite poor because of structure in the uniforms. The ratio-of-uniforms method transforms all points lying on one line through the origin into a single number. Because of the lattice structure of the uniforms from a linear congruential generator, the lines passing through the origin have regular patterns, which result in structural gaps in the numbers yielded by the ratio-of-uniforms method. Noting these distribution problems, Hörmann and Derflinger (1994) make some comparisons of the ratio-of-uniforms method with the transformed rejection method (Algorithm 4.9, page 121), and based on their empirical study, they recommend the transformed rejection method over the ratio-of-uniforms method. The quality of the output of the ratio-of-uniforms method, however, is more a function of the quality of the uniform generator and would usually not be of any concern if a good uniform generator is used. The relative computational efficiencies of the two methods depend on the majorizing functions used. The polygonal majorizing functions used by Leydold (2000) in the ratio-of-uniforms method apparently alleviate some of the problems found by Hörmann (1994b).

### Ratio-of-Uniforms for Multivariate Distributions

Stefănescu and Văduva (1987) and Wakefield, Gelfand, and Smith (1991) extend the ratio-of-uniforms method to multivariate distributions.

As we mentioned in discussing simple acceptance/rejection methods, the probability of rejection may be quite high for multivariate distributions. High correlations in the target distribution can also reduce the efficiency of the ratio-of-uniforms method even further.

## 4.8  Alias Method

Walker (1977) shows that a discrete distribution with $k$ mass points can be represented as an equally weighted mixture of $k$ two-point distributions; that is, distributions with only two mass points. Consider the random variable $X$ such that

$$\Pr(X = x_i) = p_i, \quad i = 1, \ldots, k,$$

and $\sum_{i=1}^{k} p_i = 1$. Walker constructed $k$ two-point distributions,

$$\Pr(Y_i = y_{ij}) = q_{ij}, \quad j = 1, 2; \quad i = 1, \ldots, k$$

(with $q_{i1} + q_{i2} = 1$) in such a way that any $p_i$ can be represented as $k^{-1}$ times a sum of $q_{i,j}$s. (It is easy to prove that this can be done; use induction, starting with $k = 1$.)

A setup procedure for the alias method is shown in Algorithm 4.15. The setup phase associates with each $i = 1$ to $k$ a value $P_i$ that will determine whether the original mass point or an "alias" mass point, indexed by $a_i$, will be delivered when $i$ is chosen with equal probability, $\frac{1}{k}$. Two lists, $L$ and $H$, are maintained to determine which points or point pairs have probabilities less than or greater than $\frac{1}{k}$. At termination of the setup phase, all points or point pairs have probabilities equal to $\frac{1}{k}$. Marsaglia calls the setup phase "leveling the histogram". The outputs of the setup phase are two lists, $P$ and $a$, each of length $k$.

**Algorithm 4.15 Alias Method Setup to Initialize the Lists $a$ and $P$**

0. For $i = 1$ to $k$,
   set $a_i = i$;
   set $P_i = 0$;
   set $b_i = p_i - \frac{1}{k}$;
   and if $b_i < 0$, put $i$ in the list $L$;
   otherwise, put $i$ in the list $H$.

1. If $\max(b_i) = 0$, then stop.

2. Select $l \in L$ and $h \in H$.

3. Set $c = b_l$ and $d = b_h$.

4. Set $b_l = 0$ and $b_h = c + d$.

5. Remove $l$ from $L$.

6. If $b_h \leq 0$, then remove $h$ from $H$; and if $b_h < 0$, then put $h$ in $L$.

7. Set $a_l = h$ and $P_l = 1 + kc$.

8. Go to step 1.  ∎

Notice that $\sum b_i = 0$ during every step. The steps are illustrated in Figure 4.13 for a distribution such that

$$
\begin{aligned}
\Pr(X = 1) &= .30, \\
\Pr(X = 2) &= .05, \\
\Pr(X = 3) &= .20, \\
\Pr(X = 4) &= .40, \\
\Pr(X = 5) &= .05.
\end{aligned}
$$

At the beginning, $L = \{2, 5\}$ and $H = \{1, 4\}$. In the first step, the values corresponding to 2 and 4 are adjusted.

The steps to generate deviates, after the values of $P_i$ and $a_i$ are computed by the setup, are shown in Algorithm 4.16.

**Algorithm 4.16 Generation Using the Alias Method Following the Setup in Algorithm 4.15**

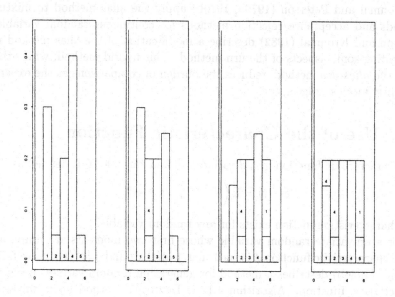

Figure 4.13: Setup for the Alias Method; Leveling the Histogram

1. Generate $u$ from a $U(0,1)$ distribution.

2. Generate $i$ from a discrete uniform over $1, 2, \ldots, k$.

3. If $u \leq P_i$, then
    3.a.  deliver $x_i$;
  otherwise,
    3.b.  deliver $x_{a_i}$.                                              ■

It is clear that the setup time for Algorithm 4.15 is $O(k)$ because the total number of items in the lists $L$ and $H$ goes down by at least one at each step. If, in step 2, the minimum and maximum values of $b$ are found, as in the original algorithm of Walker (1977), the algorithm may proceed slightly faster in some cases, but then the algorithm is $O(k \log k)$. The setup method given in Algorithm 4.15 is from Kronmal and Peterson (1979a). Vose (1991) also describes a setup procedure that is $O(k)$. Once the setup is finished by whatever method, the generation time is constant, or $O(1)$. The alias method is always at least as fast as the guide table method of Chen and Asau (Algorithm 4.3 on page 107). Its speed relative to the table-lookup method of Marsaglia as implemented by Norman and Cannon (Algorithm 4.2 on page 106) depends on the distribution. Variates from distributions with a substantial proportion of mass points whose base $b$ representations (equation (4.5)) have many zeros can be generated very rapidly by the table-lookup method.

In the IMSL Libraries, the routine **rngda** performs both the setup and the generation of discrete random deviates using an alias method.

Kronmal and Peterson (1979a, 1979b) apply the alias method to mixture methods and acceptance/rejection methods for continuous random variables. Peterson and Kronmal (1982) describe a modification of the alias method incorporating some aspects of the urn method. This hybrid method, which they called the *alias-urn* method, reduces the burden of comparisons at the expense of slightly more storage space.

## 4.9    Use of the Characteristic Function

The characteristic function of a $d$-variate random variable $X$ is defined as

$$\phi_X(t) = \mathrm{E}\left(e^{it^T X}\right), \quad t \in \mathbb{R}^d. \tag{4.14}$$

The characteristic function exists for any random variable.

For a univariate random variable whose first two moments are finite, and whose characteristic function $\phi$ is such that $\int |\phi(t)|\, dt$ and $\int |\phi''(t)|\, dt$ are finite, Devroye (1986b) describes a method for generating random variates using the characteristic function. Algorithm 4.17 is Devroye's method for a univariate continuous random variable with probability density function $p(\cdot)$ and characteristic function $\phi(\cdot)$.

**Algorithm 4.17 Conversion of Uniform Random Numbers Using the Characteristic Function**

0. Set $a = \sqrt{\frac{1}{2\pi} \int |\phi(t)|\, dt}$ and $b = \sqrt{\frac{1}{2\pi} \int |\phi''(t)|\, dt}$.

1. Generate $u$ and $v$ independently from a U$(-1, 1)$ distribution.

2. If $u < 0$, then
    2.a. set $y = bv/a$ and $t = a^2|u|$;
   otherwise,
    2.b. set $y = b/(va)$ and $t = a^2 v^2 |u|$.

3. If $t \leq p(y)$, then
    3.a. take $y$ as the desired realization;
   otherwise,
    3.b. return to step 1.                                     ∎

This method relies on the facts that, under the existence conditions,

$$p(y) \leq \frac{1}{2\pi} \int |\phi(t)|\, dt \quad \text{for all } y$$

and

$$p(y) \leq \frac{1}{2\pi x^2} \int |\phi''(t)|\, dt \quad \text{for all } y.$$

Both of these facts are easily established by use of the inverse characteristic function transform, which exists by the integrability conditions on $\phi(t)$.

The method requires evaluation of the density at each step. Devroye (1996b) also discusses variations that depend on Taylor expansion coefficients.

Devroye (1991) describes a related method for the case of a discrete random variable.

The characteristic function allows evaluation of all moments that exist. If only some of the moments are known, an approximate method described by Devroye (1989) can be used.

# 4.10 Use of Stationary Distributions of Markov Chains

Many of the methods for generating deviates from a given distribution are based on a representation of the density that allows the use of some simple transformation or some selection rule for deviates generated from a different density. In the univariate ratio-of-uniforms method, for example, we identify a bivariate uniform random variable with a region of support such that one of the marginal distributions is the distribution of interest. We then generate bivariate uniform random deviates over the region, and then, by a very simple transformation and selection, we get univariate deviates from the distribution of interest.

Another approach is to look for a stochastic process that can be easily simulated and such that the distribution of interest can be identified as a distribution at some point in the stochastic process. The simplest useful stochastic process is a Markov chain with a stationary distribution corresponding to the distribution of interest.

### Markov Chains: Basic Definitions

A *Markov chain* is a sequence of random variables, $X_1, X_2, \ldots$, such that the distribution of $X_{t+1}$ given $X_t$ is independent of $X_{t-1}, X_{t-2}, \ldots$. A sequence of realizations of such random variables is also called a Markov chain (that is, the term Markov chain can be used to refer either to a random sequence or to a fixed sequence of realizations). In this section, we will briefly discuss some types of Markov chains and their properties. The main purpose is to introduce the terms that are used to characterize the Markov chains in the applications that we describe later. See Meyn and Tweedie (1993) for extensive discussions of Markov chains. Tierney (1996) discusses the aspects of Markov chains that are particularly relevant for the applications that we consider in later sections.

The union of the supports of the random variables is called the *state space* of the Markov chain. Whether or not the state space is countable is an important characteristic of a Markov chain. A Markov chain with a countable or "discrete" state space is easier to work with and can be used to approximate a Markov chain with a continuous state space. Another important characteristic of a Markov chain is the nature of the indexing. As we have written the sequence above, we have implied that the index is discrete. We can generalize this to a continuous index, in which case we usually use the notation $X(t)$.

A Markov chain is *time homogeneous* if the distribution of $X_t$ is independent of $t$. For our purposes, we can usually restrict attention to a time-homogeneous discrete-state Markov chain with a discrete index, and this is what we assume in the following discussion in this section.

For the random variable $X_t$ in a discrete-state Markov chain with state space $S$, let $I$ index the states; that is, $i$ in $I$ implies that $s_i$ is in $S$. For $s_i \in S$, let

$$\Pr(X_t = s_i) = p_{ti}.$$

The Markov chain can be characterized by an initial distribution and a square *transition matrix* or *transition kernel* $K = (k_{ij})$, where

$$k_{ij} = \Pr(X_{t+1} = s_i | X_t = s_j).$$

The distribution at time $t$ is characterized by a vector of probabilities $p_t = (p_{t1}, p_{t2}, \ldots)$, so the vector itself is called a *distribution*. The initial distribution is $p_0 = (p_{01}, p_{02}, \ldots)$, and the distribution at time $t = 1$ is $Kp_0$. We sometimes refer to a Markov chain by the doubleton $(K, p_0)$ or just $(K, p)$.

In general, we have

$$\begin{aligned} p_t &= Kp_{t-1} \\ &= K^t p_0. \end{aligned}$$

We denote the elements of $K^t$ as $k_{ij}^{(t)}$. The relationships above require that $\sum_j k_{ij} = 1$. (A matrix with this property is called a stochastic matrix.)

A distribution $p$ such that

$$Kp = p$$

is said to be *invariant* or *stationary*. From that definition, we see that an invariant is an eigenvector of the transition matrix corresponding to an eigenvalue of 1. (Notice the unusual usage of the word "distribution"; in this context, it means a vector.) For a given Markov chain, it is of interest to know whether the chain has an invariant (that is, whether the transition matrix has an eigenvalue equal to 1) and if so, whether the invariant can be reached from the starting distribution $p_0$.

Some Markov chains oscillate among a set of distributions. (For example, think of a two-state Markov chain whose transition matrix has elements $k_{11} = k_{22} = 0$ and $k_{12} = k_{21} = 1$.) We will be interested in chains that do not oscillate; that is, chains that are *aperiodic*. A chain is guaranteed to be aperiodic if, for some $t$ sufficiently large, $k_{ii}^{(t)} > 0$ for all $i$ in $I$.

A Markov chain is *reversible* if, for any $t$, the conditional probability of $X_t$ given $X_{t+1}$ is the same as the conditional probability of $X_t$ given $X_{t-1}$. A discrete-space Markov chain obviously is reversible if and only if its transition matrix is symmetric. A Markov chain defined by $(K, p)$ is said to be in *detailed balance* if $K_{ij} p_i = K_{ji} p_j$.

A Markov chain is *irreducible* if, for all $i, j$ in $I$, there exists a $t > 0$ such that $k_{ij}^{(t)} > 0$. If the chain is irreducible, detailed balance and reversibility are equivalent.

Another property of interest is when a Markov chain first takes on a given state. This is called the *first passage time* for that state. Given that the chain is in a particular state, the first passage time to that state is the *first return time* for that state. Let $T_{ii}$ be the first return time to state $i$; that is, for a discrete time chain, let $T_{ii} = \min\{t, \text{ s.t. } X_t = s_i | X_0 = s_i\}$. ($T_{ii}$ is a random variable.) An irreducible Markov chain is *recurrent* if, for some $i$, $\Pr(T_{ii} < \infty) = 1$. (For an irreducible chain, this implies the condition for all $i$.) An irreducible Markov chain is *positive recurrent* if, for some $i$, $\mathrm{E}(T_{ii}) < \infty$. (For an irreducible chain, this implies the condition for all $i$.)

An aperiodic, irreducible, positive recurrent Markov chain is associated with a stationary distribution or invariant distribution, which is the *limiting distribution* of the chain. In applications of Markov chains, the question of whether the chain has converged to this limiting distribution is one of the primary concerns.

Applications that we discuss in later sections have uncountable state spaces, but the basic concepts extend to those. For a continuous state space, instead of a vector specifying the distribution at any given time, we have a probability density at that time, $K$ is a conditional probability density for $X_{t+1} | X_t$, and we have a similar expression for the density at $t + 1$ formed by integrating over the conditional density weighted by the unconditional density at $t$. Tierney (1996) carefully discusses the generalization to an uncountable state space and a continuous index.

## Markov Chain Monte Carlo

There are various ways of using a Markov chain to generate random variates from some distribution related to the chain. Such methods are called *Markov chain Monte Carlo*, or *MCMC*.

An algorithm based on a stationary distribution of a Markov chain is an *iterative method* because a sequence of operations must be performed until they *converge*. A Markov chain is the basis for several schemes for generating random numbers. The interest is not in the sequence of the Markov chain itself. The elements of the chain are accepted or rejected in such a way as to form a different chain whose stationary distribution is the distribution of interest.

Following engineering terminology for sampling sequences, the techniques based on these chains are generally called "samplers". The static sample, and not the sequence, is what is used. The objective in the Markov chain samplers is to generate a sequence of autocorrelated points with a given stationary distribution.

## The Metropolis Random Walk

For a distribution with density $p_X$, the Metropolis algorithm, introduced by Metropolis et al. (1953), generates a random walk and performs an acceptance/rejection based on $p$ evaluated at successive steps in the walk. In the simplest version, the walk moves from the point $y_i$ to a candidate point $y_{i+1} = y_i + s$, where $s$ is a realization from $U(-a, a)$, if

$$\frac{p_X(y_{i+1})}{p_X(y_i)} \geq u, \qquad (4.15)$$

where $u$ is an independent realization from $U(0, 1)$. If the new point is at least as probable (that is, if $p_X(y_{i+1}) \geq p_X(y_i)$), the condition (4.15) implies acceptance without the need to generate $u$. The random walk of Metropolis et al. is the basic algorithm of *simulated annealing*, which is currently widely used in optimization problems. It is also used in simulations of models in statistical mechanics (see Section 7.9). The algorithm is described in Exercise 7.16 on page 277.

If the range of the distribution is finite, the random walk is not allowed to go outside of the range. Consider, for example, the von Mises distribution, with density

$$p(x) = \frac{1}{2\pi I_0(c)} e^{c\cos(x)} \quad \text{for } -\pi \leq x \leq \pi, \qquad (4.16)$$

where $I_0$ is the modified Bessel function of the first kind and of order zero. Notice, however, that it is not necessary to know this normalizing constant because it is canceled in the ratio. The fact that all we need is a nonnegative function that is proportional to the density of interest is an important property of this method. In the ordinary acceptance/rejection methods, we need to know the constant.

If $c = 3$, after a quick inspection of the amount of fluctuation in $p$, we may choose $a = 1$. The output for $n = 1000$ and a starting value of $y_0 = 1$ is shown in Figure 4.14. The output is a Markov chain. A histogram, which is not affected by the sequence of the output in a large sample, is shown in Figure 4.15.

The von Mises distribution is an easy one to simulate by the Metropolis algorithm. This distribution is often used by physicists in simulations of lattice gauge and spin models, and the Metropolis method is widely used in these simulations. Notice the simplicity of the algorithm: we do not need to determine a majorizing density nor even evaluate the Bessel function that is the normalizing constant for the von Mises density.

The Markov chain samplers generally require a "burn-in" period (that is, a number of iterations before the stationary distribution is achieved). In practice, the variates generated during the burn-in period are discarded. The number of iterations needed varies with the distribution and can be quite large sometimes thousands. The von Mises example shown in Figure 4.14 is unusual; no burn-in

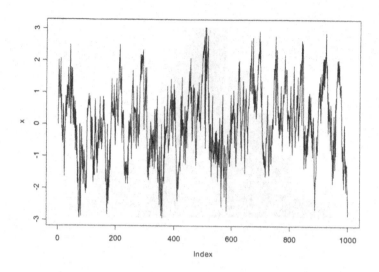

Figure 4.14: Sequential Output from the Metropolis Algorithm for a Von Mises Distribution

is required. In general, convergence is much quicker for univariate distributions with finite ranges such as this one.

It is important to remember what convergence means; it does *not* mean that the sequence is independent from the point of convergence forward. The deviates are still from a Markov chain.

The Metropolis acceptance/rejection sequence is illustrated in Figure 4.16. Compare this with the acceptance/rejection method based on independent variables, as illustrated in Figure 4.5.

### The Metropolis–Hastings Method

Hastings (1970) describes an algorithm that uses a more general chain for the acceptance/rejection step. Instead of just basing the decision on the probability density $p_X$ as in the inequality (4.15), the *Metropolis–Hastings sampler* to generate deviates from a distribution with a probability density $p_X$ uses deviates from a Markov chain with density $g_{Y_{t+1}|Y_t}$. The method is shown in Algorithm 4.18. The conditional density $g_{Y_{t+1}|Y_t}$ is chosen so that it is easy to generate deviates from it.

### Algorithm 4.18 Metropolis–Hastings Algorithm

0. Set $k = 0$.

1. Choose $x^{(k)}$ in the range of $p_X$. (The choice can be arbitrary.)

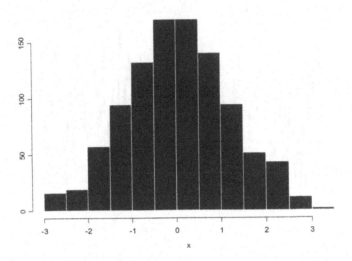

Figure 4.15: Histogram of the Output from the Metropolis Algorithm for a Von Mises Distribution

2. Generate $y$ from the density $g_{Y_{t+1}|Y_t}(y|x^{(k)})$.

3. Set $r$:
$$r = p_X(y)\frac{g_{Y_{t+1}|Y_t}(x^{(k)}|y)}{p_X(x^{(k)})g_{Y_{t+1}|Y_t}(y|x^{(k)})}.$$

4. If $r \geq 1$, then
   4.a. set $x^{(k+1)} = y$;
   otherwise,
       4.b. generate $u$ from $U(0,1)$ and
           if $u < r$, then
               4.b.i. set $x^{(k+1)} = y$;
           otherwise,
               4.b.ii. set $x^{(k+1)} = x^{(k)}$.

5. If convergence has occurred, then

| random walk | $y_i$ | $y_{i+1} =$ | $y_{i+3} =$ | $y_{i+2} =$ | |
|---|---|---|---|---|---|
| | | $y_i + s_{i+1}$ | $y_{i+1} + s_{i+2}$ | $y_{i+2} + s_{i+3}$ | $\cdots$ |
| accept? | no | yes | no | yes | $\cdots$ |
| i.i.d. from $p_X$ | | $x_j$ | | $x_{j+1}$ | $\cdots$ |

Figure 4.16: Metropolis Acceptance/Rejection

5.a. deliver $x = x^{(k+1)}$;

otherwise,

5.b. set $k = k + 1$, and go to step 2. ∎

Compare Algorithm 4.18 with the basic acceptance/rejection method in Algorithm 4.6, page 114. The analog to the majorizing function in the Metropolis–Hastings algorithm is the reference function

$$\frac{g_{Y_{t+1}|Y_t}(x|y)}{p_X(x)\, g_{Y_{t+1}|Y_t}(y|x)}.$$

In Algorithm 4.18, $r$ is called the "Hastings ratio", and step 4 is called the "Metropolis rejection". The conditional density $g_{Y_{t+1}|Y_t}(\cdot|\cdot)$ is called the "proposal density" or the "candidate generating density". Notice that because the reference function contains $p_X$ as a factor, we only need to know $p_X$ to within a constant of proportionality. As we have mentioned already, this is an important characteristic of the Metropolis algorithms.

We can see that this algorithm delivers realizations from the density $p_X$ by using the same method suggested in Exercise 4.2 (page 160); that is, determine the CDF and differentiate. The CDF is the probability-weighted sum of the two components corresponding to whether the chain moved or not. In the case in which the chain does move (that is, in the case of acceptance), for the random variable $Z$ whose realization is $y$ in Algorithm 4.18, we have

$$
\begin{aligned}
\Pr(Z \le x) &= \Pr\left(Y \le x \,\middle|\, U \le p(Y)\frac{g(x_i|Y)}{p(x_i)g(Y|x_i)}\right) \\
&= \frac{\int_{-\infty}^{x}\int_{0}^{p(t)g(x_i|t)/(p(x_i)g(t|x_i))} g(t|x_i)\,\mathrm{d}s\,\mathrm{d}t}{\int_{-\infty}^{\infty}\int_{0}^{p(t)g(x_i|t)/(p(x_i)g(t|x_i))} g(t|x_i)\,\mathrm{d}s\,\mathrm{d}t} \\
&= \int_{-\infty}^{x} p_X(t)\,\mathrm{d}t.
\end{aligned}
$$

We can illustrate the use of the Metropolis–Hastings algorithm using a Markov chain in which the density of $X_{t+1}$ is normal with a mean of $X_t$ and a variance of $\sigma^2$. Let us use this density to generate a sample from a standard normal distribution (that is, a normal with a mean of 0 and a variance of 1). We start with $x_0$ chosen arbitrarily. We take logs and cancel terms in the expression for $r$ in Algorithm 4.18. The sequential output for $n = 1000$, a starting value of $x_0 = 10$ and a variance of $\sigma^2 = 9$ is shown in Figure 4.17. Notice that the values descend very quickly from the starting value, which would be a very unusual realization of a standard normal. This example is also special. In practice, we generally cannot expect such a short burn-in period. Notice also in Figure 4.17 the horizontal line segments where the underlying Markov chain did not advance.

There are several variations of the basic Metropolis–Hastings algorithm. See Bhanot (1988) and Chib and Greenberg (1995) for descriptions of modifications

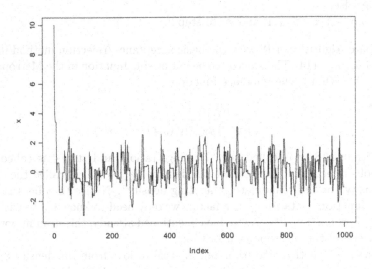

Figure 4.17: Sequential Output from a Standard Normal Distribution Using a Markov Chain, $N(X_t, \sigma^2)$

and generalizations. Also see Section 4.14 for two related methods: Gibbs sampling and hit-and-run sampling. Because those methods are particularly useful in multivariate simulation, we defer the discussion to that section.

The Markov chain Monte Carlo method has become one of the most important tools in statistics in recent years. Its applications pervade Bayesian analysis as well as Monte Carlo procedures in many settings. See Gilks, Richardson, and Spiegelhalter (1996) for several examples.

Whenever a correlated sequence such as a Markov chain is used, variance estimation must be performed with some care. In the more common cases of positive autocorrelation, the ordinary variance estimators are negatively biased. The method of batch means or some other method that attempts to account for the autocorrelation should be used. See Section 7.4 for discussions of these methods.

Tierney (1991, 1994) describes an *independence sampler*, a Metropolis–Hastings sampler for which the proposal density does not depend on $Y_t$; that is, $g_{Y_{t+1}|Y_t}(\cdot|\cdot) = g_{Y_{t+1}}(\cdot)$. For this type of proposal density, it is more critical that $g_{Y_{t+1}}(\cdot)$ approximates $p_X(\cdot)$ fairly well and that it can be scaled to majorize $p_X(\cdot)$ in the tails. Liu (1996) and Roberts (1996) discuss some of the properties of the independence sampler and its relationship to other Metropolis–Hastings methods.

As with the acceptance/rejection methods using independent sequences, the acceptance/rejection methods based on Markov chains apply immediately to multivariate random variables. As mentioned above, however, convergence generally becomes slower as the number of elements in the random vector increases.

As an example of MCMC in higher dimensions, consider an example similar to that shown in Figure 4.17 except for a multivariate normal distribution instead of a univariate one. We use a $d$-dimensional normal with a mean vector $x_t$ and a variance-covariance matrix $\Sigma$ to generate $x_{t+1}$ for use in the Metropolis–Hastings method of Algorithm 4.18. Taking $d = 3$,

$$\Sigma = \begin{bmatrix} 9 & 0 & 0 \\ 0 & 9 & 0 \\ 0 & 0 & 9 \end{bmatrix},$$

and starting with $x_0 = (10, 10, 10)$, the first 1000 values of the first element (which should be a realization from a standard univariate normal) are shown in Figure 4.18.

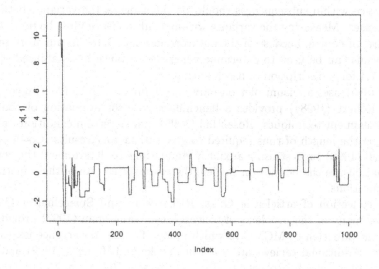

Figure 4.18: Sequential Output of $x_1$ from a Trivariate Standard Normal Distribution Using a Markov Chain, $N(X_t, \Sigma)$

## Convergence

Two of the most important issues in MCMC concern the rate of convergence (that is, the length of the burn-in) and the frequency with which the chain advances. In many applications of simulation, such as studies of waiting times in queues, there is more interest in transient behavior than in stationary behavior. This is not the case in random number generation using an iterative method. For general use in random number generation, the stationary distribution is the only thing of interest. (We often use the terms "Monte Carlo" and "simulation" rather synonymously; stationarity and transience, however, are often the key distinctions between Monte Carlo applications and simulation applications. In simulation in practice, the interest is rarely in the stationary behavior, but it is in these Monte Carlo applications.)

The issue of convergence is more difficult to address in multivariate distributions. It is for multivariate distributions, however, that the MCMC method is most useful. This is because the Metropolis–Hastings algorithm does not require knowledge of the normalizing constants, and the computation of a normalizing constant may be more difficult for multivariate distributions.

Gelman and Rubin (1992b) give examples in which the burn-in is much longer than might be expected. Various diagnostics have been proposed to assess convergence. Cowles and Carlin (1996) discuss and compare thirteen different ones. Most of these diagnostics use multiple chains in one way or another; see, for example, Gelman and Rubin (1992a), Roberts (1992), and Johnson (1996). Multiple chains or separate subsequences within a chain can be compared using analysis-of-variance methods. Once convergence has occurred, the variance within subsequences should be the same as the variance between subsequences. Measuring the variance within a subsequence must be done with some care, of course, because of the autocorrelations. Batch means from separate streams can be used to determine when the variance has stabilized. (See Section 7.4 for a description of batch means.)

Yu (1995) uses a cusum plot on only one chain to help to identify convergence. Robert (1998a) provides a benchmark case for evaluation of convergence assessment techniques. Rosenthal (1995), under certain conditions, gives bounds on the length of runs required to give satisfactory results. Cowles and Rosenthal (1998) suggest using auxiliary simulations to determine if the conditions that ensure the bounds on the lengths are satisfied. All of these methods have limitations.

The collection of articles in Gilks, Richardson, and Spiegelhalter (1996) addresses many of the problems of convergence. Gamerman (1997) provides a general introduction to MCMC in which many of these convergence issues are explored. Additional reviews are given in Brooks and Roberts (1999) and the collection of articles in Robert (1998b). Mengersen, Robert, and Guihenneuc-Jouyaux (1999) give a classification of methods and review their performance. Methods of assessing convergence are currently an area of active research.

Use of any method that indicates that convergence has occurred based on the generated data can introduce bias into the results, unless somehow the probability of making the decision that convergence has occurred can be accounted for in any subsequent inference. This is the basic problem in any adaptive statistical procedure. Cowles, Roberts, and Rosenthal (1999) discuss how bias may be introduced in inferences made using an MCMC method after a convergence diagnostic has been used in the sampling. The main point in this section is that there are many subtle issues, and MCMC must be used with some care.

Various methods have been proposed to speed up the convergence; see Gelfand and Sahu (1994), for example. Frigessi, Martinelli, and Stander (1997) discuss general issues of convergence and acceleration of convergence. How quickly convergence occurs is obviously an important consideration for the efficiency of the method. The effects of slow convergence, however, are not as disastrous as the effects of prematurely assuming that convergence has occurred.

### Coupled Markov Chains and "Perfect" Sampling

Convergence is an issue because we want to sample from the stationary distribution. The approach discussed above is to start at some arbitrary point, $t = 0$, and proceed until we think convergence has occurred. Propp and Wilson (1996, 1998) suggested another approach for aperiodic, irreducible, positive recurrent chains with finite state spaces. Their method is based on starting multiple chains at an earlier point.

The method is to generate chains that are *coupled* by the same underlying element of the sample space. The coupling can be accomplished by generating a single realization of some random variable and then letting that realization determine the updating for each of the chains. This can be done in several ways. The simplest, perhaps, is to choose the coupling random variable to be $U(0,1)$ and use the inverse CDF method. At the point $t$, we generate $u_{t+1}$ and update each chain with $X_{t+1}|u_{t+1}, x_t$ by the method of equation (4.3),

$$x_{t+1} = \min\{v, \text{ s.t. } u_{t+1} \leq P_{X_{t+1}|x_t}(v)\}, \tag{4.17}$$

where $P_{X_{t+1}|x_t}(\cdot)$ is the conditional CDF for $X_{t+1}$, given $X_t = x_t$. With this setup for coupled chains, any one of the chains may be represented in a "stochastic recursive sequence",

$$X_t = \phi(X_{t-1}, U_t), \tag{4.18}$$

where $\phi$ is called the *transition rule*. The transition rule also allows us to generate $U_{t+1}|x_{t+1}, x_t$, as

$$U\left(P_{X_{t+1}|x_t}(x_{t+1} - \epsilon), P_{X_{t+1}|x_t}(x_{t+1})\right), \tag{4.19}$$

where $\epsilon$ is vanishingly small.

The idea in the method of Propp and Wilson is to start coupled chains at $t_s = -1$ at each of the states and advance them all to $t = 0$. If they *coalesce* (that is, if they all take the same value), they are the same chain from then on, and $X_0$ has the stationary distribution. If they do not coalesce, then we can start the chains at $t_s = -2$ and maintain exactly the same coupling; that is, we generate a $u_{-1}$, but we use *the same $u_0$ as before*. If these chains coalesce at $t = 0$, then we accept the common value as a realization of the stationary distribution. If they do not coalesce, we back up the starting points of the chains even further. Propp and Wilson (1996) suggested doubling the starting point each time, but any point further back in time would work. The important thing is that each time chains are started that the realizations of the uniform random variable from previous runs be used. This method is called coupling from the past (CFTP). Propp and Wilson (1996) called this method of sampling "exact sampling". Note that if we do the same thing starting at a fixed point and proceeding *forward* with parallel chains, the value to which they coalesce is *not* a realization of the stationary distribution.

If the state space is large, checking for coalescence can be computationally intensive. There are various ways of reducing the burden of checking for coalescence. Propp and Wilson (1996) discussed the special case of a monotone chain (one for which the transition matrix stochastically preserves orderings of state vectors) that has two starting state vectors $x_0^-$ and $x_0^+$ such that, for all $x \in S$, $x_0^- \leq x \leq x_0^+$. In that case, they show that if the sequence beginning with $x_0^-$ and the sequence beginning with $x_0^+$ coalesce, the sequence from that point on is a sample from the stationary distribution. This is interesting, but of limited relevance.

Because CFTP depends on fixed values of $u_0, u_{-1}, \ldots$, for certain of these values, coalescence may occur with very small probability. (This is similar to the modified acceptance/rejection method described in Exercise 4.7b.) In these cases, the $t_s$ that will eventually result in coalescence may be very large in absolute value. An "impatient user" may decide just to start over. Doing so, however, biases the procedure. Fill (1998) described a method for sampling directly from the invariant distribution that uses coupled Markov chains of a fixed length. It is an acceptance/rejection method based on whether coalescence has occurred. This method can be restarted without biasing the results; the method is "interruptible". In this method, an ending time and a state corresponding to that time are chosen arbitrarily. Then, we generate backwards from $t_s$ as follows.

1. Select a time $t_s > 0$ and a state $x_{t_s}$.

2. Generate $x_{t_s-1}|x_{t_s}, x_{t_s-2}|x_{t_s-1}, \ldots, x_0|x_1$.

3. Generate $u_1|x_0, x_1, u_2|x_1, x_2, \ldots, u_{t_s}|x_{t_s-1}, x_{t_s}$ using, perhaps, the distribution (4.19).

4. Start chains at $t = 0$ at each of the states, and advance them to $t = t_s$ using the common $us$.

5. If the chains have coalesced by time $t_s$, then
    accept $x_0$;
   otherwise,
    return to step 1.

Fill gives a simple proof that this method indeed samples from the invariant distribution.

Methods that attempt to sample directly from the invariant distribution of a Markov chain, such as CFTP and interruptible coupled chains, are sometimes called "perfect sampling" methods.

The requirement of these methods of a finite state space obviously limits their usefulness. Møller and Schladitz (1999) extended the method to a class of continuous-state Markov chains. Fill et al. (2000) also discussed the problem of continuous-state Markov chains and considered ways of increasing the computational efficiency.

## 4.11 Use of Conditional Distributions

If the density of interest, $p_X$, can be represented as a marginal density of some joint density $p_{XY}$, observations on $X$ can be generated as a Markov chain with elements having densities

$$p_{Y_i|X_{i-1}}, \quad p_{X_i|Y_i}, \quad p_{Y_{i+1}|X_i}, \quad p_{X_{i+1}|Y_{i+1}}, \dots.$$

This is a simple instance of the Gibbs algorithm, which we discuss beginning on page 156. Casella and George (1992) explain this method in general.

The usefulness of this method depends on identifying a joint density with conditionals that are easy to simulate. For example, if the distribution of interest is a standard normal, the joint density

$$p_{XY}(x, y) = \frac{1}{\sqrt{2\pi}}, \quad \text{for } -\infty < x < \infty, \ 0 < y < e^{-x^2/2},$$

has a marginal density corresponding to the distribution of interest, and it has simple conditionals. The conditional distribution of $Y|X$ is $U(0, e^{-X^2/2})$, and the conditional of $X|Y$ is $U(-\sqrt{-2\log Y}, \sqrt{-2\log Y})$. Starting with $x_0$ in the range of $X$, we generate $y_1$ as a uniform conditional on $x_0$, then $x_1$ as a uniform conditional on $y_1$, and so on.

The auxiliary variable $Y$ that we introduce just to simulate $X$ is called a "latent variable".

## 4.12 Weighted Resampling

To obtain a sample $x_1, x_2, \dots, x_m$ that has an approximate distribution with density $p_X$, a sample $y_1, y_2, \dots, y_n$ from another distribution with density $g_Y$ can be resampled using weights or probabilities

$$w_i = \frac{p_X(y_i)/g_Y(x_i)}{\sum_{j=1}^{n} p_X(y_j)/g_Y(x_j)}, \quad \text{for } i = 1, 2, \dots n.$$

The method was suggested by Rubin (1987, 1988), who called it SIR for *sampling/importance resampling*. The method is also called importance-weighted resampling. The resampling should be done without replacement to give points with low probabilities a chance to be represented. Methods for sampling from a given set with given probabilities are discussed in Section 6.1, page 217. Generally, in SIR, $n$ is much larger than $m$. This method can work reasonably well if the density $g_Y$ is very close to the target density $p_X$.

This method, like the Markov chain methods above, has the advantage that the normalizing constant of the target density is not needed. Instead of the density $p_X(\cdot)$, any nonnegative proportional function $cp_X(\cdot)$ could be used. Gelman (1992) describes an iterative variation in which $n$ is allowed to increase as $m$ increases; that is, as the sampling continues, more variates are generated from the distribution with density $g_Y$.

## 4.13   Methods for Distributions with Certain Special Properties

Because of the analytical and implementation burden involved in building a random number generator, a general rule is that a single algorithm that works in two settings is better than two different algorithms, one for each setting. This is true, of course, unless the individual algorithms perform better in the respective special cases, and then the question is how much better. In random number generation from nonuniform distributions, it is desirable to have "universal algorithms" that use general methods that we have discussed above but are optimized for certain broad classes of distributions.

For distributions with certain special properties, general algorithms using mixtures and rejection can be optimized for broad classes of distributions. We have already discussed densities that are nearly linear (Algorithm 4.7, page 118) and densities that are nearly flat (Algorithm 4.11, page 126).

Another broad class of distributions are those that are infinitely divisible. Damien, Laud, and Smith (1995) give general methods for generation of random deviates from distributions that are infinitely divisible.

### Distributions with Densities that Can Be Transformed to Concave Functions

An important special property of some distributions is *concavity* of the density or of some transformation of the density. On page 119, we discuss how easy it is to form polygonal majorizing and squeeze functions for concave densities. Similar ideas can be employed for cases in which the density can be invertibly transformed into a concave function.

In some applications, especially in reliability or survival analysis, the logarithm is a standard transformation and *log-concavity* is an important property. A distribution is log-concave if its density (or probability function) has the property

$$\log p(x_1) - 2\log p\left(\frac{x_1 + x_2}{2}\right) + \log p(x_2) < 0,$$

wherever the densities are positive. If the density is twice-differentiable, this condition is satisfied if the negative of the Hessian is positive definite. Many of the commonly used distributions, such as the normal, the gamma with shape parameter greater than 1, and the beta with parameters greater than 1, are log-concave. See Pratt (1981) for discussion of these properties, and see Dellaportas and Smith (1993) for some examples in generalized linear models. Devroye (1984b) describes general methods for a log-concave distribution, and Devroye (1987) describes a method for a discrete distribution that is log-concave.

The methods of forming polygonal majorizing and squeeze functions for concave densities can also be applied to convex densities or to densities that can be invertibly transformed into concave functions by reversing the role of the majorizing and squeeze functions.

### Incremental Formation of Majorizing and Squeeze Functions: "Adaptive" Rejection

Gilks (1992) and Gilks and Wild (1992) describe a method that they call *adaptive rejection sampling* or *ARS* for a continuous log-concave distribution.

The adaptive rejection method described by Gilks (1992) begins with a set $S_k$ consisting of the points $x_0 < x_1 < \ldots < x_k < x_{k+1}$ from the range of the distribution of interest. Define $L_i$ as the straight line determined by the points $(x_i, \log p(x_i))$ and $(x_{i+1}, \log p(x_{i+1}))$; then, for $i = 1, 2, \ldots, k$, define the piecewise linear function $h_k(x)$ as

$$h_k(x) = \min\left(L_{i-1}(x), L_{i+1}(x)\right) \quad \text{for } x_i \leq x < x_{i+1}.$$

This piecewise linear function is a majorizing function for the log of the density, as shown in Figure 4.19.

The chords formed by the continuation of the line segments form functions that can be used as a squeeze function, $m_k(x)$, which is also piecewise linear.

For the density itself, the majorizing function and the squeeze function are piecewise exponentials. The majorizing function is

$$cg_k(x) = \exp h_k(x),$$

where each piece of $g_k(x)$ is an exponential density function truncated to the appropriate range. The density is shown in Figure 4.20.

In each step of the acceptance/rejection algorithm, the set $S_k$ is augmented by the point generated from the majorizing distribution, and $k$ is increased by 1. The method is shown in Algorithm 4.19. In Exercise 4.14, page 162, you are asked to write a program for performing adaptive rejection sampling for the density shown in Figure 4.20, which is the same one as in Figure 4.4, and to compare the efficiency of this method with the standard acceptance/rejection method.

### Algorithm 4.19 Adaptive Acceptance/Rejection Sampling

    0. Initialize $k$ and $S_k$.

    1. Generate $y$ from $g_k$.

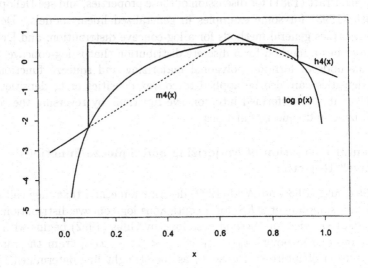

Figure 4.19: Adaptive Majorizing Function with the Log-Density (Same Density as in Figure 4.4)

2. Generate $u$ from a U(0,1) distribution.

3. If $u \leq \dfrac{\exp m_k(y)}{c g_k(y)}$, then
    3.a. deliver $y$;
  otherwise,
    3.b. if $u \leq \dfrac{p(y)}{c g_k(y)}$, then deliver $y$;
    3.c. set $k = k + 1$, add $y$ to $S_k$, and update $h_k$, $g_k$, and $m_k$.

4. Go to step 1.                                            ∎

After an update step, the new piecewise linear majorizing function for the log of the density is as shown in Figure 4.21.

Gilks and Wild (1992) describe a similar method, but instead of using secants as the piecewise linear majorizing function, they use tangents of the log of the density. This requires computation of numerical derivatives of the log density.

Hörmann (1994a) adapts the methods of Gilks (1992) and Gilks and Wild (1992) to discrete distributions.

### $T$-concave Distributions

Hörmann (1995) extends the methods for a distribution with a log-concave density to a distribution whose density $p$ can be transformed by a strictly increasing

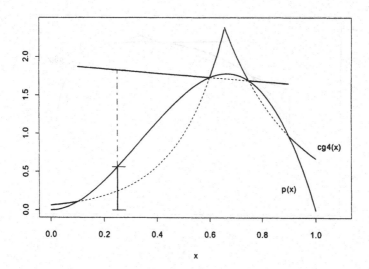

Figure 4.20: Exponential Adaptive Majorizing Function with the Density in Figure 4.4

operator $T$ such that $T(p(x))$ is concave. In that case, the density $p$ is said to be "$T$-concave". Often, a good choice is $T(s) = -1/\sqrt{s}$. A density that is $T$-concave with respect to this transformation is log-concave.

Many of the standard distributions have $T$-concave densities, and in those cases we refer to the distribution itself as $T$-concave. The normal distribution (equation (5.6)), for example, is $T$-concave for all values of its parameters. The gamma distribution (equation (5.13)) is $T$-concave for $\alpha \geq 1$ and $\beta > 0$. The beta distribution (equation (5.14)) is $T$-concave for $\alpha \geq 1$ and $\beta \geq 1$.

The transformation $T(s) = -1/\sqrt{s}$ allows construction of a table mountain majorizing function (reminiscent of a majorizing function in the ratio-of-uniforms method) that is then used in an acceptance/rejection method. Hörmann calls this method *transformed density rejection*.

Leydold (2001) describes an algorithm for $T$-concave distributions based on a ratio-of-uniforms type of acceptance/rejection method. The advantage of Leydold's method is that it requires less setup time than Hörmann's method, and so would be useful in applications in which the parameters of the distribution change relatively often compared to the number of variates generated at each fixed value. Gilks, Best, and Tan (1995; corrigendum, Gilks, Neal, Best, and Tan, 1997) develop an adaptive rejection method that does not require the density to be log-concave. They call the method adaptive rejection Metropolis sampling.

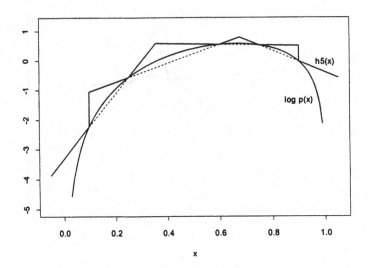

Figure 4.21: Adaptive Majorizing Function with an Additional Point

## Unimodal Densities

Many densities of interest are unimodal, and some simple methods of random number generation take advantage of that property. The ziggurat and Monty Python decomposition methods of Marsaglia and Tsang (1984, 1998) are most effective for unimodal distributions, in which the first decomposition can involve forming two regions, one on each side of the mode. Devroye (1984a) describes general methods for generating variates from such distributions. If a distribution is not unimodal, it is sometimes useful to decompose the distribution into a mixture of unimodal distributions to use the techniques on them separately.

Methods for sampling from unimodal discrete distributions, which often involve linear searches, can be more efficient if the search begins at the mode. (See, for example, the method for the Poisson distribution on page 188.)

## Multimodal Densities

For simulating densities with multiple modes, it is generally best to express the distribution as a mixture and use different methods in different regions. MCMC methods can become trapped around a local mode. There are various ways of dealing with this problem. One way to do this is to modify the target density $p_X(\cdot)$ in the Hastings ratio so that it becomes flatter, and therefore it is more likely that the sequence will move away from a local mode. Geyer and Thompson (1995) describe a method of "simulated tempering", in which a "temperature" parameter, which controls how likely it is that the sequence will move away from a current state, is varied randomly. This is similar to methods used in simulated annealing (see Section 7.9). Neal (1996) describes a systematic method of alternating between the target density and a flatter one. He called the method "tempered transition".

## 4.14    General Methods for Multivariate Distributions

Two simple methods of generating multivariate random variates make use of variates from univariate distributions. One way is to generate a vector of i.i.d. variates and then apply a transformation to yield a vector from the desired multivariate distribution. Another way is to use the representation of the distribution function or density function as a product of the form

$$p_{X_1 X_2 X_3 \cdots X_d} = p_{X_1|X_2 X_3 \cdots X_d} \cdot p_{X_2|X_3 \cdots X_d} \cdot p_{X_3|\cdots X_d} \cdots p_{X_d}.$$

In this method, we generate a marginal $x_d$ from $p_{X_d}$, then a conditional $x_{d-1}$ from $p_{X_{d-1}|X_d}$, and continue in this way until we have the full realization $x_1, x_2, \ldots, x_d$. We see two simple examples of these methods at the beginning of Section 5.3, page 197. In the first example in that section, we generate a $d$-variate normal with variance-covariance matrix $\Sigma$ either by the transformation $x = T^T z$, where $T$ is a $d \times d$ matrix such that $T^T T = \Sigma$ and $z$ is a $d$-vector of i.i.d. $N(0, 1)$ variates. In the second case, we generate $x_1$ from $N_1(0, \sigma_{11})$, then generate $x_2$ conditionally on $x_1$, then generate $x_3$ conditionally on $x_1$ and $x_2$, and so on.

As mentioned in discussing acceptance/rejection methods in Sections 4.5 and 4.10, these methods are directly applicable to multivariate distributions, so acceptance/rejection is a third general way of generating multivariate observations. As in the example of the bivariate gamma on page 123, however, this usually involves a multivariate majorizing function, so we are still faced with the basic problem of generating from some multivariate distribution.

For higher dimensions, the major problem in using acceptance/rejection methods for generating multivariate deviates results from one of the effects of the so-called "curse of dimensionality". The proportion of the volume of a closed geometrical figure that is in the outer regions of that figure increases with increasing dimensionality. (See Section 10.7 of Gentle, 2002, and Exercise 4.4f at the end of this chapter.)

An iterative method somewhat similar to the use of marginals and conditionals can also be used to generate multivariate observations. This method was used by Geman and Geman (1984) for generating observations from a Gibbs distribution (Boltzmann distribution) and so is called the *Gibbs method*. In the Gibbs method, after choosing a starting point, the components of the $d$-vector variate are generated one at a time conditionally on all others. If $p_X$ is the density of the $d$-variate random variable $X$, we use the conditional densities $p_{X_1|X_2 X_3 \cdots X_d}$, $p_{X_2|X_1 X_3 \cdots X_d}$, and so on. At each stage, the conditional distribution uses the most recent values of all of the other components. Obviously, it may require a number of iterations before the choice of the initial starting point is washed out.

The method is shown in Algorithm 4.20. (In the algorithms to follow, we represent the support of the density of interest by $S$, where $S \subseteq \mathbb{R}^d$.)

**Algorithm 4.20  Gibbs Method**

0. Set $k = 0$.

1. Choose $x^{(k)} \in S$.

2. Generate $x_1^{(k+1)}$ conditionally on $x_2^{(k)}, x_3^{(k)}, \ldots, x_d^{(k)}$,
   Generate $x_2^{(k+1)}$ conditionally on $x_1^{(k+1)}, x_3^{(k)}, \ldots, x_d^{(k)}$,

   . . .

   Generate $x_{d-1}^{(k+1)}$ conditionally on $x_1^{(k+1)}, x_2^{(k+1)}, \ldots, x_d^{(k)}$,
   Generate $x_d^{(k+1)}$ conditionally on $x_1^{(k+1)}, x_2^{(k+1)}, \ldots, x_{d-1}^{(k+1)}$.

3. If convergence has occurred, then
      3.a.  deliver $x = x^{(k+1)}$;
     otherwise,
      3.b.  set $k = k + 1$, and go to step 2. ∎

Casella and George (1992) give a simple proof that this iterative method converges; that is, as $k \to \infty$, the density of the realizations approaches $p_X$. The question of whether convergence has practically occurred in a finite number of iterations in the Gibbs method is similar to the same question in the Metropolis–Hastings method discussed in Section 4.10. In either case, to determine that convergence has occurred is not a simple problem.

Once a realization is delivered in Algorithm 4.20 (that is, once convergence has been deemed to have occurred), subsequent realizations can be generated either by starting a new iteration with $k = 0$ in step 0 or by continuing at step 1 with the current value of $x^{(k)}$. If the chain is continued at the current value of $x^{(k)}$, we must remember that the subsequent realizations are not independent. This affects variance estimates (second-order sample moments) but not means (first-order moments). In order to get variance estimates, we may use means of batches of subsequences or use just every $m^{\text{th}}$ (for some $m > 1$) deviate in step 3. (The idea is that this separation in the sequence will yield subsequences or a systematic subsample with correlations nearer 0. See Section 7.4 for a description of batch means.) If we just want estimates of means, however, it is best not to subsample the sequence; that is, the variances of the estimates of means (first-order sample moments) using the full sequence are smaller than the variances of the estimates of the same means using a systematic (or any other) subsample (as long as the Markov chain is stationary).

To see this, let $\bar{x}_i$ be the mean of a systematic subsample of size $n$ consisting of every $m^{\text{th}}$ realization beginning with the $i^{\text{th}}$ realization of the converged sequence. Now, following MacEachern and Berliner (1994), we observe that

$$|\mathrm{Cov}(\bar{x}_i, \bar{x}_j)| \le \mathrm{V}(\bar{x}_l)$$

for any positive $i$, $j$, and $l$ less than or equal to $m$. Hence, if $\bar{x}$ is the sample mean of a full sequence of length $nm$, then

$$
\begin{aligned}
\mathrm{V}(\bar{x}) &= \mathrm{V}(\bar{x}_l)/m + \sum_{i \ne j; i,j=1}^{m} \mathrm{Cov}(\bar{x}_i, \bar{x}_j)/m^2 \\
&\le \mathrm{V}(\bar{x}_l)/m + m(m-1)\mathrm{V}(\bar{x}_l)/m^2 \\
&= \mathrm{V}(\bar{x}_l).
\end{aligned}
$$

See also Geyer (1992) for a discussion of subsampling in the chain.

The paper by Gelfand and Smith (1990) was very important in popularizing the Gibbs method. Gelfand and Smith also describe a related method of Tanner and Wong (1987), called *data augmentation*, which Gelfand and Smith call *substitution sampling*. In this method, a single component of the $d$-vector is chosen (in step 1), and then multivariate subvectors are generated conditional on just one component. This method requires $d(d-1)$ conditional distributions. The reader is referred to their article and to Schervish and Carlin (1992) for descriptions and comparisons with different methods. Tanner (1996) defines a *chained data augmentation*, which is the Gibbs method described above.

In the Gibbs method, the components of the $d$-vector are changed systematically, one at a time. The method is sometimes called *alternating conditional sampling* to reflect this systematic traversal of the components of the vector.

Another type of Metropolis method is the hit-and-run sampler. In this method, all components of the vector are updated at once. The method is shown in Algorithm 4.21 in the general version described by Chen and Schmeiser (1996).

### Algorithm 4.21 Hit-and-Run Sampling

0. Set $k = 0$.

1. Choose $x^{(k)} \in S$.

2. Generate a random normalized direction $v^{(k)}$ in $\mathbb{R}^d$. (This is equivalent to a random point on a sphere, as discussed on page 201.)

3. Determine the set $S^{(k)} \subseteq \mathbb{R}$ consisting of all $\lambda \ni (x^{(k)} + \lambda v^{(k)}) \in S$. ($S^{(k)}$ is one-dimensional; $S$ is $d$-dimensional.)

4. Generate $\lambda^{(k)}$ from the density $g^{(k)}$, which has support $S^{(k)}$.

5. With probability $a^{(k)}$,
     5.a. set $x^{(k+1)} = x^{(k)} + \lambda^{(k)} v^{(k)}$;
   otherwise,
     5.b. set $x^{(k+1)} = x^{(k)}$.

6. If convergence has occurred, then

      6.a. deliver $x = x^{(k+1)}$;

   otherwise,

      6.b. set $k = k + 1$, and go to step 2. ∎

Chen and Schmeiser (1996) discuss various choices for $g^{(k)}$ and $a^{(k)}$. One choice is

$$
g^{(k)}(\lambda) = \begin{cases} \dfrac{p(x^{(k)} + \lambda v^{(k)})}{\int_{S^{(k)}} p(x^{(k)} + u v^{(k)})\, du} & \text{for } \lambda \in S^{(k)}, \\[2ex] 0 & \text{otherwise,} \end{cases}
$$

and

$$
a^{(k)} = 1.
$$

Another choice is $g$ uniform over $S^{(k)}$ if $S^{(k)}$ is bounded, or else some symmetric distribution centered on 0 (such as a normal or Cauchy distribution), together with

$$
a^{(k)} = \min\left(1, \frac{p(x^{(k)} + \lambda^{(k)} v^{(k)})}{p(x^{(k)})}\right).
$$

Smith (1984) uses the hit-and-run sampler for generating uniform points over bounded regions, and Bélisle, Romeijn, and Smith (1993) use it for generating random variates from general multivariate distributions. Proofs of the convergence of the method can be found in Bélisle, Romeijn, and Smith (1993) and Chen and Schmeiser (1996).

Gilks, Roberts, and George (1994) describe a generalization of the hit-and-run algorithm called *adaptive direction sampling*. In this method, a set of current points is maintained, and only one, chosen at random from the set, is updated at each iteration (see Gilks and Roberts, 1996).

Both the Gibbs and hit-and-run methods are special cases of the Metropolis–Hastings method in which the $r$ of step 2 in Algorithm 4.18 (page 141) is exactly 1, so there is never a rejection.

The same issues of convergence that we encountered in discussing the Metropolis–Hastings method must be addressed when using the Gibbs or hit-and-run methods. The need to run long chains can increase the number of computations to unacceptable levels. Schervish and Carlin (1992) and Cowles and Carlin (1996) discuss general conditions for convergence of the Gibbs sampler.

Dellaportas (1995) discusses some issues in the efficiency of random number generation using the Gibbs method. Berbee et al. (1987) compare the efficiency of hit-and-run methods with acceptance/rejection methods and find the hit-and-run methods to be more efficient in higher dimensions. Chen and Schmeiser (1993) give some general comparisons of Gibbs, hit-and-run, and variations. Generalizations about the performance of the methods are difficult; the best method often depends on the problem.

**Multivariate Densities with Special Properties**

We have seen that certain properties of univariate densities can be used to develop efficient algorithms for general distributions that possess those special properties. For example, adaptive rejection sampling and other special accep-tance/rejection methods can be used for distributions having concave densities or concave transformed densities, as discussed on page 150. Hörmann (2000) describes a method for log-concave bivariate distributions that uses adaptive rejection sampling to develop the majorizing function. Leydold (1998) shows that while the methods for univariate $T$-concave distributions would work for multivariate $T$-concave distributions, such methods are unacceptably slow. He splits the $T$-concave multivariate density into a set of simple cones and con-structs the majorizing function from piecewise hyperplanes that are tangent to the cones. He reports favorably on the performance of his method for as many as eight dimensions.

As we have seen, unimodal distributions are generally easier to work with than multimodal distributions. A product multivariate density having uni-modal factors will of course be unimodal. Devroye (1997) described general acceptance/rejection methods for multivariate distributions with the slightly weaker property of being orthounimodal; that is, each marginal density is uni-modal.

## 4.15   Generating Samples from a Given Distribution

Usually, in applications, rather than just generating a single random deviate, we generate a random sample of deviates from the distribution of interest. A random sample of size $n$ from a discrete distribution with probability function

$$p(X = m_i) = p_i$$

has a vector of counts of the mass points that has a multinomial $(n, p_1, \ldots, p_k)$ distribution.

If the sample is to be used as a set, rather than as a sequence, and if $n$ is large relative to $k$, it obviously makes more sense to generate a single multinomial $(x_1, x_2, \ldots, x_k)$ and use these values as counts of occurrences of the respective mass points $m_1, m_2, \ldots, m_k$. (Methods for generating multinomials are discussed in Section 5.3.2, page 198.)

This same idea can be applied to continuous distributions with a modifica-tion to discretize the range (see Kemp and Kemp, 1987).

## Exercises

4.1. The inverse CDF method.

(a) Prove that if $X$ is a random variable with an absolutely continuous distribution function $P_X$, the random variable $P_X(X)$ has a U$(0, 1)$ distribution.

(b) Prove that the inverse CDF method for discrete random variables as specified in the relationship in expression (4.2) on page 104 is correct.

4.2. Formally prove that the random variable delivered in Algorithm 4.6 on page 114 has the density $p_X$. *Hint:* For the delivered variable, $Z$, determine the distribution function $\Pr(Z \leq x)$ and differentiate.

4.3. Write a Fortran or C function to implement the acceptance/rejection method for generating a beta(3, 2) random deviate. Use the majorizing function shown in Figure 4.4 on page 115. The value of $c$ is 1.2. Use the inverse CDF method to generate a deviate from $g$. (This will involve taking a square root.)

4.4. Acceptance/rejection methods.

(a) Give an algorithm to generate a normal random deviate using the basic acceptance/rejection method with the double exponential density (see equation (5.11), page 177) as the majorizing density.

(b) What is the acceptance proportion of this method?

(c) After you have obtained the basic acceptance/rejection test, try to simplify it.

(d) Develop an algorithm to generate bivariate normal deviates with mean $(0, 0)$, variance $(1, 1)$, and correlation $\rho$ using a bivariate product double exponential density as the majorizing density. For $\rho = 0$, what is the acceptance probability?

(e) Write a program to generate bivariate normal deviates with mean $(0, 0)$, variance $(1, 1)$, and correlation $\rho$. Use a bivariate product double exponential density as the majorizing density. Now, set $\rho = 0.5$ and generate a sample of 1000 bivariate normals. Compare the sample statistics with the parameters of the simulated distribution.

(f) What is the acceptance probability for a basic acceptance/rejection method to generate $d$-variate normal deviates with mean 0 and diagonal variance-covariance matrix with all elements equal to 1 using a $d$-variate product double exponential density as the majorizing density?

4.5. What would be the problem with using a normal density to make a majorizing function for the double exponential distribution (or using a half-normal for an exponential)?

4.6. (a) Write a Fortran or C function to implement the acceptance/rejection method for a bivariate gamma distribution whose density is given in equation (4.10) on page 123 using the method described in the text. (You must develop a method for determining the mode.)

(b) Now, instead of the bivariate uniform in the rectangle near the origin, devise a pyramidal distribution to use as a majorizing density.

(c) Use Monte Carlo methods to compare efficiency of the method using the bivariate uniform and the method using a pyramidal density.

4.7. Consider the acceptance/rejection method given in Algorithm 4.6 to generate a realization of a random variable $X$ with density function $p_X$ using a density function $g_Y$.

(a) Let $T$ be the number of passes through the three steps until the desired variate is delivered. Determine the mean and variance of $T$ (in terms of $p_X$ and $g_Y$).

(b) Now, consider a modification of the rejection method in which steps 1 and 2 are reversed, and the branch in step 3 is back to the new step 2; that is:

    1. Generate $u$ from a uniform (0,1) distribution.

    2. Generate $y$ from the distribution with density function $g_Y$.

    3. If $u \leq p_X(y)/cg_Y(y)$, then take $y$ as the desired realization; otherwise return to step 2.

Is this a better method? Let $Q$ be the number of passes through these three steps until the desired variate is delivered. Determine the mean and variance of $Q$. (This method was suggested by Sibuya, 1961, and analyzed by Greenwood, 1976c.)

4.8. Formally prove that the random variable delivered in Algorithm 4.7 on page 118 has the density $p$.

4.9. Write a Fortran or C function to implement the ratio-of-uniforms method (page 130) to generate deviates from a gamma distribution with shape parameter $\alpha$. Generate a sample of size 1000 and perform a chi-squared goodness-of-fit test (see Cheng and Feast, 1979).

4.10. Use the Metropolis–Hastings algorithm (page 141) to generate a sample of standard normal random variables. Use as the candidate generating density $g(x|y)$, a normal density in $x$ with mean $y$. Experiment with different burn-in periods and different starting values. Plot the sequences generated. Test your samples for goodness-of-fit to a normal distribution. (Remember that they are correlated.) Experiment with different sample sizes.

4.11. Let $\Pi$ have a beta distribution with parameters $\alpha$ and $\beta$, and let $X$ have a conditional distribution given $\Pi = \pi$ of a binomial with parameters $n$ and $\pi$. Let $\Pi$ conditional on $X = x$ have a beta distribution with parameters

$\alpha + x$ and $n + \beta - x$. (This leads to the "beta-binomial" distribution; see page 187.) Consider a bivariate Markov chain, $(\Pi_0, X_0), (\Pi_1, X_1), \ldots,$ with an uncountable state space (see Casella and George, 1992).

(a) What is the transition kernel? That is, what is the conditional density of $(\Pi_t, X_t)$ given $(\pi_{t-1}, x_{t-1})$?

(b) Consider just the Markov chain of the beta-binomial random variable $X$. What is the $(i, j)$ element of the transition matrix?

4.12. Obtain a sample of size 100 from the beta(3,2) distribution using the SIR method of Section 4.12 and using a sample of size 1000 from the density $g_Y$ that is proportional to the triangular majorizing function used in Exercise 4.3. (Use Algorithm 6.1, page 218, to generate the sample without replacement.) Compare the efficiency of the program that you have written with the one that you wrote in Exercise 4.3.

4.13. Formally prove that the random variable delivered in Algorithm 4.19 on page 151 has the density $p_X$. (Compare Exercise 4.2.)

4.14. Write a computer program to implement the adaptive acceptance/rejection method for generating a beta(3,2) random deviate. Use the majorizing function shown in Figure 4.19 on page 152. The initial value of $k$ is 4, and $S_k = \{0.00, 0.10, 0.60, 0.75, 0.90, 1.00\}$. Compare the efficiency of the program that you have written with the ones that you wrote in Exercises 4.3 and 4.12.

4.15. Consider the trivariate normal distribution used as the example in Figure 4.18 (page 145).

(a) Use the Gibbs method to generate and plot 1000 realizations of $X_1$ (including any burn-in). Explain any choices that you make on how to proceed with the method.

(b) Use the hit-and-run method to generate and plot 1000 realizations of $X_1$ (including any burn-in). Explain any choices that you make on how to proceed with the method.

(c) Compare the Metropolis–Hastings method (page 145) and the Gibbs and hit-and-run methods for this problem.

4.16. Consider a probability model in which the random variable $X$ has a binomial distribution with parameters $n$ and $y$, which are, respectively, realizations of a conditional shifted Poisson distribution and a conditional beta distribution. For fixed $\lambda$, $\alpha$, and $\beta$, let the joint density of $X$, $N$, and $Y$ be proportional to

$$\frac{\lambda^n}{x!(n-x)!} y^{x+\alpha-1}(1-y)^{n-x+\beta-1}e^{-\lambda} \quad \text{for} \quad x = 0, 1, \ldots, n;$$

$$0 \le y \le 1;$$

$$n = 1, 2, \ldots.$$

First, determine the conditional densities for $X|y, n$, $Y|x, n$, and $N|x, y$. Next, write a Fortran or C program to sample $X$ from the multivariate distribution for given $\lambda$, $\alpha$, and $\beta$. Now, set $\lambda = 16$, $\alpha = 2$, and $\beta = 4$, run 500 independent Gibbs sequences of length $k = 10$, taking only the final variate, and plot a histogram of the observed $x$. (Use a random starting point.) Now repeat the procedure, except using only one Gibbs sequence of length 5000, and plot a histogram of all observed $x$s after the ninth one (see Casella and George, 1992).

4.17. Generate a random sample of 1000 Bernoulli variates with $\pi = 0.3$. Do not use Algorithm 4.1; instead, use the method of Section 4.15.

# Chapter 5

# Simulating Random Numbers from Specific Distributions

For the important distributions, specialized algorithms based on the general methods discussed in the previous chapter are available. The important difference in the algorithms is their speed. A secondary difference is the size and complexity of the program to implement the algorithm. Because all of the algorithms for generating from nonuniform distributions rely on programs to generate from uniform distributions, an algorithm that uses only a small number of uniforms to yield a variate of the target distribution may be faster on a computer system on which the generation of the uniform is very fast. As we have mentioned, on a given computer system, there may be more than one program available to generate uniform deviates. Often, a portable generator is slower than a nonportable one, so for portable generators of nonuniform distributions, those that require a small number of uniform deviates may be better. If evaluation of elementary functions is a part of the algorithm for generating random deviates, then the speed of the overall algorithm depends on the speed of the evaluation of the functions. The relative speed of elementary function evaluation is different on different computer systems.

The algorithm for a given distribution is some specialized version of those methods discussed in the previous chapter. Often, the algorithm uses some combination of these general techniques.

Many algorithms require some setup steps to compute various constants and to store tables; therefore, there are two considerations for the speed: the setup time and the generation time. In some applications, many random numbers from the same distribution are required. In those cases, the setup time may not be too important. In other applications, the random numbers come from different distributions—probably the same family of distributions but with changing values of the parameters. In those cases, the setup time may be very significant. If the best algorithm for a given distribution has a long setup time, it may be desirable to identify another algorithm for use when the parameters vary. Any computation that results in a quantity that is constant with respect to the parameters of the distribution should of course be performed as part of the setup computations in order to avoid performing the computation in every pass through the main part of the algorithm.

The efficiency of an algorithm may depend on the values of the parameters of the distribution. Many of the best algorithms therefore switch from one method to another, depending on the values of the parameters. In some cases, the speed of the algorithm is independent of the parameters of the distribution. Such an algorithm is called a *uniform time* algorithm. In many cases, the most efficient algorithm in one range of the distribution is not the most efficient in other regions. Many of the best algorithms therefore use mixtures of the distribution.

Sometimes, it is necessary to generate random numbers from some subrange of a given distribution, such as the tail region. In some cases, there are efficient algorithms for such truncated distributions. (If there is no specialized algorithm for a truncated distribution, acceptance/rejection applied to the full distribution will always work, of course.)

Methods for generating random variates from specific distributions are an area in which there have been literally hundreds of papers, each proposing some wrinkle (not always new or significant). Because the relative efficiencies ("efficiency" here means "speed") of the individual operations in the algorithms vary from one computing system to another, and also because these individual operations can be programmed in various ways, it is very difficult to compare the relative efficiencies of the algorithms. This provides fertile ground for a proliferation of "research" papers. Two other things contribute to the large number of insignificant papers in this area. It is easy to look at some algorithm, modify some step, and then offer the new algorithm. Thus, the intellectual capitalization required to enter the field is small. (In business and economics, this is the same reason that so many restaurants are started; only a relatively small capitalization is required.)

Another reason for the large number of papers purporting to give new and better algorithms is the diversity of the substantive and application areas that constitute the backgrounds of the authors. Monte Carlo simulation is widely used throughout both the hard and the soft sciences. Research workers in one field often are not aware of the research published in another field.

Although, of course, it is important to seek efficient algorithms, it is also necessary to consider a problem in its proper context. In Monte Carlo simulation applications, literally millions of random numbers may be generated, but the time required to generate them is likely to be only a very small fraction of the total computing time. In fact, it is probably the case that the fraction of time required for the generation of the random numbers is somehow negatively correlated with the importance of the problem. The importance of the time required to perform some task usually depends more on its proportion of the overall time of the job rather than on its total time.

Another consideration is whether the algorithm is portable; that is, whether it yields the same stream on different computer systems. As we mention in Section 4.5, methods that accept or reject a candidate variate based on a floating-point comparison may not yield the same streams on different systems.

The descriptions of the algorithms in this chapter are written with an emphasis on clarity, so they should not be incorporated directly into program code without considerations of efficiency. These considerations generally involve avoiding unnecessary computations. This may mean defining a variable not mentioned in the algorithm description or reordering the steps slightly.

# 5.1 Modifications of Standard Distributions

For many of the common distributions, there are variations that are useful either for computational or other practical reasons or because they model some stochastic process well. A distribution can sometimes be simplified by transformations of the random variable that effectively remove certain parameters that characterize the distribution. In many cases, the algorithms for generating random deviates address the simplified version of the distribution. An appropriate transformation is then applied to yield deviates from the distribution with the given parameters.

## Standard Distributions

A linear transformation, $Y = aX + b$, is simple to apply and is one of the most useful. The multiplier affects the *scale*, and the addend affects the *location*. For example, a "three-parameter" gamma distribution with density

$$p(y) = \frac{1}{\Gamma(\alpha)\beta^\alpha}(y - \gamma)^{\alpha-1}e^{-(y-\gamma)/\beta}, \quad \text{for } \gamma \leq y \leq \infty,$$

can be formed from the simpler distribution with density

$$g(x) = \frac{1}{\Gamma(\alpha)}x^{\alpha-1}e^{-x}, \quad \text{for } 0 \leq x \leq \infty,$$

using the transformation $Y = \beta X + \gamma$. *(Here, and elsewhere, when we give an expression for a probability density function, we imply that the density is equal to 0 outside of the range specified.)* The $\beta$ parameter is a scale parameter, and $\gamma$ is a location parameter. (The remaining $\alpha$ parameter is called the "shape parameter", and it is the essential parameter of the family of gamma distributions.) The simpler form is called the *standard* gamma distribution. Other distributions have similar standard forms. Standard distributions (or standardized random variables) allow us to develop simpler algorithms and more compact tables of values that can be used for a range of parameter values.

**Truncated Distributions**

In many stochastic processes, the realizations of the random variable are constrained to be within a given region of the support of the random variable. Over the allowable region, the random variable has a probability density (or probability function) that is proportional to the density (or probability) of the unconstrained random variable. If the random variable $Y$ has probability density $p(y)$ over a domain $S$, and if $Y$ is constrained to $R \subset S$, the probability density of the constrained random variable is

$$\begin{aligned} p_c(x) \; &= \; \frac{1}{\Pr(Y \in R)} \, p(x) \quad \text{for } x \in R; \\ &= \; 0, \quad \text{elsewhere.} \end{aligned} \tag{5.1}$$

The most common types of constraint are truncations, either left or right. In a left truncation at $\tau$, say, the random variable $Y$ is constrained by $\tau \leq Y$, and in a right truncation, it is constrained by $Y \leq \tau$.

Truncated distributions are useful models in applications in which the observations are *censored*. Such observations often arise in studies where a variable of interest is the time until a particular event occurs. At the end of the study, there may be a number of observational units that have not experienced the event. The corresponding times for these units are said to be censored, or right-censored; it is known only that the times for these units would be greater than some fixed value. In a similar fashion, left-censoring occurs when the exact times are not recorded early in the study. There are many issues to consider in the analysis of censored data, but it is not our purpose here to discuss the analysis.

Generation of random variates with constraints can be handled by the general methods discussed in the previous chapter. The use of acceptance/rejection is obvious; merely generate from the full distribution and reject any realizations outside of the acceptable region. Of course, choosing a majorizing density with no support in the truncated region is a better approach. Modification of the inverse CDF method to handle truncated distributions is simple. For a right truncation at $\tau$ of a distribution with CDF $P_Y$, for example, instead of the basic transformation (4.1), page 102, we use

$$X = P_Y^{-1}(U P_Y(\tau)), \tag{5.2}$$

where $U$ is a random variable from $U(0,1)$.

The method using a sequence of conditional distributions described on page 149 can often be modified easily to generate variates from truncated distributions. In some simple applications, the truncated distribution is simulated by a conditional uniform distribution, the range of which is the intersection of the full conditional range and the truncated range. See Damien and Walker (2001) for some examples.

There are usually more efficient ways of generating variates from constrained distributions. We describe a few of the more common ones (which are invariably truncations) in the following sections.

## "Inverse" Distributions

In Bayesian applications, joint probability densities of interest often involve a product of the density of some well-known random variable and what might be considered the density of the multiplicative inverse of another well-known random variable. Common examples of this are the statistics used in studentization: the chi-squared and the Wishart. Many authors refer to the distribution of such a random variable as the "inverse distribution"; for example, an "inverse chi-squared distribution" is the distribution of $X^{-1}$, where $X$ has a chi-squared distribution. Other distributions with this interpretation are the inverse gamma distribution and the inverse Wishart distribution. This interpretation of "inverse" is not the same as for that word in the inverse Gaussian distribution with density given in equation (5.30) on page 193. In the cases of the inverse gamma, chi-squared, and Wishart distributions, the method for generating random variates is the obvious one: generate from the regular distribution and then obtain the inverse.

## Folded Symmetric Distributions

For symmetric distributions, a useful nonlinear transformation is the absolute value. The distribution of the absolute value is often called a "folded" distribution. The exponential distribution, for example, is the folded double exponential distribution (see page 176). The halfnormal distribution, which is the distribution of the absolute value of a normal random variable, is a folded normal distribution.

## Mixture Distributions

In Chapter 4, we discussed general methods for generating random deviates by decomposing the density $p(\cdot)$ into a mixture of other densities,

$$p(x) = \sum_i w_i p_i(x), \qquad (5.3)$$

where $\sum_i w_i = 1$. Mixture distributions are useful in their own right. It was noticed as early as the nineteenth century by Francis Galton and Karl Pearson that certain observational data correspond very well to a mixture of two normal distributions, whereas a single normal does not fit the data well at all. Often, a simple mixture distribution can be used to model outliers or aberrant observations. This kind of mixture, in which a substantial proportion of the total probability follows one distribution and a small proportion follow another distribution, is called a "contaminated distribution". Mixture distributions are often used in robustness studies because the interest is in how well a standard procedure holds up when the data are subject to contamination by a different population or by incorrect measurements.

A very simple extension of a finite (or countable) mixture, as in equation (5.3), is one in which the parameter of the individual is used to weight the densities continuously. Let the individual densities be indexed continuously by $\theta$; that is, the density corresponding to $\theta$ is $p(\cdot; \theta)$. Now, let $w(\cdot)$ be a weight (density) associated with $\theta$ such that $w(\theta) \geq 0$ and $\int w(\theta) \mathrm{d}\theta = 1$. Then, form the mixture density $p(\cdot)$ as

$$p(x) = \int w(\theta)p(x; \theta)\mathrm{d}\theta. \qquad (5.4)$$

An example of this kind of mixture is the beta-binomial distribution, the density of which is given in equation (5.18).

**Probability-Skewed Distributions**

A special type of mixture distribution is a *probability-skewed distribution*, in which the mixing weights are the values of a CDF. The skew-normal distribution is a good example.

The (standard) skew-normal distribution has density

$$g(x) = \frac{2}{\sqrt{2\pi}} e^{-x^2/2} \Phi(\lambda x) \quad \text{for } -\infty \leq x \leq \infty, \qquad (5.5)$$

where $\Phi(\cdot)$ is the standard normal CDF, and $\lambda$ is a constant such that $-\infty < \lambda < \infty$. For $\lambda = 0$, the skew-normal distribution is the normal distribution, and in general, if $|\lambda|$ is relatively small, the distribution is close to the normal. For larger $|\lambda|$, the distribution is more skewed, either positively or negatively. This distribution is an appropriate distribution for variables that would otherwise have a normal distribution but have been screened on the basis of a correlated normal random variable. See Arnold et al. (1993) for discussions.

Other distributions symmetric about 0 can also be skewed by a CDF in this manner. The general form of the probability density is

$$g(x) \propto p(x)P(\lambda x),$$

where $p(\cdot)$ is the density of the underlying symmetric distribution, and $P(\cdot)$ is a CDF (not necessarily the corresponding one). The idea also extends to multivariate distributions. Arnold and Beaver (2000) discuss definitions and applications of such densities, specifically a skew-Cauchy density.

In most cases, if $|\lambda|$ is relatively small, generation of random variables from a probability-skewed symmetric distribution using an acceptance/rejection method with the underlying symmetric distribution as the majorizing density is entirely adequate. For larger values of $|\lambda|$, it is necessary to divide the support into two or more intervals. It is still generally possible to use the same majorizing density, but the multiplicative constant can be different in different intervals.

## 5.2   Some Specific Univariate Distributions

In this section, we consider several of the more common univariate distributions and indicate methods for simulating them. The methods discussed are generally among the better ones, at least according to some criteria, but the discussion is not exhaustive. We give the details for some simpler algorithms, but in many cases the best algorithm involves many lines of a program with several constants that optimize a majorizing function or a squeeze function or the breakpoints of mixtures. We sometimes do not describe the best method in detail but rather refer the interested reader to the relevant literature. Devroye (1986a) has given a comprehensive treatment of methods for generating deviates from various distributions, and more information on many of the algorithms in this section can be found in that reference.

The descriptions of the algorithms that we give indicate the computations, but if the reader develops a program from the algorithm, issues of computational efficiency should be considered. For example, in the descriptions, we do not identify the computations that should be removed from the main body of the algorithm and made part of some setup computations.

Two variations of a distribution are often of interest. In one variation, the distribution is truncated. In this case, as we mentioned above, the range of the original distribution is restricted to a subrange and the probability measure adjusted accordingly. In another variation, the role of the random variable and the parameter of the distribution are interchanged. In some cases, these quantities have a natural association, and the corresponding distributions are said to be conjugate. An example of two such distributions are the binomial and the beta. What is a realization of a random variable in one distribution is a parameter in the other distribution. For many distributions, we may want to generate samples of a parameter, given realizations of the random variable (the data).

## 5.2.1 Normal Distribution

The normal distribution, which we denote by $N(\mu, \sigma^2)$, has the probability density

$$p(x) = \frac{1}{\sqrt{2\pi}\sigma} e^{-(x-\mu)^2/(2\sigma^2)} \quad \text{for } -\infty \leq x \leq \infty. \tag{5.6}$$

If $Z \sim N(0, 1)$ and $X = \sigma Z + \mu$, then $X \sim N(\mu, \sigma^2)$. Because of this simple relationship, it is sufficient to develop methods to generate deviates from the *standard* normal distribution, $N(0, 1)$, so there is no setup involved. All constants necessary in any algorithm can be precomputed and stored.

There are several methods for transforming uniform random variates into normal random variates.

One transformation *not* to use is:

1. Generate $u_i$ for $i = 1, \ldots, 12$ as i.i.d. $U(0, 1)$.
2. Deliver $x = \sum u_i - 6$.

This method is the Central Limit Theorem applied to a sample of size 12. Not only is the method approximate (and based on a poor approximation!), but it is also slower than better methods.

A simple method is the Box–Muller method arising from a polar transformation: If $U_1$ and $U_2$ are independently distributed as $U(0,1)$, and

$$X_1 = \sqrt{-2\log(U_1)}\cos(2\pi U_2),$$
$$X_2 = \sqrt{-2\log(U_1)}\sin(2\pi U_2), \qquad (5.7)$$

then $X_1$ and $X_2$ are independently distributed as $N(0,1)$ (see Exercises 5.1a and 5.1b on page 213).

The Box–Muller transformation is rather slow. It requires evaluation of one square root and two trigonometric functions for every two deviates generated.

As noted by Neave (1973), if the uniform deviates used in the Box–Muller transformation are generated by a congruential generator with small multiplier, the resulting normals are deficient in the tails. Golder and Settle (1976) under similar conditions demonstrate that the density of the generated normal variates has a jagged shape, especially in the tails. Of course, if they had analyzed their small-multiplier congruential generator, they would have found that generator lacking. (See the discussion about Figure 1.3, page 16.) It is easy to see that the largest and smallest numbers generated by the Box–Muller transformation occur when a value of $u_1$ from the uniform generator is close to 0. A bound on the absolute value of the numbers generated is $\sqrt{-2\log(e)}$, where $e$ is the smallest floating-point that can be generated by the uniform generator. (In the "minimal standard" congruential generator and many similar generators, the smallest number is approximately $2^{-31}$, so the bound on the absolute value is approximately 6.56.) How close the results of the transformation come to the bound depends on whether $u_2$ is close to 0, 1/4, 1/2, 3/4, or 1 when $u_1$ is close to 0. (In the "minimal standard" generator, when $u_1$ is at the smallest value possible, $\cos(2\pi u_2)$ is close to 1 because $u_2$ is relatively close to 0. The maximum number, therefore, is very close to the upper bound of 6.56, which has a p-value of the same order of magnitude as the reciprocal of the period of the generator. On the other hand, when $u_1$ is close to 0, the value of $u_2$ is never close enough to 1/2 or 3/4 to yield a value of one of the trigonometric functions close to $-1$. The minimum value that results from the Box–Muller transformation, therefore, does not approach the lower bound of $-6.56$. The p-value of the minimum value is three to four orders of magnitude greater than the reciprocal of the period of the generator.) If the Box–Muller transformation is used with a congruential generator, especially one with a relatively small multiplier, the roles of $u_1$ and $u_2$ should be exchanged periodically. This ensures that the lower and upper bounds are approximately symmetric if the generator has full period.

Tezuka (1991) shows that similar effects also are noticeable if a poor Tausworthe generator is used. These studies emphasize the importance of using a good uniform generator for whatever distribution is to be simulated. It is especially important to be wary of the effects of a poor uniform generator in algorithms that require more than one uniform deviate (see Section 4.3).

Bratley, Fox, and Schrage (1987) show that normal variates generated by the Box–Muller transformation lie pairwise on spirals. The spirals are of exactly the same origin as the lattice of the congruential generator itself, so a solution would be to use a better uniform generator.

To alleviate potential problems of patterns in the output of a polar method such as the Box–Muller transformation, some authors have advocated that, for each pair of uniforms, only one of the resulting pair of normals be used. If there is any marginal gain in quality, it is generally not noticeable, especially if the roles of $u_1$ and $u_2$ are exchanged periodically as recommended.

The Box–Muller transformation is one of several polar methods. All of them have similar properties, but the Box–Muller transformation generally requires slower computations. Although most currently available computing systems can evaluate the necessary trigonometric functions extremely rapidly, the Box–Muller transformation can often be performed more efficiently using an acceptance/rejection algorithm, as we indicated in the general discussion of acceptance/rejection methods (see Exercise 5.1d on page 213). The Box–Muller transformation is implemented via rejection in Algorithm 5.1.

**Algorithm 5.1 A Rejection Polar Method for Normal Variates**

1. Generate $v_1$ and $v_2$ independently from U$(-1, 1)$, and set $r^2 = v_1^2 + v_2^2$.

2. If $r^2 \geq 1$, then
    go to step 1;
   otherwise,
    deliver
    $$x_1 = v_1 \sqrt{-2 \log r^2 / r^2}$$
    $$x_2 = v_2 \sqrt{-2 \log r^2 / r^2}.$$ ∎

Ahrens and Dieter (1988) describe fast polar methods for the Cauchy and exponential distributions in addition to the normal distribution.

The fastest algorithms for generating normal deviates use either a ratio-of-uniforms method or a mixture with acceptance/rejection. One of the best algorithms, called the rectangle/wedge/tail method, is described by Marsaglia, MacLaren, and Bray (1964). In that method, the normal density is decomposed into a mixture of densities with shapes as shown in Figure 5.1. It is easy to generate a variate from one of the rectangular densities, so the decomposition is done to give a high probability of being able to use a rectangular density. That, of course, means lots of rectangles, which brings some inefficiencies. The optimal decomposition must address those tradeoffs. The wedges are nearly linear densities (see Algorithm 4.7), so generating from them is relatively fast. The tail region takes the longest time, so the decomposition is such as to give a small probability to the tail. Ahrens and Dieter (1972) give an implementation of the rectangle/wedge/tail method that can be optimized at the bit level.

Kinderman and Ramage (1976) represent the normal density as a mixture and apply a variety of acceptance/rejection and table-lookup techniques for the components. The individual techniques for various regions have been developed by Marsaglia (1964), Marsaglia and Bray (1964), and Marsaglia, MacLaren, and

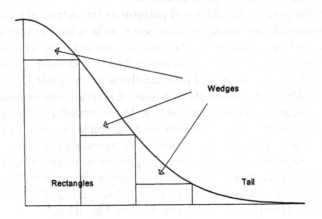

Figure 5.1: Rectangle/Wedge/Tail Decomposition

Bray (1964). Marsaglia and Tsang (1984) also give a decomposition, resulting in what they call the "ziggurat method".

Leva (1992a) describes a ratio-of-uniforms method with very tight bounding curves for generating normal deviates. (The 15-line Fortran program implementing Leva's method is Algorithm 712 of *CALGO*; see Leva, 1992b.)

Given the current speed of the standard methods of evaluating the inverse normal CDF, the inverse CDF method is often useful, especially if order statistics are of interest. Even with the speed of the standard algorithms for the inverse normal CDF, specialized versions, possibly to a slightly lower accuracy, have been suggested, for example by Marsaglia (1991) and Marsaglia, Zaman, and Marsaglia (1994). (The latter reference gives two algorithms for inverting the normal CDF: one very accurate, and one faster but slightly less accurate.)

Wallace (1996) describes an interesting method of generating normals from other normals rather than by making explicit transformations of uniforms. The method begins with a set of $kp$ normal deviates generated by some standard method. The deviates are normalized so that their sum of squares is 1024. Let $X$ be a $k \times p$ array containing those deviates, and let $A_i$ be a $k \times k$ orthogonal matrix. New normal deviates are formed by multiplication of an orthogonal matrix and the columns of $X$. A random column from the array and a random method of moving from one column to another are chosen. In Wallace's implementation, $k$ is 4 and $p$ is 256. Four orthogonal $4 \times 4$ matrices are chosen to make the matrix/vector multiplication fast:

$$A_1 = \frac{1}{2} \begin{bmatrix} 1 & 1 & -1 & 1 \\ 1 & -1 & 1 & 1 \\ 1 & -1 & -1 & -1 \\ -1 & -1 & -1 & 1 \end{bmatrix} \quad A_2 = \frac{1}{2} \begin{bmatrix} 1 & -1 & -1 & -1 \\ 1 & -1 & 1 & 1 \\ 1 & 1 & -1 & 1 \\ -1 & -1 & -1 & 1 \end{bmatrix}$$

$$A_3 = \frac{1}{2} \begin{bmatrix} 1 & -1 & 1 & 1 \\ -1 & -1 & 1 & -1 \\ -1 & -1 & -1 & 1 \\ -1 & 1 & 1 & 1 \end{bmatrix} \quad A_4 = \frac{1}{2} \begin{bmatrix} -1 & 1 & -1 & -1 \\ -1 & -1 & 1 & -1 \\ -1 & 1 & 1 & 1 \\ 1 & 1 & 1 & -1 \end{bmatrix} ;$$

hence, the matrix multiplication is usually just the addition of two elements of the vector. After a random column of $X$ is chosen (that is, a random integer between 1 and 256), a random odd number between 1 and 255 is chosen as a stride (that is, as an increment for the column number) to allow movement from one column to another. The first column chosen is multiplied by $A_1$, the next by $A_2$, the next by $A_3$, the next by $A_4$, and then the next by $A_1$, and so on. The elements of the vectors that result from these multiplications constitute both the normal deviates output in this pass and the elements of a new $k \times p$ array. Except for rounding errors, the elements in the new array should have a sum of squares of 1024 also. Just to avoid any problems from rounding, however, the last element generated is not delivered as a normal deviate but instead is used to generate a chi-squared deviate, $y$, with 1024 degrees of freedom via a Wilson–Hilferty approximation, and the 1023 other values are normalized by $\sqrt{y/1024}$. (The Wilson–Hilferty approximation relates the chi-squared random variable $Y$ with $\nu$ degrees of freedom to the standard normal random variable $X$ by

$$X \approx \frac{\left(\frac{Y}{\nu}\right)^{\frac{1}{3}} - \left(1 - \frac{2}{9\nu}\right)}{\sqrt{\frac{2}{9\nu}}}.$$

The approximation is fairly good for $\nu > 30$. See Abramowitz and Stegun, 1964.)

## Truncated Normal Distribution

In Monte Carlo studies, the tail behavior is often of interest. Variates from the tail of a distribution can always be formed by selecting variates generated from the full distribution, of course, but this can be a very slow process. Marsaglia (1964), Geweke (1991a), Robert (1995), and Damien and Walker (2001) give methods for generating variates directly from a truncated normal distribution. The truncated normal with left truncation point $\tau$ has density

$$p(x) = \frac{e^{-(x-\mu)^2/(2\sigma^2)}}{\sqrt{2\pi}\sigma \left(1 - \Phi\left(\frac{\tau-\mu}{\sigma}\right)\right)} \quad \text{for } \tau \le x \le \infty,$$

where $\Phi(\cdot)$ is the standard normal CDF.

The method of Robert uses an acceptance/rejection method with a translated exponential as the majorizing density; that is,

$$g(y) = \lambda^* e^{-\lambda^*(y-\tau)} \quad \text{for } \tau \le y \le \infty,$$

where

$$\lambda^* = \frac{\tau + \sqrt{\tau^2 + 4}}{2}. \tag{5.8}$$

(See the next section for methods to generate exponential random variates.)

The method of Damien and Walker uses conditional distributions. The range of the conditional uniform that yields the normal is taken as the intersection of the truncated range and the full conditional range $(-\sqrt{-2\log Y}, \sqrt{-2\log Y})$ in the example on page 149.

**Lognormal and Halfnormal Distributions**

Two distributions closely related to the normal are the lognormal and the halfnormal. The lognormal is the distribution of a random variable whose logarithm has a normal distribution. A very good way to generate lognormal variates is just to generate normal variates and exponentiate. The halfnormal is the folded normal distribution. The best way to generate deviates from the halfnormal is just to take the absolute value of normal deviates.

## 5.2.2   Exponential, Double Exponential, and Exponential Power Distributions

The exponential distribution with parameter $\lambda > 0$ has the probability density

$$p(x) = \lambda e^{-\lambda x} \quad \text{for } 0 \le x \le \infty. \tag{5.9}$$

If $Z$ has the *standard* exponential distribution (that is, with parameter equal to 1), and $X = Z/\lambda$, then $X$ has the exponential distribution with parameter $\lambda$ (called the "rate"). Because of this simple relationship, it is sufficient to develop methods to generate deviates from the standard exponential distribution. The exponential distribution is a special case of the gamma distribution, the density of which is given in equation (5.13). The parameters of the gamma distribution are $\alpha = 1$ and $\beta = \frac{1}{\lambda}$.

The inverse CDF method is very easy to implement and is generally satisfactory for the exponential distribution. The method is to generate $u$ from $U(0,1)$ and then take

$$x = -\frac{\log(u)}{\lambda}. \tag{5.10}$$

(This and similar computations are why we require that the simulated uniform not include its endpoints.)

Many other algorithms for generating exponential random numbers have been proposed over the years. Marsaglia, MacLaren, and Bray (1964) apply the rectangle/wedge/tail method to the exponential distribution. Ahrens and Dieter (1972) give a method that can be highly optimized at the bit level.

Ahrens and Dieter also provide a catalog of other methods for generating exponentials. These other algorithms seek greater speed by avoiding the computation of the logarithm. Many simple algorithms for random number generation involve evaluation of elementary functions. As we have indicated, evaluation of an elementary function at a random point can often be performed equivalently by acceptance/rejection, and Ahrens and Dieter (1988) describe a method for the exponential that does that. (See Hamilton, 1998, for some corrections to their algorithm.) As the software for evaluating elementary functions has become faster, the need to avoid their evaluation has decreased.

A common use of the exponential distribution is as the model of the interarrival times in a Poisson process. A (homogeneous) Poisson process,

$$T_1 < T_2 < \ldots,$$

with rate parameter $\lambda$ can be generated by taking the output of an exponential random number generator with parameter $\lambda$ as the times,

$$t_1, \ t_2 - t_1, \ \ldots.$$

We consider nonhomogeneous Poisson processes in Section 6.5.2, page 225.

### Truncated Exponential Distribution

The interarrival process is memoryless, and the tail of the exponential distribution has an exponential distribution; that is, if $X$ has the density (5.9), and $Y = X + \tau$, then $Y$ has the density

$$\lambda e^{\lambda \tau} e^{-\lambda y} \quad \text{for } \tau \leq y \leq \infty.$$

This fact provides a very simple process for generating from the tail of an exponential distribution.

### Double Exponential Distribution

The double exponential distribution, also called the Laplace distribution, with parameter $\lambda > 0$ has the probability density

$$p(x) = \frac{\lambda}{2} e^{-\lambda |x|} \quad \text{for } -\infty \leq x \leq \infty. \tag{5.11}$$

The double exponential distribution is often used in Monte Carlo studies of robust procedures because it has a heavier tail than the normal distribution yet corresponds well with observed distributions.

If $Z$ has the standard exponential distribution and $X = SZ/\lambda$, where $S$ is a random variable with probability mass $\frac{1}{2}$ at $-1$ and at $+1$, then $X$ has the double exponential distribution with parameter $\lambda$. This fact is the basis for the method of generating double exponential variates; generate an exponential, and change the sign with probability $\frac{1}{2}$. The method of bit stripping (see page 10) can be used to do this as long as the lower-order bits are the ones used and assuming that the basic uniform generator is a very good one.

### Exponential Power Distribution

A generalization of the double exponential distribution is the exponential power distribution, having density

$$p(x) \propto e^{-\lambda|x|^{\alpha}} \quad \text{for } -\infty \leq x \leq \infty. \tag{5.12}$$

For $\alpha = 2$, the exponential power distribution is the normal distribution. The members of this family with $1 \leq \alpha < 2$ are often used to model distributions with slightly heavier tails than the normal distribution. Either the double exponential or the normal distribution, depending on the value of $\alpha$, works well as a majorizing density to generate exponential power variates by acceptance/rejection (see Tadikamalla, 1980a).

## 5.2.3   Gamma Distribution

The gamma distribution with parameters $\alpha > 0$ and $\beta > 0$ has the probability density

$$p(x) = \frac{1}{\Gamma(\alpha)\beta^{\alpha}}x^{\alpha-1}e^{-x/\beta} \quad \text{for } 0 \leq x \leq \infty, \tag{5.13}$$

where $\Gamma(\alpha)$ is the complete gamma function. The $\alpha$ parameter is called the shape parameter, and $\beta$ is called the scale parameter. If the random variable $Z$ has the *standard* gamma distribution with shape parameter $\alpha$ and scale parameter 1, and $X = \beta Z$, then $X$ has a gamma distribution with parameters $\alpha$ and $\beta$. (Notice that the exponential is a gamma with $\alpha = 1$ and $\beta = 1/\lambda$.)

Of the special distributions that we have considered thus far, this is the first one that has a parameter that cannot be handled by simple translations and scalings. Hence, the best algorithms for the gamma distribution may be different depending on the value of $\alpha$ and on how many deviates are to be generated for a given value of $\alpha$.

Cheng and Feast (1979) use a ratio-of-uniforms method, as shown in Algorithm 5.2, for a gamma distribution with $\alpha > 1$. The mean time of this algorithm is $O(\alpha^{\frac{1}{2}})$, so for larger values of $\alpha$ it is less efficient. Cheng and Feast (1980) also gave an acceptance/rejection method that was better for large values of the shape parameter. Schmeiser and Lal (1980) use a composition of ten densities, some of the rectangle/wedge/tail type, followed by the acceptance/rejection method. The Schmeiser/Lal method is the algorithm used in the IMSL Libraries for values of the shape parameter greater than 1. The speed of the Schmeiser/Lal method does not depend on the value of the shape parameter. Sarkar (1996) gives a modification of the Schmeiser/Lal method that has greater efficiency because of using more intervals, resulting in tighter majorizing and squeeze functions, and because of using an alias method to help speed the process.

**Algorithm 5.2 The Cheng/Feast (1979) Algorithm for Generating Gamma Random Variates when the Shape Parameter is Greater than 1**

1. Generate $u_1$ and $u_2$ independently from $U(0,1)$, and set

$$v = \frac{\left(\alpha - \frac{1}{6\alpha}\right)u_1}{(\alpha-1)u_2}.$$

2. If

$$\frac{2(u_2-1)}{\alpha-1} + v + \frac{1}{v} \le 2,$$

then deliver $x = (\alpha-1)v$;
otherwise,
   if

$$\frac{2\log u_2}{\alpha-1} - \log v + v \le 1,$$

   then deliver $x = (\alpha-1)v$.

3. Go to step 1. ∎

An efficient algorithm for values of the shape parameter less than 1 is the acceptance/rejection method described in Ahrens and Dieter (1974) and modified by Best (1983), as shown in Algorithm 5.3. That method is the algorithm used in the IMSL Libraries for values of the shape parameter less than 1.

**Algorithm 5.3 The Best/Ahrens/Dieter Algorithm for Generating Gamma Random Variates when the Shape Parameter Is Less than 1**

0. Set $t = 0.07 + 0.75\sqrt{1-\alpha}$ and
   $$b = 1 + \frac{e^{-t}\alpha}{t}.$$

1. Generate $u_1$ and $u_2$ independently from $U(0,1)$, and set $v = bu_1$.

2. If $v \le 1$, then
      set $x = tv^{\frac{1}{\alpha}}$;
      if $u_2 \le \dfrac{2-x}{2+x}$, then deliver $x$;
      otherwise,
         if $u_2 \le e^{-x}$, then deliver $x$;
      otherwise,
      set $x = -\log\left(\dfrac{t(b-v)}{\alpha}\right)$ and $y = \dfrac{x}{t}$;

> if $u_2(\alpha + y(1 - \alpha)) \leq 1$, then deliver $x$;
> otherwise,
>> if $u_2 \leq y^{a-1}$, then deliver $x$.

3. Go to step 1.                                                    ∎

There are two cases of the gamma distribution that are of particular interest. The shape parameter $\alpha$ often is a positive integer. In that case, the distribution is sometimes called the *Erlang distribution*. If $Y_1, Y_2, \ldots, Y_\alpha$ are independently distributed as exponentials with parameter $1/\beta$, then $X = \sum Y_i$ has a gamma (Erlang) distribution with parameters $\alpha$ and $\beta$. Using the inverse CDF method (equation (5.10)) with the independent realizations $u_1, u_2, \ldots, u_\alpha$, we generate an Erlang deviate as

$$x = -\beta \log \left( \prod_{i=1}^{\alpha} u_i \right).$$

The general algorithms for gammas work better for the Erlang distribution if $\alpha$ is large.

The other special case of the gamma is the chi-squared distribution in which the scale parameter $\beta$ is 2. Twice the shape parameter $\alpha$ is called the degrees of freedom. For large or nonintegral degrees of freedom, the general methods for generating gamma random deviates are best for generating chi-squared deviates; otherwise, special methods described below are used.

An important property of the gamma distribution is:

> If $X$ and $Y$ are independent gamma variates with common scale parameter $\beta$ and shape parameters $\alpha_1$ and $\alpha_2$, then $X + Y$ has a gamma distribution with scale parameter $\beta$ and shape parameter $\alpha_1 + \alpha_2$.

This fact may be used in developing schemes for generating gammas. For example, any gamma can be represented as the sum of an Erlang variate, which is the sum of exponential variates, and a gamma variate with shape parameter less than 1. This representation may effectively be used in a method of generating gamma variates (see Atkinson and Pearce, 1976).

### Truncated Gamma Distribution

In some applications, especially ones involving censored observations, a truncated gamma is a useful model. In the case of left-censored data, we need to sample from the tail of a gamma distribution. The relevant distribution has the density

$$p(x) = \frac{1}{(\Gamma(\alpha) - \Gamma_{\tau/\beta}(\alpha))\beta^\alpha} x^{\alpha-1} e^{-x/\beta} \quad \text{for } \tau \leq x \leq \infty,$$

where $\Gamma_{\tau/\beta}(\cdot)$ is the incomplete gamma function (see page 321). As above, we consider the standard gamma.

Dagpunar (1978) describes a method of sampling from the tail of a standard ($\beta = 1$) gamma distribution with shape parameter $\alpha > 1$. The method makes use of the fact mentioned above that an exponential distribution is memoryless. A truncated exponential is used as a majorizing density in an acceptance/rejection method. The value of the exponential scale parameter that will maximize the probability of acceptance is the saddlepoint in the ratio of the truncated gamma density to the truncated exponential (both truncated at the same point, $\tau$),

$$\lambda = \frac{\tau - \alpha + \sqrt{(\tau - \alpha)^2 + 4\tau}}{2\tau}.$$

The procedure therefore is:

1. Generate $y$ from the truncated exponential and $u$ independently as U(0, 1). ($y$ can be generated by generating $u_1$ as U(0, 1) and taking $y = \frac{-\log u_1}{\lambda} + \tau$.)

2. If $(1 - \lambda)y - (\alpha - 1)(1 + \log y + \log \frac{1-\lambda}{\alpha-1}) \leq -\log u$, then deliver $x = y$.

Philippe (1997) also describes methods for a left-truncated gamma distribution for general values of the shape parameter. The reader is referred to the paper for the details.

Many common applications require truncation on the right; that is, the observations are right censored. Philippe (1997) describes a method for generating variates from a right-truncated gamma distribution, which has density

$$p(x) = \frac{1}{\Gamma_{\tau/\beta}(\alpha)\beta^\alpha} x^{\alpha-1} e^{-x/\beta} \quad \text{for } 0 \leq x \leq \tau,$$

where $\Gamma_{\tau/\beta}(\alpha)$ is the incomplete gamma function. Philippe shows that if the random variable $X$ has this distribution, then it can be represented as an infinite mixture of beta random variables:

$$X = \sum_{k=1}^{\infty} \frac{\Gamma(\alpha)}{\Gamma(\alpha + k)\Gamma_{1/\beta}(\alpha)\beta^{\alpha+k-1}e^{1/\beta}} Y_k,$$

where $Y_k$ is a random variable with a beta random variable with parameters $\alpha$ and $k$. Philippe suggested as a majorizing density a finite series

$$g_m(y) \quad \propto \quad \sum_{k=1}^{m} \frac{1}{\beta^{k-1}\Gamma(\alpha)\Gamma(k)\sum_{i=1}^{m}\frac{1}{\beta^{i-1}\Gamma(\alpha+i)}} y^{\alpha-1}(1-y)^{k-1}$$

$$= \quad \sum_{k=1}^{m} \frac{1}{\beta^{k-1}\Gamma(\alpha+k)\sum_{i=1}^{m}\frac{1}{\beta^{i-1}\Gamma(\alpha+i)}} h_k(y),$$

where $h_k$ is a beta density (equation (5.14)) with parameters $\alpha$ and $k$. Thus, to generate a variate from a distribution with density $g_m$, we select a beta with the probability equal to the weight and then use a method described in the next section for generating a beta variate. The number of terms depends on the probability of acceptance. Obviously, we want a high probability of acceptance, but this requires a large number of terms in the series. For a probability of acceptance of at least $p^*$ (with $p^* < 1$, obviously), Philippe shows that the number of terms required in the series is approximately

$$m^* = \frac{1}{4}\left(z_p + \sqrt{z_p^2 + \frac{4}{\beta}}\right)^2,$$

where $z_p = \Phi^{-1}(p)$ and $\Phi$ is the standard normal CDF.

**Algorithm 5.4  The Philippe (1997) Algorithm for Generating Gamma Random Variates Truncated on the Right at $\tau$**

0. Determine $m^*$, and initialize quantities in $g_{m^*}$.

1. Generate $y$ from the distribution with density $g_{m^*}$.

2. Generate $u$ from $\mathrm{U}(0,1)$.

3. If

$$u \le \frac{\sum_{k=1}^{m^*} \frac{1}{\beta^{k-1}\Gamma(k)}}{e^{\frac{y}{\beta}} \sum_{k=1}^{m^*} \frac{(1-y)^{k-1}}{\beta^{k-1}\Gamma(k)}},$$

   then
       deliver $x = y$;
   otherwise,
       return to step 1.                                              ∎

Damien and Walker (2001) also give a method for generating variates directly from a truncated gamma distribution. Their method uses conditional distributions, as we discuss on page 149. The range of the conditional uniform that yields the gamma is taken as the intersection of the truncated range and the full conditional range.

**Generalized Gamma Distributions**

There are a number of generalizations of the gamma distribution. The generalizations provide more flexibility in modeling because they have more parameters. Stacy (1962) defined a generalization that has two shape parameters. It is especially useful in failure-time models. The distribution has density

$$p(x) = \frac{|\gamma|}{\Gamma(\alpha)\beta^{\alpha\gamma}} x^{\alpha\gamma-1} e^{(-x/\beta)^\gamma} \quad \text{for } 0 \le x \le \infty.$$

This distribution includes as special cases the ordinary gamma (with $\gamma = 1$), the halfnormal distribution (with $\alpha = \frac{1}{2}$ and $\gamma = 2$), and the Weibull (with

$\alpha = 1$). The best way to generate a generalized gamma distribution is to use the best method for the corresponding gamma and then exponentiate.

Everitt (1998) describes a generalized gamma distribution, which he calls the "Creedy and Martin generalized gamma", with density

$$p(x) = \theta_0 \, x^{\theta_1} \, e^{\theta_2 x + \theta_3 x^2 + \theta_4 x^3} \quad \text{for } 0 \le x \le \infty.$$

This density can of course be scaled with a $\beta$, as in the other gamma distributions that we have discussed.

Ghitany (1998) and Agarwal and Al-Saleh (2001) have described a generalized gamma distribution based on a generalized gamma function,

$$\Gamma(\alpha, \nu, \lambda) = \int_0^\infty \frac{1}{(t + \nu)^\lambda} t^{\alpha-1} e^{-t} dt,$$

introduced by Kobayashi (1991). The distribution has density

$$p(x) = \frac{1}{\Gamma(\alpha, \nu, \lambda) \beta^{\alpha-\lambda}} \frac{x^{\alpha-1}}{(x + \beta\nu)^\lambda} e^{(-x/\beta)} \quad \text{for } 0 \le x \le \infty.$$

This distribution is useful in reliability studies because of the shapes of the hazard function that are possible for various values of the parameters. It is the same as the ordinary gamma for $\lambda = 0$.

### $D$-Distributions

A class of distributions, called $D$-distributions, that arise in extended gamma processes is studied by Laud, Ramgopal, and Smith (1993). The interested reader is referred to that paper for the details.

## 5.2.4   Beta Distribution

The beta distribution with parameters $\alpha > 0$ and $\beta > 0$ has the probability density

$$p(x) = \frac{1}{B(\alpha, \beta)} x^{\alpha-1} (1 - x)^{\beta-1} \quad \text{for } 0 \le x \le 1, \tag{5.14}$$

where $B(\alpha, \beta)$ is the complete beta function.

Efficient methods for generating beta variates require different algorithms for different values of the parameters. If either parameter is equal to 1, it is very simple to generate beta variates using the inverse CDF method, which in this case would just be a root of a uniform. If both values of the parameters are less than 1, the simple acceptance/rejection method of Jöhnk (1964), given as Algorithm 5.5, is one of the best. If one parameter is less than 1 and the other is greater than 1, the method of Atkinson (1979) is useful. If both parameters are greater than 1, the method of Schmeiser and Babu (1980) is very efficient, except that it requires a lot of setup time. For the case of both parameters greater than 1, Cheng (1978) gives an algorithm that requires very little setup time. The IMSL Libraries use all five of these methods, depending on the values of the parameters and how many deviates are to be generated for a given setting of the parameters.

**Algorithm 5.5 Jöhnk's Algorithm for Generating Beta Random Variates when Both Parameters are Less than 1**

1. Generate $u_1$ and $u_2$ independently from $U(0,1)$, and set $v_1 = u_1^{1/\alpha}$ and $v_2 = u_2^{1/\beta}$.

2. Set $w = v_1 + v_2$.

3. If $w > 1$, then go to step 1.

4. Set $x = \dfrac{v_1}{w}$, and deliver $x$.                                      ∎

## 5.2.5   Chi-Squared, Student's $t$, and $F$ Distributions

The chi-squared, Student's $t$, and $F$ distributions all are derived from the normal distribution. Variates from these distributions could, of course, be generated by transforming normal deviates. In the case of the chi-squared distribution, however, this would require generating and squaring several normals for each chi-squared deviate. A more direct algorithm is much more efficient. Even in the case of the $t$ and $F$ distributions, which would require only a couple of normals or chi-squared deviates, there are better algorithms.

### Chi-Squared Distribution

The chi-squared distribution, as we have mentioned above, is a special case of the gamma distribution (see equation (5.13)) in which the scale parameter, $\beta$, is 2. Twice the shape parameter, $2\alpha$, is called the degrees of freedom and is often denoted by $\nu$. If $\nu$ is large or is not an integer, the general methods for generating gamma random deviates described above are best for generating chi-squared deviates. If the degrees of freedom value is a small integer, the chi-squared deviates can be generated by taking a logarithm of the product of some independent uniforms. If $\nu$ is an even integer, the chi-squared deviate $r$ is produced from $\nu/2$ independent uniforms, $u_i$, by

$$r = -2\log \left( \prod_{i=1}^{\nu/2} u_i \right).$$

If $\nu$ is an odd integer, this method can be used with the product going to $(\nu-1)/2$, and then the square of an independent normal deviate is added to produce $r$.

The square root of the chi-squared random variable is sometimes called a chi random variable. Although, clearly, a chi random variable could be generated as the square root of a chi-squared deviate generated as above, there are more efficient direct ways of generating a chi deviate; see Monahan (1987).

**Student's $t$ Distribution**

The standard $t$ distribution with $\nu$ degrees of freedom has density

$$p(x) = \frac{\Gamma\left(\frac{\nu+1}{2}\right)}{\Gamma\left(\frac{\nu}{2}\right)\sqrt{\nu\pi}} \left(1 + \frac{x^2}{\nu}\right)^{-\frac{\nu+1}{2}} \quad \text{for } -\infty \le x \le \infty. \tag{5.15}$$

The degrees of freedom, $\nu$, does not have to be an integer, but it must be positive.

A standard normal random variable divided by the square root of a chi-squared random variable with $\nu$ degrees of freedom is a $t$ random variable with $\nu$ degrees of freedom. Also, the square root of an $F$ random variable with 1 and $\nu$ degrees of freedom is a $t$ random variable with $\nu$ degrees of freedom. These relations could be used to generate $t$ deviates, but neither yields an efficient method.

Kinderman and Monahan (1980) describe a ratio-of-uniforms method for the $t$ distribution. The algorithm is rather complicated, but it is very efficient. Marsaglia (1980) gives a simpler procedure that is almost as fast. Either is almost twice as fast as generating a normal deviate and a chi-squared deviate and dividing by the square root of the chi-squared one. Marsaglia (1984) also gives a very fast algorithm for generating $t$ variates that is based on a transformed acceptance/rejection method that he called exact-approximation (see Section 4.5).

Bailey (1994) gives the polar method shown in Algorithm 5.6 for the Student's $t$ distribution. It is similar to the polar method for normal variates given in Algorithm 5.1.

**Algorithm 5.6 A Rejection Polar Method for $t$ Variates with $\nu$ Degrees of Freedom**

1. Generate $v_1$ and $v_2$ independently from $U(-1, 1)$, and set $r^2 = v_1^2 + v_2^2$.

2. If $r^2 \ge 1$, then
   go to step 1;
   otherwise,

   $$\text{deliver } x = v_1 \sqrt{\frac{\nu(r^{-4/\nu} - 1)}{r^2}}. \qquad \blacksquare$$

The jagged shape of the frequency curve of normals generated via a polar method based on a poor uniform generator that was observed by Neave (1973) and by Golder and Settle (1976) may also occur for $t$ variates generated by this polar method. It is important to use a good uniform generator for whatever distribution is to be simulated.

In Bayesian analysis, it is sometimes necessary to generate random variates for the degrees of freedom in a $t$ distribution conditional on the data. In the hierarchical model underlying the analysis, the $t$ random variable is interpreted as a mixture of normal random variables divided by square roots of gamma random variables. For given realizations of gammas, $\lambda_1, \lambda_2, \ldots, \lambda_n$, the density of the degrees of freedom is

$$p(x) \propto \prod_{i=1}^{n} \frac{\nu^{\nu/2}}{2^{\nu/2} \Gamma\left(\frac{\nu}{2}\right)} \lambda_i^{\nu/2} e^{-\nu\lambda_i/2}.$$

Mendoza-Blanco and Tu (1997) show that three different gamma distributions can be used to approximate this density very well for three different ranges of values of $\lambda_g e^{-\lambda_a}$, where $\lambda_g$ is the geometric mean of the $\lambda_i$, and $\lambda_a$ is the arithmetic mean. Although the approximations are very good, the gamma approximations could also be used as majorizing densities.

### F Distribution

A variate from the $F$ distribution can be generated as the ratio of two chi-squared deviates, which, of course, would be only half as fast as generating a chi-squared deviate. A better way to generate an $F$ variate is as a transformed beta. If $X$ is distributed as a beta, with parameters $\nu_1/2$ and $\nu_2/2$, and

$$Y = \frac{\nu_2}{\nu_1} \frac{X}{1 - X},$$

then $Y$ has an $F$ distribution with $\nu_1$ and $\nu_2$ degrees of freedom. Generating a beta deviate and transforming it usually takes only slightly longer than generating a single chi-squared deviate.

### 5.2.6   Weibull Distribution

The Weibull distribution with parameters $\alpha > 0$ and $\beta > 0$ has the probability density

$$p(x) = \frac{\alpha}{\beta} x^{\alpha-1} e^{-x^\alpha/\beta} \quad \text{for } 0 \leq x \leq \infty. \tag{5.16}$$

The simple inverse CDF method applied to the standard Weibull distribution (i.e., $\beta = 1$) is quite efficient. The expression is simply

$$(-\log u)^{\frac{1}{\alpha}}.$$

Of course, an acceptance/rejection method could be used to replace the evaluation of the logarithm in the inverse CDF. The standard Weibull deviates are then scaled by $\beta^{1/\alpha}$.

## 5.2.7 Binomial Distribution

The probability function for the binomial distribution with parameters $n$ and $\pi$ is

$$p(x) = \frac{n!}{x!(n-x)!}\pi^x(1-\pi)^{n-x} \quad \text{for } x = 0, 1, \ldots, n, \qquad (5.17)$$

where $n$ is a positive integer and $0 < \pi < 1$.

To generate a binomial, a simple way is to sum Bernoullis (equation (4.4), and Algorithm 4.1, page 105), which is equivalent to an inverse CDF technique. If $n$, the number of independent Bernoullis, is small, this method is adequate. The time required for this kind of algorithm is obviously O($n$). For larger values of $n$, the median of a random sample of size $n$ from a Bernoulli distribution can be generated (it has an approximate beta distribution; see Relles, 1972), and then the inverse CDF method can be applied from that point. Starting at the median allows the time required to be halved. Kemp (1986) shows that starting at the mode results in an even faster method and gives a method to approximate the modal probability quickly. If this idea is applied recursively, the time becomes O($\log n$). The time required for any method based on the CDF of the binomial is an increasing function of $n$.

Several methods whose efficiencies are not so dependent on $n$ are available, and for large values of $n$ they are to be preferred to methods based on the CDF. (The value of $\pi$ also affects the speed; the inverse CDF methods are generally competitive as long as $n\pi < 500$.) Stadlober (1991) described an algorithm based on a ratio-of-uniforms method. Kachitvichyanukul (1982) gives an efficient method using acceptance/rejection over a composition of four regions (see Schmeiser, 1983; and Kachitvichyanukul and Schmeiser, 1988a, 1990). This is the method used in the IMSL Libraries.

### Beta-Binomial Distribution

The beta-binomial distribution is the mixture distribution that is a binomial for which the parameter $\pi$ is a realization of a random variable that has a beta distribution. This distribution is useful for modeling overdispersion or "extravariation" in applications where there are clusters of separate binomial distributions.

The probability function for the beta-binomial distribution with parameters $n$, which is a positive integer, $\alpha > 0$, and $\beta > 0$ is

$$p(x) = \frac{n!}{x!(n-x)!\mathrm{B}(\alpha,\beta)} \int_0^1 \pi^{\alpha-1+x}(1-\pi)^{n+\beta-1-x}\mathrm{d}\pi \quad \text{for } x = 0, 1, \ldots, n,$$

$$(5.18)$$

where $\mathrm{B}(\alpha,\beta)$ is the complete beta function. (The integral in this expression is $\mathrm{B}(\alpha+x, n+\beta-x)$.)

The mean of the beta-binomial is in the form of the binomial mean, $n\pi$, with the beta mean, $\alpha/(\alpha + \beta)$, in place of $\pi$, but the variance is

$$\frac{n\alpha\beta}{(\alpha + \beta)^2} \frac{n + \alpha + \beta}{1 + \alpha + \beta}.$$

A simple way of generating deviates from a beta-binomial distribution is first to generate the parameter $\pi$ as the appropriate beta and then to generate the binomial (see Ahn and Chen, 1995). In this case, an inverse CDF method for the binomial may be more efficient because it does not require as much setup time as the generally more efficient ratio-of-uniforms or acceptance/rejection methods referred to above.

### 5.2.8   Poisson Distribution

The probability function for the Poisson distribution with parameter $\theta > 0$ is

$$p(x) = \frac{e^{-\theta}\theta^x}{x!} \quad \text{for } x = 0, 1, 2, \ldots. \tag{5.19}$$

A Poisson with a small mean, $\theta$, can be generated efficiently by the inverse CDF technique. Kemp and Kemp (1991) describe a method that begins at the mode of the distribution and proceeds in the appropriate direction to identify the inverse. They give a method for identifying the mode and computing the modal probability.

Many of the other methods that have been suggested for the Poisson also require longer times for distributions with larger means. Ahrens and Dieter (1980) and Schmeiser and Kachitvichyanukul give efficient methods having times that do not depend on the mean (see Schmeiser, 1983). The method of Schmeiser and Kachitvichyanukul uses acceptance/rejection over a composition of four regions. This is the method used in the IMSL Libraries.

### 5.2.9   Negative Binomial and Geometric Distributions

The probability function for the negative binomial is

$$p(x) = \binom{x + r - 1}{r - 1} \pi^r (1 - \pi)^x \quad \text{for } x = 0, 1, 2, \ldots, \tag{5.20}$$

where $r > 0$ and $0 < \pi < 1$. If $r$ is an integer, the negative binomial distribution is sometimes called the Pascal distribution. If $\pi$ is the probability of a success in a single Bernoulli trial, the random variable can be thought of as the number of failures before $r$ successes are obtained.

If $r\pi/(1 - \pi)$ is relatively small and $(1 - \pi)^r$ is not too small, the inverse CDF method works well. Otherwise, a gamma $(r, \pi/(1 - \pi))$ can be generated and used as the parameter to generate a Poisson. The Poisson variate is then delivered as the negative binomial variate.

The geometric distribution is a special case of the negative binomial with $r = 1$. The probability function is

$$p(x) = \pi(1 - \pi)^x \quad \text{for } x = 0, 1, 2, \ldots. \tag{5.21}$$

The integer part of an exponential random variable with parameter $\lambda = -\log(1 - \pi)$ has a geometric distribution with parameter $\pi$; hence, the simplest, and also one of the best, methods for the geometric distribution with parameter $\pi$ is to generate a uniform deviate $u$ and take

$$\left\lceil \frac{\log u}{\log(1 - \pi)} \right\rceil.$$

It is common to see the negative binomial and the geometric distributions defined as starting at 1 instead of 0, as above. The distributions are the same after making an adjustment of subtracting 1.

## 5.2.10 Hypergeometric Distribution

The probability function for the hypergeometric distribution is

$$p(x) = \frac{\dbinom{M}{x} \dbinom{L - M}{N - x}}{\dbinom{L}{N}} \tag{5.22}$$

$$\text{for } x = \max(0, N - L + M), \ldots, \min(N, M).$$

The usual method of developing the hypergeometric distribution is with a finite sampling model: $N$ items are to be sampled, independently with equal probability and without replacement, from a lot of $L$ items of which $M$ are special; the random variable $X$ is the number of special items in the random sample.

A good method for generating from the hypergeometric distribution is the inverse CDF method. The inverse CDF can be evaluated recursively using the simple expression for the ratio $p(x + 1)/p(x)$, so the build-up search of Algorithm 4.4 or the chop-down search of Algorithm 4.5 could be used. In either case, beginning at the mean, $MN/L$, can speed up the search.

Another simple method that is good is straightforward use of the finite sampling model that defines the distribution.

Kachitvichyanukul and Schmeiser (1985) give an algorithm based on acceptance/rejection of a probability function decomposed as a mixture, and Stadlober (1990) describes an algorithm based on a ratio-of-uniforms method. Both of these can be faster than the inverse CDF for larger values of $N$ and $M$. Kachitvichyanukul and Schmeiser (1988b) give a Fortran program for sampling from the hypergeometric distribution. The program uses either the inverse CDF or the acceptance/rejection method depending on the mode,

$$m = \left\lceil \frac{(N + 1)(M + 1)}{L + 2} \right\rceil.$$

If $m - \max(0, N + M - L) < 10$, then the inverse CDF method is used; otherwise, the composition/acceptance/rejection method is used.

**Extended Hypergeometric Distribution**

A related distribution, called the extended hypergeometric distribution, can be developed by assuming that $X$ and $Y = N - X$ are binomial random variables with parameters $M$ and $\pi_X$ and $L - M$ and $\pi_Y$, respectively. Let $\rho$ be the odds ratio,

$$\rho = \frac{\pi_X(1 - \pi_Y)}{\pi_Y(1 - \pi_X)};$$

then, the conditional distribution of $X$ given $X + Y = N$ has probability mass function

$$p(x|x + y = N) \;=\; \frac{\binom{M}{x}\binom{L - M}{N - x}\rho^x}{\sum_{j=a}^{b}\binom{M}{j}\binom{L - M}{N - j}\rho^j} \tag{5.23}$$

$$\text{for } x = a\ldots, b,$$

where $a = \max(0, N - L + M)$ and $b = \min(N, M)$.

This function can also be evaluated recursively and random numbers generated by the inverse CDF method, similarly to the hypergeometric distribution. Liao and Rosen (2001) describe methods for evaluating the probability mass functions and also for computing the mode of the distribution in order to speed up the evaluation of the inverse CDF.

Another generalization, called the noncentral hypergeometric distribution, is developed by allowing different probabilities of selecting the two types of items. If the relative probability of selecting an item of the special type to that of selecting an item of the other type (given an equal number of each type) $\omega$, the realization of $X$ can be built sequentially by Bernoulli realizations with probability $M_k/(M_k + \omega(L_k - M_k))$, where $M_k$ is the number of special items remaining. Variates from this distribution can be generated by the finite sampling model underlying the distribution.

## 5.2.11  Logarithmic Distribution

The probability function for the logarithmic distribution with parameter $\theta$ is

$$p(x) = -\frac{\theta^x}{x \log(1 - \theta)} \quad \text{for } x = 1, 2, 3, \ldots, \tag{5.24}$$

where $0 < \theta < 1$.

Kemp (1981) describes a method for generating efficiently from the inverse logarithmic CDF either using a chop-down approach (see page 108) to move rapidly down the set of CDF values or using a mixture in which highly likely values are given priority.

## 5.2.12 Other Specific Univariate Distributions

Many other interesting distributions have simple relationships to the standard distributions discussed above. When that is the case, because there are highly optimized methods for the standard distributions, it is often best just to use a very good method for the standard distribution and then apply the appropriate transformation. For some distributions, the inverse CDF method is almost as good as more complicated methods.

### Cauchy Distribution

Variates from the Cauchy or Lorentzian distribution, which has density

$$p(x) = \frac{1}{\pi a \left(1 + \left(\frac{x-b}{a}\right)^2\right)} \quad \text{for } -\infty \leq x \leq \infty, \tag{5.25}$$

can be generated easily by the inverse CDF method. For the standard Cauchy distribution (that is, with $a = 1$ and $b = 0$), given $u$ from $U(0,1)$, a Cauchy is $\tan(\pi u)$. The tangent function in the inverse CDF could be evaluated by acceptance/rejection in the manner mentioned on page 121, but if the inverse CDF is to be used, it is probably better just to use a good numerical function to evaluate the tangent. Kronmal and Peterson (1981) express the Cauchy distribution as a mixture and give an acceptance/complement method that is very fast.

### Rayleigh Distribution

For the Rayleigh distribution with density

$$p(x) = \frac{x}{\sigma^2} e^{-x^2/2\sigma^2} \quad \text{for } 0 \leq x \leq \infty \tag{5.26}$$

(which is a Weibull distribution with parameters $\alpha = 2$ and $\beta = 2\sigma^2$), variates can be generated by the inverse CDF method as

$$x = \sigma\sqrt{-2\log u}.$$

Faster acceptance/rejection methods can be constructed, but if the computing system has fast functions for exponentiation and taking logarithms, the inverse CDF is adequate.

### Pareto Distribution

For the Pareto distribution with density

$$p(x) = \frac{ab^a}{x^{a+1}} \quad \text{for } b \leq x \leq \infty, \tag{5.27}$$

variates can be generated by the inverse CDF method as

$$x = \frac{b}{u^{1/a}}.$$

There are many variations of the continuous Pareto distribution, the simplest of which is the one defined above. In addition, there are some discrete versions, including various zeta and Zipf distributions. (See Arnold, 1983, for an extensive discussion of the variations.) Variates from these distributions can usually be generated by discretizing some form of a Pareto distribution. Dagpunar (1988) describes such a method for a zeta distribution, in which the Pareto variates are first generated by an acceptance/rejection method.

## Zipf Distribution

The standard Zipf distribution assigns probabilities to the positive integers $x$ proportional to $x^{-\alpha}$, for $\alpha > 1$. The probability function is

$$p(x) = \frac{1}{\zeta(\alpha)x^\alpha} \quad \text{for } x = 1, 2, 3, \ldots, \tag{5.28}$$

where $\zeta(\alpha) = \sum_{x=1}^{\infty} x^{-\alpha}$ (the Riemann zeta function).

Variates from the simple Zipf distribution can be generated efficiently by a direct acceptance/rejection method given by Devroye (1986a). In this method, first two variates $u_1$ and $u_2$ are generated from $U(0,1)$, and then $x$ and $t$ are defined as

$$x = \lfloor u_1^{-1/(\alpha-1)} \rfloor$$

and

$$t = (1 + 1/x)^{\alpha-1}.$$

The variate $x$ is accepted if

$$x \le \frac{t}{t-1} \frac{2^{\alpha-1} - 1}{2^{\alpha-1} u_2}.$$

## Von Mises Distribution

Variates from the von Mises distribution with density

$$p(x) = \frac{1}{2\pi I_0(c)} e^{c\cos(x)} \quad \text{for } -\pi \le x \le \pi, \tag{5.29}$$

as discussed on page 140, can be generated by the acceptance/rejection method. Best and Fisher (1979) use a transformed folded Cauchy distribution as the majorizing distribution. The majorizing density is

$$g(y) = \frac{1 - \rho^2}{\pi(1 + \rho^2 - 2\rho\cos y)} \quad \text{for } 0 \le y \le \pi,$$

where $\rho$ is chosen in $[0, 1)$ to optimize the probability of acceptance for a given value of the von Mises parameter, $c$. (Simple plots of $g(\cdot)$ with different values of $\rho$ compared to a plot of $p(\cdot)$ with a given value of $c$ visually lead to a relatively good choice of $\rho$.) This is the method used in the IMSL Libraries. Dagpunar (1990) gives an acceptance/rejection method for the von Mises distribution that is often more efficient.

**Inverse Gaussian Distribution**

The inverse Gaussian distribution is widely used in reliability studies. The density, for location parameter $\mu > 0$ and scale parameter $\lambda > 0$, is

$$p(x) = \left(\frac{\lambda}{2\pi}\right)^{1/2} x^{-3/2} \exp\left(\frac{-\lambda(x - \mu)^2}{2\mu^2 x}\right) \quad \text{for } 0 \le x \le \infty. \tag{5.30}$$

The inverse Gaussian distribution with $\mu = 1$ is called the Wald distribution. It is the distribution of the first passage time in a standard Brownian motion with positive drift. Michael, Schucany, and Haas (1976) and Atkinson (1982) discussed methods for simulating inverse Gaussian random deviates. The method of Michael et al., given as Algorithm 5.7, is particularly straightforward but efficient.

**Algorithm 5.7 Michael/Schucany/Haas Method for Generating Inverse Gaussians**

1. Generate $v$ from $N(0, 1)$, and set $y = v^2$.

2. Set $x_1 = \mu + \frac{\mu^2 y}{2\lambda} - \frac{\mu}{2\lambda}\sqrt{4\mu\lambda y + \mu^2 y^2}$.

3. Generate $u$ from $U(0, 1)$.

4. If $u \le \frac{\mu}{\mu + x_1}$, then
   deliver $x = x_1$;
   otherwise,
   deliver $x = \frac{\mu^2}{x_1}$.    ∎

The generalized inverse Gaussian distribution has an additional parameter that is the exponent of $x$ in the density (5.30), which allows for a wider range of shapes. Barndorff-Nielsen and Shephard (2001) discuss the generalized inverse Gaussian distribution, and describe a method for generating random numbers from it.

## 5.2.13 General Families of Univariate Distributions

In simulation applications, one of the first questions is what is the distribution of the random variables that model observational data. Some distributions, such as Poisson or hypergeometric distributions, are sometimes obvious from first principles of the data-generating process. In other cases, there may be a "natural" distribution, such as normal or exponential, or a distribution derived from a natural distribution, such as Student's $t$ or chi-squared. Sometimes, however, we may not be willing to assume one of these parametric models. We may seek a more general approach that loosely models an observed dataset. We may want only to specify some of the moments, for example. Alternatively, we may want the model distribution to correspond to a few specified quantiles.

There are a number of general families of distributions that may be useful in simulating data that correspond well to observed data. Some commonly used ones are the Pearson family, the Johnson family, the generalized lambda family, and the Burr family. The Pearson family is probably the best known of these distributions. A specific member of the family is determined by the first four moments, so a common way of fitting a distribution to an observed set of data is by matching the moments of the distribution to those of the sample.

Another widely used general family of distribution is the Johnson family. A specific member of this family is also determined by the first four or five moments, depending on the parametrization (see Hill, Hill, and Holder, 1976, and Bacon-Shone, 1985). Specific members of the Johnson family can also be constructed from the percentiles of the observations to be modeled (see Chou et al., 1994). Devroye (1986a) describes a method for simulating Johnson variates for a given parametrization.

A generalized lambda family of distributions was described by Ramberg and Schmeiser (1974). This system, which is a generalization of a system introduced by John Tukey, has four parameters that can be chosen to fit a variety of distributional shapes. They specify the distribution in terms of the inverse of its distribution function,

$$P^{-1}(u) = \lambda_1 + \frac{u^{\lambda_3} - (1 - u)^{\lambda_4}}{\lambda_2}. \qquad (5.31)$$

The distribution function itself cannot be written in closed form, but the inverse allows deviates from this distribution to be generated easily by the inverse CDF method; just generate $u$ and apply equation (5.31).

Karian, Dudewicz, and McDonald (1996) and Karian and Dudewicz (1999, 2000) describe methods to determine values of the $\lambda$ parameters to arrive at a distribution function that fits a given set of data well. Freimer et al. (1988) give a different parametrization of the generalized Tukey lambda family, and discuss methods of fitting data under their parametrization, which seems somewhat simpler. Albert, Delampady, and Polasek (1991) defined a family of distributions that is very similar to the lambda distributions and is particularly useful in Bayesian analysis with location-scale models.

Another family of distributions that is very flexible and that can have a wide range of shapes is the Burr family of distributions (Burr, 1942). One of the common forms (Burr and Cislak, 1968) has the CDF

$$P(x) = 1 - \frac{1}{(1 + x^\alpha)^\beta} \quad \text{for } 0 \leq x \leq \infty; \ \alpha, \beta > 0, \qquad (5.32)$$

which is easily inverted. Other forms of the Burr family have more parameters, allowing modeling of a wider range of empirical distributions.

Fleishman (1978) suggested representing the random variable of interest as a polynomial in a standard normal random variable, in which the coefficients are determined so that the moments match specific values. If $Z$ has a $N(0, 1)$ distribution, then the random variable of interest, $X$, is expressed as

$$X = c_0 + c_1 Z + \cdots + c_k Z^k. \tag{5.33}$$

If $m$ moments are to be matched to prespecified values, then $k$ can be chosen as $m - 1$, and the $c$s can be determined from $m$ equations in $m$ unknowns that involve expectations of powers of a $N(0, 1)$ random variable. Fleishman used this representation to match four moments; hence, he used a third-degree polynomial in a standard normal random variable.

Tadikamalla and Johnson (1982) describe a flexible general family of distributions based on transformations of logistic variables.

The generalized gamma distributions mentioned on page 182 can also be very useful in modeling random variables with a variety of properties.

Tadikamalla (1980b) compares the use of the Pearson, Johnson, and generalized lambda families of distributions and the Fleishman normal polynomials for simulating distributions with various given shapes or given moments. As might be expected, some systems are better in one situation and others are better in other cases. The use of the polynomial of a normal random deviate, as Fleishman suggested, seems to fit a wide range of distributions, and its simplicity recommends it in many cases.

The motivation for some of the early work with general families of distributions was to use them as approximations to some standard distribution, such as a gamma, for which it is more difficult to generate deviates. As methods for the standard distributions have improved, it is more common just to generate directly from the distribution of interest. The general families, however, often provide more flexibility in choosing a distribution that better matches sample data. The distribution is fit to the sample data using either percentiles or moments. Pearson, Johnson, and Burr (1979) discuss differences in the results of using percentiles and moments. They give comparisons of the percentage points of distributions with the same first four moments that are constructed from eight different families of distributions. If a large number of moments are known or assumed, the methods of Devroye (1989, 1991) that use only the knowledge of the moments may be useful. If the percentiles are known, the inverse CDF method may be used to bracket the deviate, and then interpolation can be used to evaluate the deviate.

## Families of Distributions with Heavy Tails

Random number generation is important in studying the performance of various statistical methods, especially when the assumptions underlying the methods are not satisfied. The question is whether the statistical method is *robust*.

One concern in robust statistics is how well the method would hold up in very extreme conditions such as in the presence of a heavy-tailed distribution. The Cauchy distribution has very heavy tails; none of its moments exist. It is often used in robustness studies.

The Pareto distribution has relatively heavy tails; for some values of the parameter, the mean exists but the variance does not. A "Pareto-type" distribution is one whose distribution function satisfies the relationship

$$P(x) = 1 - x^{-\gamma}g(x),$$

where $g(x)$ is a *slowly varying function*; that is, for fixed $t > 0$,

$$\lim_{x \to \infty} \frac{g(tx)}{g(x)} = 1.$$

The Burr distribution with the CDF given in (5.32) is of the Pareto type, with $\gamma = \alpha\beta$.

The stable family of distributions is a flexible family of generally heavy-tailed distributions. This family includes the normal distribution at one extreme value of one of the parameters and the Cauchy distribution at the other extreme value. There are various parameterizations of the stable distributions; see Nolan (1998b). Depending on one of the parameters, $\alpha$, the index of stability, the characteristic function (equation (4.14), page 136) for random variables of this family of distributions has one of two forms:

$$\phi(t \mid \alpha, \sigma, \beta, \mu) = \exp\left(-\sigma^\alpha |t|^\alpha \left(1 - i\beta \text{sign}(t)\tan(\pi\alpha/2)\right) + i\mu t\right) \quad \text{if } \alpha \neq 1,$$

or

$$\phi(t \mid 1, \sigma, \beta, \mu) = \exp\left(-\sigma |t| \left(1 + 2i\beta \text{sign}(t)\log(t)/\pi\right) + i\mu t\right) \quad \text{if } \alpha = 1$$

for $0 < \alpha \leq 2$, $0 \leq \sigma$, and $-1 \leq \beta \leq 1$. For $\alpha = 2$, this is the normal distribution (in which case $\beta$ is irrelevant), and for $\alpha = 1$ and $\beta = 0$, this is the Cauchy distribution.

Chambers, Mallows, and Stuck (1976) give a method for generating deviates from stable distributions. (Watch for some errors in the constants in the auxiliary function D2 for evaluating $(e^x - 1)/x$.) Their method is used in the IMSL Libraries. For a symmetric stable distribution, Devroye (1986a) points out that a faster method can be developed by exploiting the relationship of the symmetric stable to the Fejer–de la Vallee Poussin distribution.

The member of the stable family with $\alpha = 1$ and $\beta = 1$ is called the Landau distribution, which has applications in modeling fluctuation of energy loss in a system of charged particles. Chamayou (2001) describes a method for generating variates from a Landau distribution (based on the Chambers-Mallows–Stuck method). He also gives a method for a generalization of the Landau distribution called the Vavilov distribution.

Buckle (1995) shows how to simulate the parameters of a stable distribution conditional on the data.

# 5.3 Some Specific Multivariate Distributions

Multivariate distributions can be built from univariate distributions either by a direct transformation of a vector of i.i.d. scalar variates or by a sequence of conditional scalar variates.

The use of various Markov chain methods in Monte Carlo simulation has made the conditional approaches more popular in generating multivariate deviates. The hit-and-run sampler (see page 157) is particularly useful in generating variates from multivariate distributions (see Bélisle, Romeijn, and Smith, 1993, for example). A survey of methods and applications of multivariate simulation can be found in Johnson (1987).

Elliptically contoured multivariate distributions are of special interest. The densities of these distributions have concentric ellipses with constant values. The density of an elliptically contoured distribution is of the form

$$p(x) = c \frac{1}{|\Sigma|^{\frac{1}{2}}} g\big((x - \mu)^{\mathrm{T}} \Sigma^{-1} (x - \mu)\big),$$

where $c$ is a positive constant of proportionality, and $g$ is a nonnegative real scalar-valued function. The multivariate normal distribution is an elliptically contoured distribution.

## 5.3.1 Multivariate Normal Distribution

The $d$-variate normal distribution with mean vector $\mu$ and nonsingular variance-covariance matrix $\Sigma$, which we denote by $N_d(\mu, \Sigma)$, has the probability density function

$$p(x) = \frac{1}{(2\pi)^{\frac{d}{2}} |\Sigma|^{\frac{1}{2}}} \exp\left(-\frac{(x - \mu)^{\mathrm{T}} \Sigma^{-1} (x - \mu)}{2}\right). \qquad (5.34)$$

A direct way of generating random vectors from the multivariate normal distribution is to generate a $d$-vector of i.i.d. standard normal deviates $z = (z_1, z_2, \ldots, z_d)$ and then to form the vector

$$x = T^{\mathrm{T}} z + \mu, \qquad (5.35)$$

where $T$ is a $d \times d$ matrix such that $T^{\mathrm{T}} T = \Sigma$. ($T$ could be a Cholesky factor of $\Sigma$; see Gentle, 1998, page 93, for discussion and an algorithm for a Cholesky factor.) Then, $x$ has a $N_d(\mu, \Sigma)$ distribution.

Another approach for generating the $d$-vector $x$ from $N_d(\mu, \Sigma)$ is to generate $x_1$ from $N_1(0, \sigma_{11})$, generate $x_2$ conditionally on $x_1$, generate $x_3$ conditionally on $x_1$ and $x_2$, and so on.

Deák (1990) describes a method for generating multivariate normals by using a transformation to spherical coordinates together with a random orthogonal transformation. This method is also useful in evaluating multivariate normal probabilities.

Multivariate distributions that are restricted to some subspace of the standard range may be difficult to simulate in general. Often, we must resort to generation of variates from the full distribution followed by rejection of those that do not meet the restriction. A multivariate normal distribution that is truncated by linear restrictions, however, can be handled easily. If the variate $x$ from $N_d(\mu, \Sigma)$ must satisfy the restriction

$$a \leq Cx \leq b,$$

where $C$ is a full rank matrix, the restrictions can be applied to a vector of i.i.d. standard normal deviates $z$,

$$\left(T^T\right)^{-1}(C^{-1}a - \mu) \leq z \leq \left(T^T\right)^{-1}(C^{-1}b - \mu),$$

where $T^T$ is as in equation (5.35). The standard normal deviates with these restrictions can be generated as described above using an exponential majorizing density. Geweke (1991a) describes this method and gives some timing comparisons.

### 5.3.2  Multinomial Distribution

The probability function for the $d$-variate multinomial distribution is

$$p(x) = \frac{n!}{\prod x_j!} \prod \pi_j^{x_j} \quad \text{for } x_j \geq 0, \text{ and } \Sigma x_j = n. \tag{5.36}$$

The parameters $\pi_j$ must be positive and sum to 1.

To generate a multinomial, a simple way is to work with the marginals; they are binomials. The generation is done sequentially. Each succeeding conditional marginal is binomial. For efficiency, the first marginal considered would be the one with the largest probability.

Another interesting algorithm for the multinomial, due to Brown and Bromberg (1984), is based on the fact that the conditional distribution of independent Poisson random variables, given their sum, is multinomial. The use of this relationship requires construction of extensive tables. Davis (1993) found the Brown–Bromberg method to be slightly faster than the sequential conditional marginal binomial method once the setup operations are performed. If multinomials are to be generated from distributions with different parameters, however, the sequential conditional marginal method is more efficient.

### 5.3.3  Correlation Matrices and Variance-Covariance Matrices

Generation of random correlation matrices or random variance-covariance matrices in general requires some definition of a probability measure (as generation of any random object does, of course). A probability measure for a variance-covariance matrix, for example, may be taken to correspond to that of sample variance-covariance matrices from a random sample of a given size from a $d$-variate normal distribution with variance-covariance matrix $\Sigma$. In this case, the probability distribution, aside from a constant, is a Wishart distribution.

Hartley and Harris (1963) and Odell and Feiveson (1966) give the method in Algorithm 5.8 to generate random variance-covariance matrices corresponding to a sample of size $n$ from a $d$-variate normal distribution.

**Algorithm 5.8 Sample Variance-Covariance Matrices from the Multivariate Normal with Variance-Covariance Matrix $\Sigma$**

0. Determine $T$ such that $T^{\mathrm{T}}T = \Sigma$. $T$ could be a Cholesky factor of $\Sigma$, for example.

1. Generate a sequence of i.i.d. standard normal deviates $z_{ij}$ for $i = 1, 2, \ldots, j$ and $j = 2, 3, \ldots, d$.

2. For $i = 1, 2, \ldots, d$, generate independent chi-squared variates, $y_i$, with $n + 1 - i$ degrees of freedom.

3. Compute $B = (b_{ij})$:
$$b_{11} = y_1,$$
$$b_{jj} = y_j + \sum_{i=1}^{j-1} z_{ij}^2 \quad \text{for } j = 2, 3, \ldots, d,$$
$$b_{1j} = z_{1j}\sqrt{y_1},$$
$$b_{ij} = z_{ij}\sqrt{y_i} + \sum_{k=1}^{i-1} z_{ki}z_{kj} \quad \text{for } i < j = 2, 3, \ldots, d,$$
$$b_{ij} = b_{ji} \quad \text{for } j < i = 2, 3, \ldots, d.$$

4. Deliver $V = \frac{1}{n}T^{\mathrm{T}}BT$.　∎

The advantage of this algorithm is that it does not require $n$ $d$-variate normals for each random variance-covariance matrix. The $B$ in Algorithm 5.8 is a Wishart matrix. Smith and Hocking (1972) give a Fortran program for generating Wishart matrices using this technique. The program is available from `statlib` as the *Applied Statistics* algorithm AS 53. (See the bibliography.)

Everson and Morris (2000) describe a modification of the method above for generating Wishart matrices whose eigenvalues are constrained to be less than a specified vector. The main modification is to generate right-truncated chi-squared variates, which they do by the inverse CDF method as in equation (5.2) on page 168. This yields a first principal diagonal matrix (that is, the scalar $b_{11}$) with an eigenvalue that satisfies the constraint. They then modify the computations of the $b_{ij}$ for $i, j \geq 2$ to include a rejection step that ensures that the new eigenvalue of each principal diagonal in turn is less than its constraint.

A random matrix from a noncentral Wishart distribution with noncentrality matrix having columns $c_i$ can be generated from a random Wishart matrix $B$ using the relationship

$$S = T^{\mathrm{T}}\left(\sum_{i=1}^{d}(z_i + c_i)(z_i + c_i)^{\mathrm{T}} + B\right)T,$$

where $z_i$ are i.i.d. random vectors from a $N_d(0, I_d)$ distribution, and the $z_i$ are independent of $B$. This requires $d^2$ additional scalar normal variates. Gleser (1976) gives a modification of this method that requires only $r^2$ additional normal deviates, where $r$ is the number of linearly independent vectors $c_i$.

For generating random correlation matrices, again we need some probability measure or at least some description of the space from which the matrices are to be drawn. Chalmers (1975) and Bendel and Mickey (1978) develop procedures that satisfy the definition of a correlation matrix (positive definite with 1s on the diagonal) and having eigenvalues approximately equal to a specified set. Another way of restricting the space from which random correlation matrices are to be drawn is to specify the expected values of the correlation matrices.

Marsaglia and Olkin (1984) consider the problem of generating correlation matrices with a given mean and the problem of generating correlation matrices with a given set of eigenvalues. Their algorithm for the latter is similar to that of Chalmers (1975). Starting with a diagonal matrix $\Lambda$ with elements $0 \leq \lambda_1, \lambda_2, \ldots, \lambda_d \leq 1$ and such that $\sum \lambda_j = d$, the algorithm forms a random correlation matrix of the form $P\Lambda P^T$. It is shown as Algorithm 5.9.

## Algorithm 5.9 Marsaglia–Olkin Method for Random Correlation Matrices with Given Eigenvalues

0. Set $E = I_d$ (the $d \times d$ identity matrix) and $k = 1$.

1. Generate a $d$-vector $w$ of i.i.d. standard normal deviates, form $x = Ew$, and compute $a = \sum (1 - \lambda_i) x_i^2$.

2. Generate a $d$-vector $z$ of i.i.d. standard normal deviates, form $y = Ez$, and compute $b = \sum (1 - \lambda_i) x_i y_i$, $c = \sum (1 - \lambda_i) y_i^2$, and $e^2 = b^2 - ac$.

3. If $e^2 < 0$, then go to step 2.

4. Choose a random sign, $s = -1$ or $s = 1$. Set $r = \dfrac{b + se}{a} x - y$.

5. Choose another random sign, and set $p_k = \dfrac{s}{(r^T r)^{\frac{1}{2}}} w$.

6. Set $E = E - rr^T$, and set $k = k + 1$.

7. If $k < d$, then go to step 1.

8. Generate a $d$-vector $w$ of i.i.d. standard normal deviates, form $x = Ew$, and set $p_d = \dfrac{x}{(x^T x)^{\frac{1}{2}}}$.

9. Construct the matrix $P$ using the vectors $p_k$ as its rows. Deliver $P\Lambda P^T$ as the random correlation matrix.  ∎

Ryan (1980) shows how to construct fixed correlation matrices with a given structure. Although the matrices are fixed, they have applications in Monte Carlo studies for generating other deviates.

Heiberger (1978), Stewart (1980), Anderson, Olkin, and Underhill (1987), and Fang and Li (1997) give methods for generating random orthogonal matrices. The methods use reflector or rotator matrices to transform a random matrix to the form desired. (See Gentle, 1998, for definitions and methods of computing these matrices.)

Heiberger (1978) discussed the use of random orthogonal matrices in generating random correlation matrices. (See Tanner and Thisted, 1982, for a correction of Heiberger's algorithm. This is the method used in the IMSL Libraries.) Stewart (1980) used random orthogonal matrices to study an estimator of the condition number of a matrix. One of the methods given by Fang and Li (1997) results in matrices with quasirandom elements (see Chapter 3).

Eigenvalues of random matrices are often of interest, especially the extreme eigenvalues. One way to generate them, of course, is to generate the random matrix and then extract the eigenvalues. Marasinghe and Kennedy (1982) describe direct methods for generation of the extreme eigenvalues for certain Wishart matrices. They show that the maximum eigenvalue of a random $2 \times 2$ Wishart matrix, corresponding to a sample of size $n$, can be generated by the inverse CDF method,

$$w = \left(1 + \sqrt{1 - u^{2/(n-1)}}\right)/2,$$

where $u$ is generated from $U(0, 1)$.

Using a Weibull density,

$$g(y) = (n - 3)x e^{-(n-3)y^2/2},$$

as a majorizing density, they also give the following acceptance/rejection algorithm for the $2 \times 2$ case:

1. Generate $u_1$ from $U(0, 1)$, and set $y = -2\log(u_1)/(n - 3)$.

2. If $y \geq 1$, then go to step 1.

3. Generate $u_2$ from $U(0, 1)$, and if $u_1 u_2 > (1 - y)^{(n-3)/2}$, then go to step 1;
   otherwise,
       deliver      $w = (\sqrt{y} + 1)/2$.

## 5.3.4   Points on a Sphere

Coordinates of points uniformly distributed over the surface of a sphere (here "sphere" may mean "circle" or "hypersphere") are equivalent to independent normals scaled to lie on the sphere (that is, the sum of their squares is equal to the radius of the sphere). This general method is equivalent to the polar

methods for normal random variables discussed previously. It can be used in
any dimension.

Marsaglia (1972b) describes methods for three and four dimensions that are
about twice as fast as using normal variates. For three dimensions, $u_1$ and $u_2$
are generated independently from $U(-1,1)$, and if $s_1 = u_1^2 + u_2^2 \leq 1$, then the
coordinates are delivered as

$$
\begin{aligned}
x_1 &= 2u_1\sqrt{1 - s_1}, \\
x_2 &= 2u_2\sqrt{1 - s_1}, \\
x_3 &= 1 - 2s_1.
\end{aligned}
$$

For four dimensions, $u_1$, $u_2$, and $s_1$ are generated as above (with the same
rejection step), $u_3$ and $u_4$ are generated independently from $U(-1,1)$ until
$s_2 = u_3^2 + u_4^2 \leq 1$. The rejection steps are applied independently, and for the
accepted points, the coordinates are delivered as

$$
\begin{aligned}
x_1 &= u_1 \text{ as above,} \\
x_2 &= u_2 \text{ as above,} \\
x_3 &= u_3\sqrt{\frac{1 - s_1}{s_2}}, \\
x_4 &= u_4\sqrt{\frac{1 - s_1}{s_2}}.
\end{aligned}
$$

The IMSL routine **rnsph** uses these methods for three or four dimensions and
uses scaled normals for higher dimensions.

Banerjia and Dwyer (1993) consider the related problem of generating random points in a ball, which would be equivalent to generating random points
on a sphere and then scaling the radius by a uniform deviate. They describe
a divide-and-conquer algorithm that can be used in any dimension and that
is faster than scaling normals or scaling points from Marsaglia's method, assuming that the speed of the underlying uniform generator is great relative to
square root computations.

## 5.3.5   Two-Way Tables

Boyett (1979) and Patefield (1981) consider the problem of generating random
entries in a two-way table subject to given marginal row and column totals.
The distribution is uniform over the integers that yield the given totals. Boyett
derives the joint distribution for the cell counts and then develops the conditional distribution for a given cell, given the counts in all cells in previous rows
and all cells in previous columns of the current given row. Patefield then uses
the conditional expected value of a cell count to generate a random entry for
each cell in turn.

Let $a_{ij}$ for $i = 1, 2, \ldots, r$ and $j = 1, 2, \ldots, c$ be the cell count, and use the
"dot notation" for summation: $a_{\bullet j}$ is the sum of the counts in the $j^{\text{th}}$ column,

for example, and $a_{\bullet\bullet}$ is the grand total. The conditional probability that the count in the $(l, m)^{\text{th}}$ cell is $a_{lm}$ given the counts $a_{ij}$, for $1 \leq i < l$ and $1 \leq j \leq c$ and for $i = l$ and $1 \leq j < m$, is

$$
\frac{\left(a_{l\bullet} - \sum_{j<m} a_{lj}\right)! \left(a_{\bullet\bullet} - \sum_{i \leq l} a_{i\bullet} - \sum_{j<m} a_{\bullet j} + \sum_{j<m} \sum_{i \leq l} a_{ij}\right)!}{a_{lm}! \left(a_{l\bullet} - \sum_{j \leq m} a_{lj}\right)! \left(a_{\bullet\bullet} - \sum_{i \leq l} a_{i\bullet} - \sum_{j \leq m} a_{\bullet j} + \sum_{j<m} \sum_{i \leq l} a_{ij}\right)!}
$$

$$
\times \frac{\left(a_{\bullet m} - \sum_{i \leq l} a_{im}\right)! \left(\sum_{j>m} \left(a_{\bullet j} - \sum_{i<l} a_{ij}\right)\right)!}{\left(a_{\bullet m} - \sum_{i<l} a_{im}\right)! \left(\sum_{j \geq m} \left(a_{\bullet j} - \sum_{i<l} a_{ij}\right)\right)!}.
$$

For each cell, a random uniform is generated, and the discrete inverse CDF method is used. Sequential evaluation of this expression is fairly simple, so the probability accumulation proceeds rapidly. The full expression is evaluated only once for each cell. Patefield (1981) also speeds up the process by beginning at the conditional expected value of each cell rather than accumulating the CDF from 0. The conditional expected value of the random count in the $(l, m)^{\text{th}}$ cell, $A_{lm}$, for $1 \leq i < l$ and $1 \leq j \leq c$ and for $i = l$ and $1 \leq j < m$, is

$$
\mathrm{E}(A_{lm} \mid a_{ij}) = \frac{\left(a_{\bullet m} - \sum_{i=1}^{l-1} a_{im}\right) \left(a_{l\bullet} - \sum_{j=1}^{m-1} a_{lj}\right)}{\sum_{j=m}^{c} \left(a_{\bullet j} - \sum_{i=1}^{l-1} a_{ij}\right)}
$$

unless the denominator is 0, in which case $\mathrm{E}(A_{lm}|a_{ij})$ is zero.

Patefield (1981) gives a Fortran program implementing the method described. This is the method used in the IMSL routine **rntab**.

## 5.3.6   Other Specific Multivariate Distributions

Only a few of the standard univariate distributions have standard multivariate extensions. Various applications often lead to different extensions; see Kotz, Balakrishnan, and Johnson (2000). If the density of a multivariate distribution exists and can be specified, it is usually possible to generate variates from the distribution using an acceptance/rejection method. The majorizing density can often be just the product density; that is, a multivariate density with components that are the independent univariate variables, as in the example of the bivariate gamma on page 123.

### Multivariate Bernoulli Variates and the Multivariate Binomial Distribution

A multivariate Bernoulli distribution of correlated binary random variables has applications in modeling system reliability, clinical trials with repeated measures, and genetic transmission of disease. For the multivariate Bernoulli distribution with marginal probabilities $\pi_1, \pi_2, \ldots, \pi_d$ and pairwise correlations $\rho_{ij}$, Emrich and Piedmonte (1991) propose identifying a multivariate normal distribution with similar pairwise correlations. The normal is determined by solving for normal pairwise correlations, $r_{ij}$, in a system of $d(d-1)/2$ equations involving the bivariate standard normal CDF, $\Phi_2$, evaluated at percentiles $z_\pi$ corresponding to the Bernoulli probabilities:

$$\Phi_2(z_{\pi_i}, z_{\pi_j}; r_{ij}) = \rho_{ij}\sqrt{\pi_i(1-\pi_i)\pi_j(1-\pi_j)} + \pi_i\pi_j. \qquad (5.37)$$

Once these pairwise correlations are determined, a multivariate normal $y$ is generated and transformed to a Bernoulli, $x$, by the rule

$$\begin{aligned} x_i &= 1 \quad \text{if } y_i \leq z_{\pi_i} \\ &= 0 \quad \text{otherwise.} \end{aligned}$$

Sums of multivariate Bernoulli random variables are multivariate binomial random variables. Phenomena modeled by binomial distributions, within clusters, often exhibit greater or less intracluster variation than independent binomial distributions would indicate. This behavior is called "overdispersion" or "underdispersion". Overdispersion can be simulated by the beta-binomial distribution discussed earlier. A beta-binomial cannot model underdispersion, but the method of Emrich and Piedmonte (1991) to induce correlations in the Bernoulli variates can be used to model either overdispersion or underdispersion. Ahn and Chen (1995) discuss this method and compare it with the use of a beta-binomial in the case of overdispersion. They also compared the output of the simulation models for both underdispersed and overdispersed binomials with actual data from animal litters.

Park, Park, and Shin (1996) give a method for generating correlated binary variates based on sums of Poisson random variables in which the sums have some terms in common. They let $Z_1$, $Z_2$, and $Z_3$ be independent Poisson random variables with nonnegative parameters $\alpha_{11} - \alpha_{12}$, $\alpha_{22} - \alpha_{12}$, and $\alpha_{12}$, respectively, with the convention that a Poisson with parameter 0 is a degenerate random variable equal to 0, and define the random variables $Y_1$ and $Y_2$ as

$$Y_1 = Z_1 + Z_3$$

and

$$Y_2 = Z_2 + Z_3.$$

They then define the binary random variables $X_1$ and $X_2$ by

$$\begin{aligned} X_i &= 1 \quad \text{if } Y_i = 0 \\ &= 0 \quad \text{otherwise.} \end{aligned}$$

They then determine the constants, $\alpha_{11}$, $\alpha_{22}$, and $\alpha_{12}$, so that

$$E(X_i) = \pi_i$$

and

$$\text{Corr}(X_1, X_2) = \rho_{12}.$$

It is easy to see that

$$\alpha_{ij} = \log\left(1 + \rho_{ij}\sqrt{\frac{(1-\pi_i)(1-\pi_j)}{\pi_i\pi_j}}\right) \tag{5.38}$$

yields those relations. After the $\alpha$s are computed, the procedure is as shown in Algorithm 5.10.

**Algorithm 5.10 Park/Park/Shin Method for Generating Correlated Binary Variates**

0. Set $k = 0$.

1. Set $k = k + 1$. Let $\beta_k = \alpha_{rs}$ be the smallest positive $\alpha_{ij}$.

2. If $\alpha_{rr} = 0$ or $\alpha_{ss} = 0$, then stop.

3. Let $S_k$ be the set of all indices, $i, j$, for which $\alpha_{ij} > 0$. For all $\{i, j\} \subseteq S_k$, set $\alpha_{ij} = \alpha_{ij} - \beta_k$.

4. If not all $\alpha_{ij} = 0$, then go to step 1.

5. Generate $k$ Poisson deviates, $z_j$, with parameters $\beta_j$. For $i = 1, 2, \ldots, d$, set

$$y_i = \sum_{i \in S_j} z_j.$$

6. For $i = 1, 2, \ldots, d$, set

$$\begin{aligned} x_i &= 1 \quad \text{if } y_i = 0 \\ &= 0 \quad \text{otherwise.} \end{aligned}$$

∎

Lee (1993) gives another method to generate multivariate Bernoullis that uses odds ratios. (Odds ratios and correlations uniquely determine each other for binary variables.)

**Multivariate Beta or Dirichlet Distribution**

The Dirichlet distribution is a multivariate extension of a beta distribution. The density of the $d$-variate Dirichlet is the obvious extension of the beta density (equation (5.14)).

$$p(x) = \frac{\Gamma(\Sigma_{i=1}^{d+1}\alpha_i)}{\prod_{i=1}^{d+1}\Gamma(\alpha_i)}\prod_{j=1}^{d} x_j^{\alpha_j-1}(1 - x_1 - x_2 - \cdots - x_d)^{\alpha_{d+1}-1}$$
$$\text{for } 0 \le x_j \text{ and } \sum x_j \le 1. \tag{5.39}$$

Arnason and Baniuk (1978) consider several ways to generate deviates from the Dirichlet distribution, including a sequence of conditional betas and the use of the relationship of order statistics from a uniform distribution to a Dirichlet. (The $i^{th}$ order statistic from a sample of size $n$ from a U(0,1) distribution has a beta distribution with parameters $i$ and $n-i+1$.) The most efficient method that they found was the use of a relationship between independent gamma variates and a Dirichlet variate. If $Y_1, Y_2, \ldots, Y_d, Y_{d+1}$ are independently distributed as gamma random variables with shape parameters $\alpha_1, \alpha_2, \ldots, \alpha_d, \alpha_{d+1}$ and common scale parameter, then the $d$-vector $X$ with elements

$$X_j = \frac{Y_j}{\sum_{k=1}^{d+1} Y_k}, \quad j = 1, \ldots, d,$$

has a Dirichlet distribution with parameters $\alpha_1, \alpha_2, \ldots, \alpha_d, \alpha_{d+1}$. This relationship yields the straightforward method of generating Dirichlets by generating gammas.

### Dirichlet-Multinomial Distribution

The Dirichlet-multinomial distribution is the mixture distribution that is a multinomial with parameter $\pi$ that is a realization of a random variable having a Dirichlet distribution. Just like the beta-binomial distribution (5.18), this distribution is useful for modeling overdispersion or extravariation in applications where there are clusters of separate multinomial distributions.

A simple way of generating deviates from a Dirichlet-multinomial distribution is first to generate the parameter $\pi$ as the appropriate Dirichlet and then to generate the multinomial conditionally.

There are other ways of inducing overdispersion in multinomial distributions. Morel (1992) describes a simple algorithm to generate a finite mixture of multinomials by clumping individual multinomials. This mixture distribution has the same first two moments of the Dirichlet-multinomial distribution, but it is not the same distribution.

### Multivariate Hypergeometric Distribution

The multivariate hypergeometric distribution is a generalization of the hypergeometric distribution for more than two types of outcomes. The model is an urn filled with balls of different colors. The multivariate random variable is the vector of numbers of each type of ball when $N$ balls are drawn randomly and without replacement. The probability function for the multivariate hypergeometric distribution is the same as that for the univariate hypergeometric distribution (equation (5.22), page 189) except with more classes.

To generate a multivariate hypergeometric random deviate, a simple way is to work with the marginals. The generation is done sequentially. Each succeeding conditional marginal is hypergeometric. To generate the deviate, combine all classes except the first in order to form just two classes. Next, generate a univariate hypergeometric deviate $x_1$. Then remove the first class and form two classes consisting of the second one and the third through the last combined, and generate a univariate hypergeometric deviate based on $N - x_1$ draws. This gives $x_2$, the number of the second class. Continue in this manner until the number remaining to be drawn is 0 or until the classes are exhausted. For efficiency, the first marginal used would be the one with the largest probability.

## Multivariate Uniform Distribution

Falk (1999) considers the problem of simulating a $d$-variate $U_d(0, 1)$ distribution with specified correlation matrix $R = (\rho_{ij})$. A simple approximate method is to generate $y$ from $N_d(0, R)$ and take

$$x_i = \Phi(y_i),$$

where $\Phi$ is the standard normal CDF. Falk shows that the correlation matrix of variates generated in this way is very close to the target correlation matrix $R$. He also shows that if the matrix

$$\begin{aligned} \tilde{R} &= (r_{ij}) \\ &= \left(2\sin(\pi\rho_{ij}/6)\right) \end{aligned}$$

is positive semidefinite, and if $Y \sim N_d(0, \tilde{R})$ and $X_i = \Phi(Y_i)$, then $\text{Corr}(X) = R$. Therefore, if the target correlation matrix $R$ is such that the corresponding matrix $\tilde{R}$ is positive semidefinite, then variates generated as above are from a $d$-variate $U_d(0, 1)$ distribution with exact correlation matrix $R$.

## Multivariate Exponential Distributions

A multivariate exponential distribution can be defined in terms of Poisson shocks (see Marshall and Olkin, 1967), and variates can be generated from that distribution by generating univariate Poisson variates (see Dagpunar, 1988). There are various ways to define a multivariate double exponential distribution, or a multivariate Laplace distribution. Ernst (1998) describes an elliptically contoured multivariate Laplace distribution with density

$$p(x) = \frac{\gamma\Gamma(d/2)}{2\pi^{d/2}\Gamma(d/\gamma)|\Sigma|^{\frac{1}{2}}} \exp\left(-\left((x-\mu)^{\mathrm{T}}\Sigma^{-1}(x-\mu)\right)^{\gamma/2}\right). \qquad (5.40)$$

Ernst shows that a simple way to generate a variate from this distribution is to generate a point $s$ on the $d$-dimensional sphere (see Section 5.3.4), generate a generalized univariate gamma variate $y$ (page 182) with parameters $d$, 1, and $\gamma$, and deliver

$$x = yT^{\mathrm{T}}s + \mu,$$

where $T^{\mathrm{T}}T = \Sigma$.

Kozubowski and Podgórski (2000) describe an asymmetric multivariate Laplace distribution (not elliptically contoured). They also describe a method for generating random deviates from that distribution.

**Multivariate Gamma Distributions**

The bivariate gamma distribution of Becker and Roux (1981) discussed in Section 4.5 (page 123) is only one possibility for extending the gamma. Others, motivated by different models of applications, are discussed by Mihram and Hultquist (1967), Ronning (1977), Ratnaparkhi (1981), and Jones, Lai, and Rayner (2000), for example. Ronning (1977) and London and Gennings (1999) describe specific multivariate gamma distributions and describe methods for generating variates from the multivariate gamma distribution that they considered.

The bivariate gamma distribution of Jones, Lai, and Rayner (2000) is formed from two univariate gamma distributions with fixed shape parameters and scale parameters $\zeta$ and $\xi$, each of which takes one of two values with a generalized Bernoulli distribution. For $i, j = 1, 2$, $\Pr(\zeta = \zeta_i, \xi = \xi_j) = \pi_{ij}$. The correlation between the two elements of the bivariate gamma depends on the $\pi_{ij}$, as Jones, Lai, and Rayner (2000) discuss. It is easy to generate random variates from this bivariate distribution: for each variate, generate a value for $\zeta$ and $\xi$, and then generate two of the univariate gammas.

**Multivariate Stable Distributions**

Various multivariate extensions of the stable distributions can be defined. Modarres and Nolan (1994) give a representation of a class of multivariate stable distributions in which the multivariate stable random variable is a weighted sum of a univariate stable random variable times a point on the unit sphere. The reader is referred to the paper for the description of the class of multivariate stable distributions for which the method applies. See also Nolan (1998a).

### 5.3.7   Families of Multivariate Distributions

Methods are available for generating multivariate distributions with various specific properties. Extensions have been given for multivariate versions of some of the general families of univariate distributions discussed on page 193. Parrish (1990) gives a method to generate random deviates from a multivariate Pearson family of distributions. Takahasi (1965) defines a multivariate extension of the Burr distributions. Generation of deviates from the multivariate Burr distribution can be accomplished by transformations of univariate samples.

Gange (1995) gives a method for generating general multivariate categorical variates using iterative proportional fitting to the marginals.

A useful general class of multivariate distributions are the elliptically contoured distributions. A nonsingular elliptically contoured distribution has a density of the general form

$$p(x) = \frac{c}{|\Sigma|^{\frac{1}{2}}} g\big((x - \mu)^{\mathrm{T}} \Sigma^{-1} (x - \mu)\big),$$

where $g(\cdot)$ is a nonnegative function, and $\Sigma$ is a positive definite matrix. The multivariate normal distribution is obviously of this class, as is the multivariate Laplace distribution discussed above. There are other interesting distributions in this class, including two types of multivariate Pearson distributions. Johnson (1987) discusses general methods for generating variates from the Pearson elliptically contoured distributions. The book edited by Fang and Anderson (1990) contains several papers on applications of elliptically contoured distributions.

Cook and Johnson (1981, 1986) define families of non-elliptically symmetric multivariate distributions, and consider their use in applications for modeling data. Johnson (1987) discusses methods for generating variates from those distributions.

## Distributions with Specified Correlations

Li and Hammond (1975) propose a method for a $d$-variate distribution with specified marginals and variance-covariance matrix. The Li–Hammond method uses the inverse CDF method to transform a $d$-variate normal into a multivariate distribution with specified marginals. The variance-covariance matrix of the multivariate normal is chosen to yield the specified variance-covariance matrix for the target distribution. The determination of the variance-covariance matrix for the multivariate normal to yield the desired target distribution is difficult, however, and does not always yield a positive definite variance-covariance matrix for the multivariate normal. (An approximate variance-covariance or correlation matrix that is not positive definite can be a general problem in applications of multivariate simulation. See Exercise 6.1 in Gentle, 1998, for a possible solution.)

Lurie and Goldberg (1998) modify the Li–Hammond approach by iteratively refining the correlation matrix of the underlying normal using the sample correlation matrix of the transformed variates. They begin with a fixed sample of $t$ multivariate normals with the identity matrix as the variance-covariance. These normal vectors are first linearly transformed by the matrix $T^{(k)}$ as described on page 197 and then transformed by the inverse CDF method into a sample of $t$ vectors with the specified marginal distributions. The correlation matrix of the transformed sample is computed and compared with the target correlation. A measure of the difference in the sample correlation matrix and the target correlation is minimized by iterations over $T^{(k)}$. A good starting point for $T^{(0)}$ is the $d \times d$ matrix that is the square root of the target correlation matrix $R$ (that is, the Cholesky factor) so that $T^{(0)\mathrm{T}}T^{(0)} = R$.

The measure of the difference in the sample correlation matrix and the target correlation is a sum of squares, so the minimization is a nonlinear least squares problem. The sample size $t$ to use in the determination of the optimal transformation matrix must be chosen in advance. Obviously, $t$ must be large enough to give some confidence that the sample correlation matrices reflect the target accurately. Because of the number of variables in the optimization problem, it is likely that $t$ should be chosen proportional to $d^2$.

Once the transformation matrix is chosen, to generate a variate from the target distribution, first generate a variate from $N_d(0, I)$, then apply the linear transformation, and finally apply the inverse CDF transformation. To generate $n$ variates from the target distribution, Lurie and Goldberg (1998) also suggest that the normals be adjusted so that the sample has a mean of 0 and a variance-covariance matrix exactly equal to the expected value that the transformation would yield. (This is *constrained sampling*, as discussed on page 248.)

Vale and Maurelli (1983) also generate general random variates using a multivariate normal distribution with the target correlation matrix as a starting point. They express the individual elements of the multivariate random variable of interest as polynomials in the elements of the multivariate normal random variable, similar to the method of Fleishman (1978) in equation (5.33). They then determine the coefficients in the polynomials so that the lower-order marginal moments correspond to specified values. This does not, of course, mean that the correlation matrix of the random variable determined in this way is the desired matrix. Vale and Maurelli suggest use of the first four marginal moments.

Parrish (1990), as mentioned above, gives a method for generating variates from a multivariate Pearson family of distributions. A member of the Pearson family is specified by the first four moments, which of course includes the covariances.

Kachitvichyanukul, Cheng, and Schmeiser (1988a) describe methods for inducing correlation in binomial and Poisson random variates.

# 5.4   Data-Based Random Number Generation

Often, we have a sample and need to generate random numbers from the unknown distribution that yielded it. Specifically, we have a set of observations, $\{x_1, x_2, \ldots, x_n\}$, and we wish to generate a pseudorandom sample from the same distribution as the given dataset. This kind of method is called *data-based random number generation*.

### Discrete Distributions

How we proceed depends on some assumptions about the distribution. If the distribution is discrete and we have no other information about it than what is available from the given dataset, the best way of generating a pseudorandom sample from the distribution is just to generate a random sample of indices with replacement (see Chapter 6) and then use the index set as indices for the given sample. For scalar $x$, this is equivalent to using the inverse CDF method on the ECDF (the empirical cumulative distribution function):

$$P_n(x) = \frac{1}{n}(\text{number of } x_i \leq x).$$

This method of generating a random sample using given data is what is done in nonparametric Monte Carlo bootstrapping (see Section 7.7.2).

## Parametric Families of Distributions

If it is assumed that the given sample is from some parametric family, one approach is to use the sample to estimate the parameters and then use one of the methods given above. This is what is done in the parametric Monte Carlo bootstrap.

## General Families of Distributions

If it is assumed that the distribution is continuous, but no other assumptions are made (other than perhaps some general assumptions of existence and smoothness of the probability density), the problem can be thought of as two steps. The first step is to estimate the density, perhaps using one of the general families of distributions discussed in Sections 5.2.13 and 5.3.7. The second step is to generate random deviates from that density. The method suggested above for a discrete distribution is what we would do if we first estimated the density (probability function) with a histogram with bins of zero width and then generated the random deviates from a distribution with probability function the same as the histogram.

## General Univariate Distributions

For a univariate continuous distribution, use of the ECDF directly is not acceptable because it will only yield values corresponding to the given sample. We may, however, use some smoothed version of the ECDF and then use the inverse CDF method. There are several ways that this can be done. One way, when the given sample size is small, is just to connect the jump points of the empirical distribution function with line segments to form a piecewise linear, increasing function. Another way is to bin the data and form either a histogram or a frequency polygon and then use the inverse CDF method on the corresponding distribution function. The distribution function corresponding to a histogram is a piecewise linear function, and the distribution function corresponding to a frequency polygon is a piecewise quadratic polynomial.

These methods use different nonparametric estimators of the probability density function. There are other methods of nonparametric density estimation that could be used (see Gentle, 2002, Chapter 9), but the simple methods indicated above are generally adequate.

## General Multivariate Distributions

As we have mentioned above, it is not practical to use the inverse CDF method directly for multivariate distributions.

Taylor and Thompson (1986) suggest a different way that avoids the step of estimating a density. The method has some of the flavor of density estimation; however, in fact it is essentially equivalent to fitting a density with a normal kernel. It uses the $m$ nearest neighbors of a randomly selected point; $m$ is a *smoothing parameter*. The method is particularly useful for multivariate data. Suppose that the given sample is $\{x_1, x_2, \ldots, x_n\}$ (the $x$s are vectors). A random vector deviate is generated by the steps given in Algorithm 5.11.

**Algorithm 5.11 Thompson–Taylor Data-Based Simulation**

1. Randomly choose a point, $x_j$, from the given sample.

2. Identify the $m$ nearest neighbors of $x_j$ (including $x_j$), $x_{j_1}, x_{j_2}, \ldots, x_{j_m}$, and determine their mean, $\bar{x}_j$.

3. Generate a random sample, $u_1, u_2, \ldots, u_m$, from a uniform distribution with lower bound $\frac{1}{m} - \sqrt{\frac{3(m-1)}{m^2}}$ and upper bound $\frac{1}{m} + \sqrt{\frac{3(m-1)}{m^2}}$.

4. Deliver the random variate

$$z = \sum_{k=1}^{m} u_k(x_{j_k} - \bar{x}_j) + \bar{x}_j.$$

■

The limits of the uniform weights and the linear combination for $z$ are chosen so that the expected value of the $i^{\text{th}}$ element of a random variable $Z$ that yields $z$ is the $i^{\text{th}}$ element of the sample mean of the $x$s, $\bar{x}_i$; that is,

$$E(Z_i) = \bar{x}_i.$$

(The subscripts in these expressions refer to the elements of the data vectors rather than to the element of the sample.) Likewise, the variance and covariance of elements of $Z$ are close to the sample variance and covariance of the elements of the given sample. If $m = 1$, they would be exactly the same. For $m > 1$, the variance is slightly larger because of the variation due to the random weights. The exact variance and covariance, however, depend on the distribution of the given sample because the linear combination is of nearest points. The routine rndat in the IMSL Libraries implements this method.

Prior to generating any pseudorandom deviates by the Thompson–Taylor method, the given dataset should be processed into a $k$-$d$ tree (see Friedman, Bentley, and Finkel, 1977) so that the nearest neighbors of a given observation can be identified quickly.

## 5.5  Geometric Objects

It is often of interest to generate random geometric objects (for example, in a space-filling problem). Exercise 5.8 on page 215 discusses a simple random-packing problem to fill a volume with balls of equal size. More interesting and complicated applications involve objects of variable shape such as stones to be laid down as paving or to be dispersed through a cement mixture.

Other applications are in the solution of partial differential equations and in probability density estimation, when we may wish to tessellate space using objects similar to triangles and rectangles. Rectangles are simple tessellating objects that have similar properties in various dimensions. In one dimension, they are intervals, in three dimensions they are often called rectangular solids, and in general they are called hyperrectangles. Devroye, Epstein, and Sack (1993) give efficient methods for generating random intervals and hyperrectangles.

Simplices (plural of simplex) are often used in tesselating space because of their simplicity. A simplex in $d$ dimensions is deterimined by $d + 1$ points. Simple triangles are simplices in two dimensions.

# Exercises

5.1. Polar transformations.

(a) Show that the Box–Muller transformation (equation (5.7), page 172) correctly transforms independent uniform variates into independent standard normals.

(b) Let $X_1$ and $X_2$ be independent standard normal random variables and represent the pair in polar coordinates

$$X_1 = R\cos\Theta,$$
$$X_2 = R\sin\Theta.$$

Show that $R$ and $\Theta$ are independent and that $\Theta$ has a $U(0, 2\pi)$ distribution and $R^2$ has an exponential distribution with parameter $\frac{1}{2}$; that is, $f_R(r) = re^{-r^2/2}$.

(c) Show that the rejection polar method of Algorithm 5.1 does indeed deliver independent normals.

(d) Write a Fortran or a C program implementing both the regular Box–Muller transformation using the intrinsic system functions and the rejection polar method of Algorithm 5.1. Empirically compare the efficiencies of the two methods by timing the execution of your program.

5.2. Truncated normal.

(a) Write the steps for generating a truncated normal using the method of Robert (1995) (page 176).

(b) Show that for majorizing densities that are translated exponentials, the one with scale parameter $\lambda^*$ in equation (5.8) is optimal. ("Optimal" means that it results in the greatest probability of acceptance.)

     (c) Write a program in a compiled language to generate truncated normals using the method of Robert (1995) and another program using the method of Damien and Walker (2001). Run timing comparisons for the two methods. Which appears to be more efficient?

5.3. Prove that the random variable delivered in equation (5.10) on page 176 has an exponential distribution.

5.4. Identify quantities in Algorithm 5.2, page 179, that should be computed as part of a setup step prior to the steps shown. There are four quantities that should be precomputed, including such trivial new variables as $a = \alpha - 1$.

5.5. Write a program to generate truncated gamma variates using the method of Philippe (1997) (page 182). The program should accept $\alpha$, $\beta$, $\tau$, and $n^*$ as arguments. Now, write a program that uses the Cheng/Feast (1979) algorithm, page 179, and just rejects variates greater than $\tau$. (Remember to precompute quantities such as those that are used in every pass through the accept/reject loop.) For $\alpha = 5$, $\beta = 1$, and various values of $\tau$, generate 10,000 deviates by both methods and compare the times of the two algorithms.

5.6. Correlated binary variates.

     (a) Derive equation (5.38) on page 205 (see Park, Park, and Shin, 1996).

     (b) Consider the pair of binary variates with $\pi = (0.2,\, 0.7)$ and $\rho = 0.05$. Generate 1000 pairs using the method of Emrich and Piedmonte (1991). First, compute $r = 0.0961$ from equation (5.37). Likewise, generate 1000 pairs using the method of Park, Park, and Shin (1996) and compute the sample correlation. How close are they to $\rho$?

5.7. Multivariate normal variates.

     (a) Consider a bivariate random variable $(X_1, X_2)$ with a bivariate normal distribution with means $\mu_1$ and $\mu_2$, variances $\sigma_1^2$ and $\sigma_2^2$, and correlation $\rho$. Assume you can generate a vector of $d$ i.i.d. univariate standard normal deviates by the function rnorm(d). Now, assume a given value $X_2 = x_2$. Write an expression for a random realization of $X_1$.

     (b) Consider a $d$-variate random variable $X$ with a multivariate normal distribution with mean $\mu$ and variances-covariance matrix $\Sigma$. ($X$ and $\mu$ are $d$-vectors and $\Sigma$ is a $d \times d$ symmetric positive definite matrix.) Now, consider a partition of $X$ into $d_1$ and $d_2$ elements, $(X_1, X_2)$, with a corresponding partitioning of $\mu$ into $(\mu_1, \mu_2)$, and $\Sigma$ into

$$\Sigma = \left[ \begin{array}{cc} \Sigma_{11} & \Sigma_{12} \\ \Sigma_{21} & \Sigma_{22} \end{array} \right],$$

where the $\Sigma_{ij}$ are matrices of the appropriate dimensions. Assume, as before, you can generate a vector of $d$ i.i.d. univariate standard normal deviates by the function rnorm(d). Now, assume a given value of the $d_2$ vector $X_2 = x_2$. Write an expression for a random realization of the $d_1$ vector $X_1$. Formally show that your expression has the correct conditional mean and covariance.

5.8. Consider a cube with each edge of length $s$. Imagine the cube to represent a box with no top. Develop a program to simulate the dropping of balls of diameter $r$ $(r < s)$ into the box until it is filled. ("Filled" can mean either that no more balls will stack on the others or that no ball has its top higher than the top of the box. Your program should allow either definition.) Assume that the balls will roll freely to their lowest local energy state within the constraints imposed by the sides of the box and the other balls. Let $s = 10$ and $r = 3$. How many balls go into the box? Try the simulation several times. Do you get the same number? Now, let $s \gg r$. How many balls go into the box? Try the simulation several times. Do you get the same number? Is this number in accordance with the Kepler conjecture? (The Kepler conjecture states that the densest arrangement of balls is obtained by stacking triangular layers of balls—the "cannonball arrangement".)

# Chapter 6

# Generation of Random Samples, Permutations, and Stochastic Processes

## 6.1 Random Samples

In applications of statistical techniques, as well as in Monte Carlo studies, it is often necessary to take a random sample from a given finite set. A common form of random sampling is simple random sampling without replacement, in which a sample of $n$ items is selected from a population $N$ in such a way that every subset of size $n$ from the universe of $N$ items has an equal chance of being the sample chosen. This is equivalent to a selection mechanism in which $n$ different items are selected, each with equal probability, $n/N$, and without regard to which other items are selected.

In a variation called Bernoulli sampling, each item is selected independently with a probability of $n/N$. In this variation, the sample size itself is a random variable. In another variation, called simple random sampling with replacement, a fixed sample of size $n$ is chosen with equal and independent probabilities, without the restriction that the sample items be different. Selection of a simple random sample with replacement is equivalent to generating $n$ random deviates from a discrete uniform distribution over the integers $1, 2, \ldots, N$.

Variations in sampling designs include stratified sampling and multistage sampling. In multistage sampling, *primary sampling units* are selected first. The primary sampling units consist of *secondary sampling units*, which are then sampled in a second stage. For example, the primary sampling units may be geographical districts, and the secondary sampling units may be farms. The more complicated sampling schemes can be implemented by using the same algorithms as for the simpler designs on different sampling units or at different stages.

Often, rather than each item in the population having the same probability of being included in the sample, the $i^{\text{th}}$ item in the population has a preassigned probability, $p_i$. In many cases, the $p_i$s are assigned based on some auxiliary variable, such as a measure of the size of the population units. In multistage sampling, the primary sampling units are often chosen with probability proportional to the number of secondary units that they contain.

In one variation of sampling with unequal probabilities, similar to Bernoulli sampling, each item in the population is selected independently with its preassigned probability. This is called Poisson sampling. In this case, the sample size is a random variable.

There are other operational variations. Sometimes, the sampling activity begins with the generation of a set of indices, which are then used to determine whether a given member of the population is included in the population. In a more common situation, however, the items in the population are encountered sequentially, and a decision must be made at the time of encountering the item. This is called a "draw sequential" scheme.

The method shown in Algorithm 6.1 is a draw sequential scheme for simple random sampling if the $p_i = n/N$ for each $i$. If $\sum p_i$ is known in advance (and presumably $\sum p_i = n_0$, where $n_0$ is some expected sample size), Algorithm 6.1 yields a probability sample without replacement, but the sample size is a random variable. Unless the sample size is fixed to be 1 or all of the $p_i$s are equal, the schemes to achieve a fixed sample size are very complicated. See Särndal, Swensson, and Wretman (1992) for discussions of various sampling designs.

### Algorithm 6.1 Draw Sequential Probability Sampling

0. Set $i = 0$, $s_i = 0$, $t_i = 0$, and $T = \sum p_j$.

1. Set $i = i + 1$ and generate $u$ from $\text{U}(0, 1)$.

2. If $u \leq \frac{T p_i - s_{i-1}}{T - t_{i-1}}$, then
   include the $i^{\text{th}}$ item, and set $s_i = s_{i-1} + p_i$.

3. If $i < N$, set $t_i = t_{i-1} + p_i$, and go to step 1.     ∎

This algorithm is obviously $\text{O}(N)$. For simple random sampling (that is, if the probabilities are all equal), we can improve on Algorithm 6.1 in two ways. If $N$ is known in advance, Vitter (1984) and Ahrens and Dieter (1985) give methods to obtain the simple random sample in $\text{O}(n)$ time because we can generate the random skips in the population before the next item is selected for inclusion in the sample.

The other improvement for the case of simple random sampling is to drop the requirement that $N$ be known in advance. If a sample of size $n$ is to be selected from a population of size $N$, which is not known in advance, a "reservoir sample" of size $n$ is filled and then updated by replacements until the full population has been considered; the reservoir is the sample. This process is called reservoir sampling. In a simple-minded implementation of a reservoir algorithm, a decision is made in turn, for each population item, whether to replace one of the reservoir items with it. This process is obviously $O(N)$ (McLeod and Bellhouse, 1983). This can also be improved on by generating a random number of population items to skip before including one in the sample.

Any one-pass algorithm for the case where $N$ is unknown must be a reservoir method (Vitter, 1985). Li (1994) summarizes work on reservoir sampling and gives the following algorithm, which is $O(n(1 + \log(N/n)))$. In this description, let the population items be denoted by $X_1, X_2, X_3, \ldots$.

**Algorithm 6.2 Li's Reservoir Sampling**

0. Initialize the reservoir with the first $n$ items of the population. Call the sample items $x_1, x_2, \ldots, x_n$. Set $i = n + 1$.

1. Generate $u_1$ from $U(0, 1)$, and compute $w = u_1^{\frac{1}{n}}$.

2. Generate $u_2$ from $U(0, 1)$, and compute $s = \left\lfloor \frac{\log u_2}{\log(1-w)} \right\rfloor$.

3. If $X_{i+s}$ is in the population, then
       3.a. generate $j$ uniformly from $1, 2, \ldots, n$, replace $x_j$ with $X_{i+s}$, and go to step 1;
   otherwise,
       3.b. terminate. ∎

At first, one may think that it would be unreasonable to choose an $n$ for a sample size without knowing $N$, but indeed this is a common situation, especially in sampling for proportions. Using the normal approximation for setting a confidence interval, a sample of size 1067 will yield a 95% confidence interval no wider than 0.06 for a population proportion, even if the finite population correction factor is ignored. In practice, a "random sample" of size 1100 is often used in sampling for proportions no matter what the size of the population. Often, of course, simple random samples are not used. Instead, systematic samples or cluster samples are used because of their operational simplicity.

Because databases are often organized in computer storage as trees rather than just as sequential files, it is sometimes necessary to be able to sample from trees. Knuth (1975) gives a general sampling algorithm for trees, and Rosenbaum (1993) describes a more efficient method for equal probability sampling from trees. Alonso and Schott (1995) and Wilson and Propp (1996) describe algorithms for generating random spanning trees from a directed graph. Olken and Rotem (1995a) give a survey of methods for sampling from databases organized as trees or with more complex structures.

Olken and Rotem (1995b) extend the reservoir sampling methods to the problem of sampling from a spatial database such as may be used in geographic information systems (GIS).

## 6.2  Permutations

The problem of generation of a random permutation is related to random sampling. A random permutation can be generated in one pass as shown in Algorithm 6.3. In this description, assume that the population items are indexed by $1, 2, 3, \ldots N$.

**Algorithm 6.3 Random Permutation**

  0.  Set $i = 0$.

  1.  Generate $j$ uniformly from the integers $1, 2, \ldots, N - i$.

  2.  Exchange the elements in the $(N - i)^{\text{th}}$ and the $j^{\text{th}}$ positions.

  3.  If $i < N - 2$, then
      3.a.  set $i = i + 1$ and go to step 1;
    otherwise,
      3.b.  terminate.                                     ∎

The algorithm is $O(N)$; it only makes one pass through the data, and the work at each step is constant.

## 6.3  Limitations of Random Number Generators

We have mentioned possible problems with random number generators that have short periods. As we have seen, the period of some of the most commonly used generators is of the order of $2^{31}$ (or $10^9$), although some generators may have a substantially longer period. We have indicated that a long period is desirable, and this is intuitively obvious. Generation of random samples and permutations is a simple problem that really points out a limitation of a finite period.

Both of these problems involve a classic operation for yielding large numbers: the factorial operation. Consider the seemingly simple case of generating a random sample of size 500 from a population of size 1,000,000. The cardinality of the sample space is

$$\binom{1000000}{500} \approx 10^{2000}.$$

This number is truly large by almost any scale. It is clear that no practical finite period generator can cover this sample space (see Greenwood, 1976a, 1976b).

In no other application considered in this book does the limitation of a finite period stand out so clearly as it does in the case of generation of random samples and permutations.

# 6.4 Generation of Nonindependent Samples

A sequence of random variables can be thought of either as a stochastic process or as a single multivariate random variable. Many multivariate distributions can usefully be formulated as stochastic processes, and the conditional generation algorithms that we described for the multivariate normal and the multinomial distributions are simple examples of this. General methods for inducing correlations either within a single stream of random variates or between two streams have been described by Kachitvichyanukul, Cheng, and Schmeiser (1988) and by Avramidis and Wilson (1995).

## 6.4.1 Order Statistics

In some applications, a random sample needs to be sorted into ascending or descending order, and in some cases we are interested only in the extreme values of a random set. In either case, the random numbers are *order statistics* from a random sample. In the early 1970s, several authors pointed out that order statistics could be generated directly, without simulating the full set and then sorting (Lurie and Hartley, 1972; Reeder, 1972; and Schucany, 1972). There are four possible ways of generating order statistics other than just generating the variates and sorting them:

1. *Direct transformations*: If the order statistics from a given distribution have some simple known distribution, generate the order statistics directly from that distribution.

2. *Sequential*: Generate one order statistic directly using its distribution, and then generate additional ones sequentially using their conditional distributions.

3. *Spacings*: Generate one order statistic directly using its distribution, and then generate additional ones sequentially using the distribution of the spacings between the order statistics.

4. *Multinomial groups*: Divide the range of the random variable into intervals, generate one multinomial variate for the counts in the intervals corresponding to the probability of the interval, and then generate the required number of ordered variates within each interval by any of the previous methods.

There are only a few distributions for which we can use these methods directly because the derived distributions that are necessary for the application of the methods are generally not easily worked out.

### Order Statistics from a Uniform Distribution

For the uniform distribution, all four of the methods mentioned above can be easily applied. It is straightforward to show that the $i^{\text{th}}$ order statistic from a sample of size $n$ from a $U(0,1)$ distribution has a beta distribution with parameters $i$ and $n - i + 1$ (see, e.g., David, 1981). We can therefore generate uniform order statistics directly as beta variates.

Notice that for the first or the $n^{\text{th}}$ order statistic from a uniform distribution, the beta distribution has one parameter equal to 1. For such a beta distribution, it is easy to generate a deviate because the density, and hence the distribution function, is proportional to a monomial. The inverse CDF method just involves taking the $n^{\text{th}}$ root of a uniform.

To proceed with the sequential method, we use the fact that conditional on $u_{(i)}$, for $j > i$, the $j^{\text{th}}$ order statistic from a sample of size $n$ from a $U(0,1)$ distribution has the same distribution as the $(j - i)^{\text{th}}$ order statistic from a sample of size $n - i$ from a $U(u_{(i)}, 1)$ distribution. We again use the relationship to the beta distribution, except we have to scale it into the interval $(u_{(i)}, 1)$.

The spacings method depends on an interesting relationship between ratios of independent standard exponential random variables and the difference between order statistics from a $U(0,1)$ distribution. It is easy to show that if $Y_1, Y_2, \ldots, Y_n, Y_{n+1}$ are independently distributed as exponential with parameter equal to 1, for $i = 1, 2, \ldots, n$, the ratio

$$R_i = \frac{Y_i}{Y_1 + Y_2 + \ldots + Y_n + Y_{n+1}}$$

has the same distribution as

$$U_{(i)} - U_{(i-1)},$$

where $U_{(i)}$ is the $i^{\text{th}}$ order statistic from a sample of size $n$ from a $U(0,1)$ distribution, and $U_{(0)} = 0$.

Reeder (1972), Lurie and Mason (1973), and Rabinowitz and Berenson (1974) compare the performance of these methods for generating order statistics from a uniform distribution, and Gerontidis and Smith (1982) compare the performance for some nonuniform distributions. If all $n$ order statistics are required, the multinomial groups method is usually best. Often, however, only a few order statistics from the full set of $n$ are required. In that case, either the sequential method or the spacings method is to be preferred.

Nagaraja (1979) points out an interesting anomaly: although the distribution of the order statistics using any of four methods that we have discussed is the same,

$$\Pr\left(X_{(n)}^{\text{seq}} > X_{(n)}^{\text{sort}}\right) > 0.5,$$

where $X_{(n)}^{\text{seq}}$ is the maximum uniform order statistic generated by first generating $X_{(1)}$ as the $n^{\text{th}}$ root of a uniform and then generating the sequence up to $X_{(n)}$, and $X_{(n)}^{\text{sort}}$ is the maximum uniform order statistic generated by sorting. The probability difference is small, and this fact is generally not important in Monte Carlo studies.

### Order Statistics from Other Distributions

If the target distribution is not the uniform distribution, it still may be possible to use one of these methods. If the inverse CDF method can be applied, however, it is usually better to generate the appropriate order statistics from the uniform distribution and then use the inverse CDF method to generate the order statistics from the target distribution.

## 6.4.2 Censored Data

In applications studying lifetime data, such as reliability studies and survival analysis, it is common not to observe all lifetimes of the experimental units in the study. The unobserved data are said to be "censored". There are several ways that the censoring can be performed. Often, either the very small values or the very large values are not observed. Data collected by such procedures can be simulated as truncated distributions, such as we have already discussed.

More complicated censoring may be difficult to simulate. In "progressive censoring" or "multiple censoring", more than one experimental unit is removed from the study at a time. In type-I progressive censoring, the censoring times are fixed. At the $i^{\text{th}}$ preselected time, $r_i$ items are removed from the test; that is, their lifetimes are censored. In type-II censoring, the censoring times are the random failure times of single units. At the $i^{\text{th}}$ failure, $r_i$ items are removed from the test. In this type of accelerated life testing, we observe $m$ failures (or lifetimes), and

$$n = m + r_1 + \cdots + r_m,$$

where $n$ items begin in the study. Balakrishnan and Sandhu (1995) give a method for progressive type-II censoring. They show that if $U_{(1)}, U_{(2)}, \ldots, U_{(m)}$ are progressive type-II censored observations from U(0, 1),

$$V_i = \frac{1 - U_{(m-i+1)}}{1 - U_{(m-i)}} \quad \text{for } i = 1, 2, \ldots, m-1,$$

$$V_m = 1 - U_{(1)},$$

and

$$W_i = V_i^{i + r_m + r_{m-1} + \cdots + r_{m-i+1}} \quad \text{for } i = 1, 2, \ldots, m,$$

then the $W_i$ are i.i.d. U(0, 1).

This fact (reminiscent of the relationships of order statistics from the uniform distribution) provides the basis for simulating progressive type-II censored observations. Also similar to the situation for order statistics, the uniform case can yield samples from other distributions by use of the inverse CDF method.

### Algorithm 6.4 Progressive Type-II Censored Observations from a Distribution with CDF $P$

1. Generate $m$ independent U(0, 1) deviates $w_i$.

2. Set $v_i = w_i^{1/(i+r_m+r_{m-1}+\cdots+r_{m-i+1})}$.

3. Set $u_i = 1 - v_m v_{m-1} v_{m-i+1}$.
   The $u_i$ are deviates from progressive type-II censoring from the U(0, 1) distribution.

4. Set $x_i = P^{-1}(u_i)$ (or use the appropriate modification discussed in Chapter 4 if $P$ is not easily invertible).                                           ∎

## 6.5  Generation of Nonindependent Sequences

There are many types of stochastic processes in which the distribution of a given element in the process depends on the realizations of previous random elements in the process. Examples of such processes are the position of an animal as it moves in time and the price of a financial asset. There are various types of models that are useful for stochastic processes, and in this section we will briefly describe a few and discuss their simulation.

We denote a stochastic process by a sequence $X_0, X_1, \ldots$ and an associated sequence $t_0, t_1, \ldots$. The $X$s and/or the $t$s may be random variables. The value that an $X$ assumes is called the state of the system. The set of all possible states is called the state space. The associated sequence of $t$s are indexes, so we may write the sequence of states as $X_{t_0}, X_{t_1}, \ldots$. If the indexes are an increasing sequence of scalars, the index is usually called "time". Another common type of index is location, usually in two dimensions. We distinguish types of processes based on the kinds of values that the states can assume (whether they are discrete or continuous) and the nature of the index values. When the values of the indexes are restricted to increasing integral values, the process is called a discrete-time process, and if the indexes can assume any real value in an increasing sequence, the process is called a continuous-time process. In either case, the indexes are usually restricted to be nonnegative.

Multivariate processes are often of interest. In signal processing applications, for example, we often assume an unobservable process $X_0, X_1, \ldots$ and a sequence of observations $Y_0, Y_1, \ldots$ with distributions that depend on the $X_t$s. There are many variations of this basic model.

Often, the simulation of a stochastic process is just a straightforward sequential application of the methods that we have discussed for generation of independent variates.

### 6.5.1  Markov Process

The Markov (or Markovian) property for a stochastic process is the requirement that the conditional distribution of $X_{t_i} | X_{t_{i-1}}$ be independent of $X_{t_0}, \ldots, X_{t_{i-2}}$. This is the condition that a Markov process does not have *memory*. A Markov process, whether discrete time or continuous time, in which the state space is a finite or countable set is called a Markov chain. In a discrete-time Markov chain, with the states indexed by $1, 2, \ldots$, let $p_{ij}$ be the probability that the state at time $t_k$ is $j$ given that the state at time $t_{k-1}$ is $i$. The matrix $P = (p_{ij})$ is called the transition matrix. See Meyn and Tweedie (1993) for an extensive coverage of types of Markov chains and their properties.

Simulation of Markov chains is generally straightforward. When some of the transition probabilities are small, combination of states followed by splitting can be faster than a direct method. Juneja and Shahabudding (2001) discuss some efficient methods for dealing with small transition probabilities.

A simple random walk on a grid is a Markov chain that has applications in many areas. A generalization to a random walk on a plane is also a Markov process. A self-avoiding random walk (that is, a random walk with the restriction that no path of the walk can cross itself) is not a Markov process. A self-avoiding random walk on a grid has been used to simulate the growth of polymers.

## 6.5.2   Nonhomogeneous Poisson Process

A simple Poisson process is characterized by a sequence of independent exponential random variables. It can be viewed as a continuous-time Markov chain in which the states are a fixed sequence and the indexes are the cumulative sums of the independent exponentials, which are called interarrival times.

Some important Poisson processes have interarrival times that are distributed as exponential with rate parameter $\lambda$ varying in time $\lambda(t)$. For a nonhomogeneous Poisson process with a rate function that is varying in time, Lewis and Shedler (1979) develop a method called "thinning" that uses acceptance/rejection on Poisson variates from a process with a greater rate function, $\lambda^*$ (which may be stepwise constant). The method is shown in Algorithm 6.5.

### Algorithm 6.5 Nonhomogeneous Poisson Process by Thinning

0. Set $i = 0$, and initialize $t_0$.

1. Set $d = 0$.

2. Generate $e$ from an exponential with rate function $\lambda^*$, and set $d = d + e$.

3. Generate a $u$ from $U(0, 1)$.

4. If $u \leq \lambda(t_i + d)/\lambda^*$, then
   4.a. set $i = i + 1$, set $t_i = t_{i-1} + d$, deliver $t_i$, and go to step 1; otherwise,
   4.b. go to step 2.                                          ∎

The IMSL Libraries routine rnnpp generates a nonhomogeneous Poisson process using the Lewis and Shedler thinning method.

The thinning technique also applies to other distributions that are varying in time. Chen and Schmeiser (1992) describe methods for generating a nonhomogeneous Poisson process in which the rate varies cyclically as a trigonometric function.

## 6.5.3   Other Time Series Models

One of the most widely used models for a time series is the autoregressive moving average (ARMA) model,

$$\phi(B)(X_t) = \theta_0 + \theta(B)(A_t) \quad \text{for } t = 0, 1, 2, \ldots, \tag{6.1}$$

where B is the backward shift operator defined by $B^k(X_t) = X_{t-k}$,

$$\phi(B) = 1 - \phi_1 B - \cdots - \phi_p B^p,$$

and

$$\theta(B) = 1 - \theta_1 B - \cdots - \theta_q B^q,$$

where $p, q \geq 0$. We can therefore rewrite equation (6.1) as

$$X_t = \phi_1 X_{t-1} + \cdots + \phi_p X_{t-p} + \theta_0 + A_t - \theta_1 A_{t-1} - \cdots - \theta_q A_{t-q}.$$

To begin the process, we usually set any $X$ or $A$ with a negative subscript as 0.

The $A_t$ are i.i.d. random variables called innovations. In most cases, we assume that the $A_t$s have a normal distribution.

There are many variations on this basic model. Some variations are quite amenable for analysis, but in some cases the model is best studied by simulation. This is especially true if the innovations do not have a normal distribution.

The model is fairly easy to simulate. Numerical errors can accumulate for a very long simulation, but this is not generally a problem. The IMSL routine rnarm generates random realizations of ARMA sequences, and the S-Plus function arima.sim generates random realizations of ARIMA sequences (ARMA sequences with differencing).

If the innovations in the time series model are not i.i.d., various other models may be more appropriate. One approach that seems to work well in some situations is the generalized autoregressive conditional heteroscedastic (GARCH) model,

$$X_t = E_t \sigma_t, \tag{6.2}$$

where the $E_t$ are i.i.d. $N(0,1)$,

$$\sigma_t^2 = \sigma^2 + \beta_1 \sigma_{t-1}^2 + \cdots + \beta_p \sigma_{t-p}^2 + \alpha_1 X_{t-1}^2 + \cdots + \alpha_p X_{t-q}^2,$$

$$\sigma > 0, \ \beta_i \geq 0, \ \alpha_i \geq 0,$$

and

$$\sum_{i=1}^{p} \beta_i + \sum_{j=1}^{q} \alpha_j < 1.$$

A GARCH(1,1) model is often adequate. This model is widely used in modeling stock price movements. The assumption of normality for the $E_t$ can easily be changed to some other distribution. Because normality is usually assumed in a GARCH model, if some other distribution is assumed, the name is often changed. For example, if a Student's $t$ distribution is used, the model is sometimes called TGARCH, and if a stable distribution (other than the normal) is used, the model is called SGARCH.

# Exercises

6.1. One scheme for sampling with replacement and with probability proportional to size is called Lahiri's method (see Särndal, Swensson, and Wretman, 1992). Suppose that a population consists of $N$ items, with sizes $M_1, M_2, \ldots, M_N$, and a sample of size $n$ is to be selected with replacement in such a way that the probability of the $i^{\text{th}}$ item is proportional to $M_i$. (The $N$ population items may be primary sampling units, and the $M_i$s may be numbers of secondary sampling units in the primary units.) Lahiri's method uses acceptance/rejection:

0. Set $i = 0$, $j = 0$, and $M = \max(M_1, M_2, \ldots, M_N)$.

1. Set $i = i + 1$, and generate $k$ from a discrete uniform over $1, 2, \ldots, N$.

2. Generate $m$ from a discrete uniform over $1, 2, \ldots, M$.

3. If $m \leq M_k$, then
include the $k^{\text{th}}$ item and set $j = j + 1$.

4. If $j < n$, then go to step 1.

Prove that this method results in a probability proportional-to-size sample with replacement.

6.2. Consider the problem of generating a simple random sample of size $n$ from a population of size $N$. Suppose that $N = 10^6$, as in the discussion on page 220, and suppose that you have a random number generator with period $10^9$.

(a) With this generator, what is the largest sample that you can draw in accordance with the definition of a simple random sample? Discuss. (A simple answer involves a major assumption. What is it? Without this assumption, it is difficult even to interpret the question.)

(b) Suppose that you wish to use the generator to draw a random sample of size 500. How serious do you believe the problem to be? Discuss.

6.3. Show that the $i^{\text{th}}$ order statistic from a sample of size $n$ from a U$(0, 1)$ distribution has a beta distribution with parameters $i$ and $n - i + 1$. *Hint: Just write out the density (as in David, 1981, for example), plug in the distribution function of the uniform, and identify the terms.*

6.4. Write a Fortran or C function to accept $a$, $b$, $n$, $n_1$, and $n_2$ (with $1 \leq n_1 \leq n_2 \leq n$) and to return the $n_1^{\text{th}}$ through the $n_2^{\text{th}}$ order statistics from a two-parameter Weibull distribution with parameters $a$ and $b$. Use the random spacings method to generate uniform order statistics, and then use the inverse CDF to generate order statistics from a Weibull distribution.

6.5. Derive two different algorithms, and write programs implementing them to generate a deal of hands for the game of bridge. (That is, form

four random mutually exclusive sets each containing 13 of the integers
$1, 2, \ldots, 52$.) Base one algorithm on the techniques of random sampling
and the other on methods of random permutations. Which is more efficient?

6.6. Random walks.

(a) Write a program to simulate a symmetric random walk on a two-
dimensional grid. (That is, for integers $i$ and $j$, if the state at time
$t$ is $(i, j)$, then the probability at time $t + 1$ is $1/4$ for each of the
states $(i - 1, j)$, $(i + 1, j)$, $(i, j - 1)$, and $(i, j + 1)$, and, of course,
0 for all other states.) Use your program to estimate the mean
and variance of $D_t = \sqrt{(i_t - i_0)^2 + (j_t - j_0)^2}$. Also determine these
values analytically.

(b) Write a program to simulate a symmetric self-avoiding random walk
on a two-dimensional grid. Notice that often the self-avoiding walk
reaches a point from which no further movement is possible. In some
applications, a good model is a self-avoiding random walk that does
not reach a point from which further movement is not possible. To
simulate such a process, we generate self-avoiding random walks for
the time period of interest and discard all that stop early. Use your
program to estimate the mean and variance of $D_t$ for walks that do
not stop for $t$ up to 200.

(c) Again, as in Exercise 6.6a, simulate a symmetric random walk on a
two-dimensional grid, except this time assuming that the state goes
from $(i, j)$ to $(i\pm 1, j)$ and to $(i, j\pm 1)$ with probability $1/(4(1+e^{-1/2}))$
and goes to $(i \pm 1, j\pm)$ with probability $e^{-1/2}/(4(1 + e^{-1/2}))$. Use
your program to estimate the mean and variance of $D_t$.

(d) Now, simulate a symmetric random walk on a two-dimensional grid,
this time assuming that the state goes from $(x_{t-1}, y_{t-1})$ to the point
$(x_t, y_t)$ with a probability density proportional to $\exp(((x_t - x_{t-1})^2 + (y_t - y_{t-1})^2)/2)$. Use your program to estimate the mean and variance of $D = \sqrt{(x_t - x_0)^2 + (y_t - y_0)^2}$.

6.7. Consider a symmetric random walk on a grid that is known to be at position $(0, 0)$ at time $t = 0$ and to be at position $(i_t, j_t)$ at time $t$. Develop an
algorithm and write a program to simulate a constrained symmetric random walk on a grid. (This is not an easy problem!) Develop a statistical
test of your algorithm, and test it empirically.

# Chapter 7

# Monte Carlo Methods

Monte Carlo simulation is the use of experiments with random numbers to evaluate mathematical expressions. The experimental units are the random numbers. The expressions may be definite integrals, systems of equations, or more complicated mathematical models. In most cases, when mathematical expressions are to be evaluated, the standard approximations from numerical analysis are to be preferred, but Monte Carlo methods provide an alternative that is sometimes the only tractable approach. Monte Carlo is often the preferred method for evaluating integrals over high-dimensional domains, for example. Very large and sparse systems of equations can sometimes be solved effectively by Monte Carlo methods. (See Chapter 7 of Hammersley and Handscomb, 1964.) In these applications, there is no inherent randomness. Random variables are defined and then simulated in order to solve a problem that is strictly deterministic.

Sometimes, an ensemble of physical units is described by a density, which can be normalized to be a probability density, and Monte Carlo methods are used to simulate properties of the ensemble. Evaluating models of large numbers of particles is a common application in statistical physics (see Newman and Barkema, 1999, for example).

In some cases, there is an inherent randomness in the model, and the objective of the Monte Carlo study is to discover the distribution of some variable. We may think of the problem as having a set of inputs with known or assumed random distributions and a set of outputs with unknown random distributions. As a simple example, consider the question of the distribution of the midrange of a gamma distribution. (Analytic methods should be used when practical, of course. The distribution in this example can and should be worked out analytically.) The input is a variable with a gamma distribution, and the output is the midrange statistic. Instead of estimating a finite-dimensional parameter such as the mean and variance of the midrange, the objective of the Monte Carlo study may be to estimate the probability density of the midrange. The object to be estimated is a continuous function. There are various methods of probability density estimation that could be used with a sample of midranges generated by Monte Carlo methods. The results of the Monte Carlo study generally are presented as graphs representing the probability density of the statistic of interest.

There are many practical issues that must be taken into account in conducting a Monte Carlo study. These considerations include what software to use and how the computer time is to be used. Often, a Monte Carlo study must be conducted as a set of separate computer runs. In that case, it is necessary that the separate runs not use overlapping sequences of random numbers. After the underlying sequence of random numbers is interrupted, the study should resume at the same point in the sequence.

The results of any scientific experiment should be reproducible. In Monte Carlo experimentation, we have a stronger requirement: the experimental units should be *strictly reproducible*. Strict reproducibility of the experimental units means that any stream of random numbers can be repeated (to within machine precision on each element), given the same initial conditions. As we have indicated, in very rare instances, methods that make decisions based on comparisons of floating-point computations that affect the number of uniform numbers used may not yield strictly reproducible streams of random numbers.

Scientific experiments involving large problems or many variables must be designed to yield information efficiently. Estimators and other statistics in a well-designed experiment have small variance. Monte Carlo methods should also be designed to be efficient. In this chapter, we discuss some of the ways of reducing the variance of Monte Carlo estimators. Additional discussion and other strategies can be found in Liu (2001).

In all applications, we should be aware of the cardinality of the sample space. As pointed out in Section 6.3, even in some seemingly simple problems, the sample space is so large that a random number generator may not be able to generate all points in it.

A Monte Carlo method begins with the identification of a random variable such that the expected value of some function of the random variable is a parameter in the problem to be solved. To use the Monte Carlo method, we must be able to simulate samples from the random variable. The problem being addressed may be strictly deterministic. The evaluation of a definite integral, which is a deterministic quantity, provides a good example of a Monte Carlo method.

## 7.1 Evaluating an Integral

In its simplest form, Monte Carlo simulation is the evaluation of a definite integral

$$\theta = \int_D f(x)\,dx \tag{7.1}$$

by identifying a random variable $Y$ with support on $D$ and density $p(y)$ and a function $g$ such that the expected value of $g(Y)$ is $\theta$:

$$
\begin{aligned}
\mathrm{E}(g(Y)) &= \int_D g(y)p(y)\,dy \\
&= \int_D f(y)\,dy \\
&= \theta.
\end{aligned}
$$

Let us first consider the case in which $D$ is the interval $[a, b]$, $Y$ is taken to be a random variable with a uniform density over $[a, b]$, and $g$ is taken to be $f$. In this case,

$$ \theta = (b - a)\mathrm{E}(f(Y)). $$

The problem of evaluating the integral becomes the familiar statistical problem of estimating a mean, $\mathrm{E}(f(Y))$.

The statistician quite naturally takes a random sample and uses the sample mean. For a sample of size $m$, an estimate of $\theta$ is

$$ \widehat{\theta} = (b - a)\frac{\sum_{i=1}^{m} f(y_i)}{m}, \tag{7.2} $$

where the $y_i$ are values of a random sample from a uniform distribution over $(a, b)$. The estimate is unbiased:

$$
\begin{aligned}
\mathrm{E}(\widehat{\theta}) &= (b - a)\frac{\sum \mathrm{E}(f(Y_i))}{m} \\
&= (b - a)\mathrm{E}(f(Y)) \\
&= \int_a^b f(x)\,dx.
\end{aligned}
$$

The variance is

$$
\begin{aligned}
\mathrm{V}(\widehat{\theta}) &= (b - a)^2 \frac{\sum \mathrm{V}(f(Y_i))}{m^2} \\
&= \frac{(b - a)^2}{m} \mathrm{V}(f(Y)) \\
&= \frac{(b - a)}{m} \int_a^b \left( f(x) - \int_a^b f(t)\,dt \right)^2 dx. \tag{7.3}
\end{aligned}
$$

The integral in equation (7.3) is a measure of the *roughness* of the function. (There are various ways of defining roughness. Most definitions involve derivatives. The more derivatives that exist, the less rough the function. Other definitions, such as the one here, are based on a norm of a function. The $L_2$ norm of the difference of the function from its integrated value is a very natural measure of roughness of the function. Another measure is just the $L_2$ norm of the function itself, which, of course, is not translation-invariant.)

The method of estimating an integral just described is sometimes called "crude Monte Carlo". In Exercise 7.2, page 271, we describe another method, called "hit-or-miss", and ask the reader to show that the crude method is superior to hit-or-miss.

Suppose that the original integral can be written as

$$\theta = \int_D f(x)\,dx$$
$$= \int_D g(x)p(x)\,dx, \tag{7.4}$$

where $p(x)$ is a probability density over $D$. As with the uniform example considered earlier, it may require some scaling to get the density to be over the interval $D$. (In the uniform case, $D = (a, b)$, both $a$ and $b$ must be finite, and $p(x) = 1/(b - a)$.) Now, suppose that we can generate $m$ random variates $y_i$ from the distribution with density $p$. Then, our estimate of $\theta$ is just

$$\widehat{\theta} = \frac{\sum g(y_i)}{m}. \tag{7.5}$$

Compare this estimator with the estimator in equation (7.2).

The use of a probability density as a weighting function allows us to apply the Monte Carlo method to improper integrals (that is, integrals with infinite ranges of integration; see Exercise 7.9). The first thing to note, therefore, is that the estimator (7.5) applies to integrals over general domains, while the estimator (7.2) applies only to integrals over finite intervals. Another important difference is that the variance of the estimator in equation (7.5) is likely to be smaller than that of the estimator in equation (7.2) (see Exercise 7.7 on page 273 and see Section 7.5).

Quadrature is an important topic in numerical analysis, and a number of quadrature methods are available. They are generally classified as Newton–Cotes methods, extrapolation or Romberg methods, and Gaussian quadrature. These methods involve various approximations to the integrand over various subdomains of the range of integration. The use of these methods involves consideration of error bounds, which are often stated in terms of some function of a derivative of the integrand evaluated at some unknown point. Monte Carlo quadrature differs from these numerical methods in a fundamental way: Monte Carlo methods involve random (or pseudorandom) sampling. The expressions in the Monte Carlo quadrature formulas do not involve any approximations, so questions of bounds of the error of approximation do not arise. Instead of error bounds or order of the error as some function of the integrand, we use the variance of the random estimator to indicate the extent of the uncertainty in the solution.

The square root of the variance (that is, the standard deviation of the estimator) is a good measure of the range within which different realizations of the estimator of the integral may fall. Under certain assumptions, using the standard deviation of the estimator, we can define statistical "confidence intervals" for the true value of the integral $\theta$. Loosely speaking, a confidence interval is an interval about an estimator $\widehat{\theta}$ that in repeated sampling would include the true value $\theta$ a specified portion of the time. (The specified portion is the "level" of the confidence interval and is often chosen to be 90% or 95%. Obviously, all other things being equal, the higher the level of confidence, the wider the interval must be.)

Because of the dependence of the confidence interval on the standard deviation, the standard deviation is sometimes called a "probabilistic error bound". The word "bound" is misused here, of course, but in any event, the standard deviation does provide some measure of a sampling "error".

The important thing to note from equation (7.3) is the order of error in terms of the Monte Carlo sample size; it is $O(m^{-\frac{1}{2}})$. This results in the usual diminished returns of ordinary statistical estimators; to halve the error, the sample size must be quadrupled.

We should be aware of a very important aspect of a discussion of error bounds for the Monte Carlo estimators. It applies to random numbers. The pseudorandom numbers that we actually use only simulate the random numbers, so "unbiasedness" and "variance" must be interpreted carefully.

The Monte Carlo quadrature methods extend directly to multivariate integrals, although, obviously, it takes larger samples to fill the space. It is, in fact, only for multivariate integrals that Monte Carlo quadrature should ordinarily be used. The preference for Monte Carlo methods in multivariate quadrature results from the independence of the pseudoprobabilistic error bounds and the dimensionality mentioned above. See Evans and Swartz (2000) for further discussion of techniques (both Monte Carlo and deterministic ones) for evaluation of definite integrals, including integrals over multivariate spaces. Ogata (1990) discusses Monte Carlo quadrature in the context of Bayesian analysis, especially with multiple parameters. The compendium of articles in Flournoy and Tsutakawa (1991) covers use of Monte Carlo methods in the evaluation of a number of specialized integrals. Some of the articles also review standard numerical methods for the approximation of integrals. Gelman and Meng (1998) review various methods for Monte Carlo quadrature, and demonstrate relationships among various methods, including sequential ones discussed in the next section.

# 7.2  Sequential Monte Carlo Methods

In many applications, random behavior can be modeled by an ensemble of i.i.d. random variables. Also, in many applications of Monte Carlo methods, such as the evaluation of an integral, i.i.d. sampling is the best way to approach the problem. In other cases, however, the distributions of the elements in the process being studied depend on the realizations of the elements up to that point. Such a stochastic process could be studied by use of a single nested integral, of course, but it is often more efficient, both in developing the model and in performing the simulation, to use sequential Monte Carlo methods.

In sequential Monte Carlo simulation, we generate variates from conditional distributions with densities $p_{X_0}, p_{X_1|X_0}, p_{X_2|X_1X_0}, \ldots$. In applications, the index often represents time; that is, $X_0$ is the random variable at time 0, $X_1$ at time 1, and so on. If the process is continuous in time (that is, if the index in the subscript is continuous), in practice we assume that observations are taken at discrete points in time.

A realization of the stochastic process is called a *trajectory* or a *particle*. Sequential Monte Carlo methods generally require generation of many realizations of the stochastic process. To simplify the notation, let us use $X_{i:j}$ for integers $i$ and $j$ with $i < j$ to represent the trajectory from $X_i$ to $X_j$, that is, the vector $(X_i, \ldots, X_j)$. Properties of the process are estimated by averaging the trajectories.

A special kind of stochastic process is a *Markov process*, or Markov chain, as we used in Section 4.10. A Markov process is characterized by densities with the property that $p_{X_i|X_{i-1}\cdots X_0} = p_{X_i|X_{i-1}}$. A simple example of such a process is a random walk. Two types of modifications of a random walk are a self-avoiding random walk and a constrained random walk (see Exercises 7.8b and 7.8c).

We often assume an underlying Markov process, $X_0, X_1, \ldots$, that is not observable ("hidden") and a sequence of observations $Y_0, Y_1, \ldots$ that are conditionally independent given the hidden Markov process, and the distribution of $Y_t$ depends only on $X_t$. This type of model is useful in signal processing applications, for example. The objective is to estimate recursively the posterior (conditional) density,

$$p_{X_{0:t}|Y_{1:t}}(x_{0:t}|y_{1:t}), \tag{7.6}$$

the *filtering density*,

$$p_{X_t|Y_{1:t}}(x_t|y_{1:t}),$$

or the expectation of a function of the signal,

$$\mathrm{E}(f_t(X_{0:t})) = \int f_t(x_{0:t}) p_{X_{0:t}|Y_{1:t}}(x_{0:t}|y_{1:t}), \mathrm{d}x_0\mathrm{d}x_1 \cdots \mathrm{d}x_t. \tag{7.7}$$

In some areas of application this is called "particle filtering". A straightforward Monte Carlo estimate of the expectation in equation (7.7) can be computed by generating $m$ trajectories, $x_{0:t}^{(i)}$ for $i = 1, \ldots, m$, and averaging the $f_t(x_{0:t}^{(i)})$. Unfortunately, because of the high dimensions of the integral in equation (7.7) and of the normalizing constants in the marginal and conditional distributions, a straightforward application of Monte Carlo techniques would be overwhelmed by variance. We will briefly discuss this problem later in this chapter. Extensive discussion of the problem is in the book edited by Doucet, de Freitas, and Gordon (2001).

# 7.3   Experimental Error in Monte Carlo Methods

Monte Carlo methods are sampling methods; therefore, the estimates that result from Monte Carlo procedures have associated *sampling errors*. The fact that the estimate is not equal to its expected value (assuming that the estimator is unbiased) is not an "error" or a "mistake"; it is just a result of the variance of the random (or pseudorandom) data. Monte Carlo methods are experiments using random data. The variability of the random data results in *experimental error*, just as in other scientific experiments in which randomness is a recognized component.

As in any statistical estimation problem, an estimate should be accompanied by an estimate of its variance. The estimate of the variance of the estimator of interest is usually just the sample variance of computed values of the estimator of interest.

Following standard practice, we could use the square root of the variance (that is, the standard deviation) of the Monte Carlo estimator to form an approximate confidence interval for the integral being estimated. Of course, the confidence limits would include the unknown terms in the variance. We could, however, estimate the variance of the estimator using the same sample that we use to estimate the integral.

The standard deviation in the approximate confidence limits is sometimes called a "probabilistic error bound". The word "bound" is misused here, of course, but in any event, the standard deviation does provide some measure of a sampling "error". The important thing to note from equation (7.3) is the order of error; it is $O(m^{-\frac{1}{2}})$.

An important property of the standard deviation of a Monte Carlo estimate of a definite integral is that the order in terms of the number of function evaluations is independent of the dimensionality of the integral. On the other hand, the usual error bounds for numerical quadrature are $O(m^{-\frac{2}{d}})$, where $d$ is the dimensionality.

This discussion of error bounds for the Monte Carlo estimator applies to random numbers. The pseudorandom numbers that we actually use only simulate random numbers, so "unbiasedness" and "variance" must be interpreted carefully.

In Monte Carlo applications, a major reason for being interested in the variance of the estimator is to determine whether to increase the Monte Carlo sample size. The sample size is often determined so that the length of a confidence interval for the estimator meets a given maximum length requirement. For example, we may wish that the length of a 95% confidence interval for the parameter of interest, $\theta$, be no more than $d$. The confidence interval is of the form

$$\left( \widehat{\theta} - I_1, \quad \widehat{\theta} + I_2 \right).$$

Without knowing the distribution of $\widehat{\theta}$, of course, we cannot determine $I_1$ and

$I_2$. The usual approach is to approximate the confidence interval using the normal distribution. This leads to $I_1$ and $I_2$ having the forms $t_{\nu 1}v_\nu$ and $t_{\nu 2}v_\nu$, where $t_{\nu 1}$ and $t_{\nu 2}$ are quantiles of a Student's $t$ distribution with $\nu$ degrees of freedom, and $v_\nu$ is the square root of an estimator of the variance of $\hat{\theta}$ based on a sample size related to $\nu$. Because $v_\nu$ decreases approximately as $1/\sqrt{\nu}$, increasing the sample size ultimately results in the restriction on the length being satisfied:

$$(t_{\nu 1} + t_{\nu 2})v_\nu \leq d$$

(assuming the variance is finite).

The experimental error of Monte Carlo experiments should be treated just as carefully as the experimental error or measurement error in other scientific experimentation. The error determines a bound on the number of significant digits in numerical results. The error is propagated through any subsequent computations, and thus bounds on the number of significant digits are propagated.

In reporting numerical results from Monte Carlo simulations, it is mandatory to give some statement of the level of the experimental error. An effective way of doing this is by giving the sample standard deviation. When a number of results are reported, and the standard deviations vary from one to the other, a good way of presenting the results is to write the standard deviation in parentheses beside the result itself, for example,

$$3.147\,(0.0051).$$

Notice that if the standard deviation is of order $10^{-3}$, the precision of the main result is not greater than $10^{-3}$. Just because the computations are done at a higher precision is no reason to write the number as if it had more significant digits.

## 7.4  Variance of Monte Carlo Estimators

The variance of a Monte Carlo estimator has important uses in assessing the quality of the estimate of the integral. The expression for the variance, as in equation (7.3), is likely to be very complicated and to contain terms that are unknown. We therefore need methods for estimating the variance of the Monte Carlo estimator.

### Estimating the Variance

A Monte Carlo estimate usually has the form of the estimator of $\theta$ in equation (7.2):

$$\hat{\theta} = c\frac{\sum f_i}{m}.$$

The variance of the estimator has the form of equation (7.3):

$$V(\hat{\theta})c \int \left( f(x) - \int f(t)\, dt \right)^2 dx.$$

An estimator of the variance is

$$\hat{V}(\hat{\theta}) = c^2 \frac{\sum (f_i - \bar{f})^2}{m-1}. \tag{7.8}$$

This estimator is appropriate only if the elements of the set of random variables $\{F_i\}$, on which we have observations $\{f_i\}$, are (assumed to be) independent and thus have zero correlations.

## Estimating the Variance Using Batch Means

If the $F_i$ do not have zero correlations, the estimator (7.8) has an expected value that includes the correlations; that is, it is biased for estimating $V(\hat{\theta})$. This situation arises often in simulation. In many processes of interest, however, observations are "more independent" of observations farther removed within the sequence than they are of observations closer to them in the sequence. A common method for estimating the variance in a sequence of nonindependent observations therefore is to use the means of successive subsequences that are long enough that the observations in one subsequence are almost independent of the observations in another subsequence. The means of the subsequences are called "batch means".

If $F_1, \ldots, F_b, F_{b+1}, \ldots, F_{2b}, F_{2b+1}, \ldots, F_{kb}$ is a sequence of random variables such that the correlation of $F_i$ and $F_{i+b}$ is approximately zero, an estimate of the variance of the mean, $\bar{F}$, of the $m = kb$ random variables can be developed by observing that

$$
\begin{aligned}
V(\bar{F}) &= V\left( \frac{1}{m} \sum F_i \right) \\
&= V\left( \frac{1}{k} \sum_{j=1}^{k} \left( \frac{1}{b} \sum_{i=(j-1)b+1}^{jb} F_i \right) \right) \\
&\approx \frac{1}{k^2} \sum_{j=1}^{k} V\left( \frac{1}{b} \sum_{i=(j-1)b+1}^{jb} F_i \right) \\
&\approx \frac{1}{k} V(\bar{F}_b),
\end{aligned}
$$

where $\bar{F}_b$ is the mean of a batch of length $b$. If the batches are long enough, it may be reasonable to assume that the means have a common variance. An estimator of the variance of $\bar{F}_b$ is the standard sample variance from $k$ observations, $\bar{f}_1, \ldots, \bar{f}_k$:

$$\frac{\sum (\bar{f}_j - \bar{f})^2}{k-1}.$$

Hence, the batch-means estimator of the variance of $\bar{F}$ is

$$\widehat{V}(\bar{F}) = \frac{\sum (\bar{f}_j - \bar{f})^2}{k(k-1)}. \tag{7.9}$$

This batch-means variance estimator should be used if the Monte Carlo study yields a stream of nonindependent observations, such as in a time series or when the simulation uses a Markov chain. The size of the subsamples should be as small as possible and still have means that are independent. A test of the independence of the $\bar{F}_b$ may be appropriate to help in choosing the size of the batches. Batch means are useful in variance estimation whenever a Markov chain is used in the generation of the random deviates, as discussed in Section 4.10.

### Analysis of Variance

The variance can also be estimated from the results of completely different experiments because those results should be uncorrelated. Often, it is worth the extra trouble of running separate experiments in order to get a reliable estimate of the variance. There may also be other reasons to run separate experiments.

In Chapter 2, we advocated use of more than one random number generator and/or use of more than one seed in applications using random numbers. The different generators or different seeds result in a block design for a simulation experiment. We can represent the classical linear model for the observations $y_{ijk}$ in the experiment as

$$y_{ijk} = \mu + g_i + s_{i(j)} + e_{ij(k)},$$

where $\mu$ is the quantity about which we wish to make inference, $g_i$ is the effect on the observations of the $i^{\text{th}}$ generator, $s_{i(j)}$ is the effect of the $j^{\text{th}}$ seed within the $i^{\text{th}}$ generator (this notation for nesting is not the most commonly used notation), and $e_{ij(k)}$ is the random error of the $k^{\text{th}}$ observation within the $ij$ pair of generator and seed. If the generators are all "perfect", then $g_i = s_{i(j)} = 0$ for all $i$ and $j$. The means of all observations using the $ij$ pairs of generator and seed, $\bar{y}_{ij\bullet}$, should exhibit no systematic variation. The sample variance of the $\bar{y}_{ij\bullet}$ should be consistent with the means of appropriate size from a distribution with the variance of the underlying error, $e_{ij(k)}$. This can be formally checked by the standard statistical techniques of analysis of variance. (The standard statistical techniques rely on an assumption of normality of the $e_{ij(k)}$. If this is not reasonable, nonparametric procedures could be used in place of the standard methods.)

### Quasirandom Sampling

Our discussion of variance in Monte Carlo methods that are based on pseudorandom numbers follows the pretense that the numbers are realizations of random variables, and the main concern in pseudorandom number generation is the simulation of a sequence of i.i.d. random variables. In quasirandom number generation, the attempt is to get a sample that is spread out over the sample space more evenly than could be expected from a random sample. Monte Carlo methods based on quasirandom numbers, or "quasi-Monte Carlo" methods, do not admit discussion of variance in the technical sense.

The approximation of an integral such as in equation (7.1) using quasirandom numbers, however, involves a quantity (in one dimension) of the same form as that of an estimate using pseudorandom numbers, equation (7.2). Although the approximation of the integral does not have a variance, how good the approximation is depends on a quantity similar to the variance of an estimator based on random variables, equation (7.3). The integral in equation (7.3) is a measure of the roughness of the integrand and thus is an appropriate measure of how good the quasirandom approximation (or, indeed, any similar approximation) is. The roughness can be approximated in the same way that the variance is estimated; that is, by a quantity of the form (7.8).

Sometimes, it may be desirable to treat the results of a quasi-Monte Carlo experiment as if they had variances. This may be to form confidence intervals (rather than asymptotic error bounds). One approach to this is to use a hybrid generator (that is, a combination of a quasirandom generator and a pseudorandom generator). Braaten and Weller (1979) describe a hybrid method in which a pseudorandom generator is used to scramble a quasirandom sequence. Ökten (1998) suggests use of random samples from quasirandom sequences in order to estimate a confidence interval. Owen (1995, 1997) discusses how hybrid methods can actually improve both the variance of a pseudorandom method and the precision of a quasirandom method.

In Section 7.5 we discuss ways of reducing the variance in Monte Carlo estimation using pseudorandom numbers. One of these general methods is stratification. In some uses of stratification, the objective is to ensure that the observations are spread somewhat evenly over the sample space. We should recall that this objective is also the basic motivation for use of quasirandom sampling.

## 7.5 Variance Reduction

An objective in sampling is to reduce the variance of the estimators while preserving other good qualities, such as unbiasedness. Variance reduction results in statistically efficient estimators. The emphasis on efficient Monte Carlo sampling goes back to the early days of digital computing (Kahn and Marshall, 1953), but the issues are just as important today (or tomorrow) because, presumably, we are solving bigger problems. The general techniques used in statistical sampling apply to Monte Carlo sampling, and there is a mature theory for sampling designs that yield efficient estimators (see, for example, Särndal, Swensson, and Wretman, 1992).

Except for straightforward analytic reduction, discussed in the next section, techniques for reducing the variance of a Monte Carlo estimator are called "swindles" (especially if they are thought to be particularly clever). The common thread in variance reduction is to use additional information about the problem in order to reduce the effect of random sampling on the variance of the observations. This is one of the fundamental principles of all statistical design.

## 7.5.1 Analytic Reduction

The first principle in estimation is to use any known quantity to improve the estimate. For example, suppose that the problem is to evaluate the integral

$$\theta = \int_D f(x) \, dx$$

by Monte Carlo methods. Now, suppose that $D_1$ and $D_2$ are such that $D_1 \cup D_2 = D$ and $D_1 \cap D_2 = \emptyset$, and consider the representation of the integral

$$\theta = \int_{D_1} f(x)\,dx + \int_{D_2} f(x)\,dx$$
$$= \theta_1 + \theta_2.$$

Now, suppose that a part of this decomposition of the original problem is known (that is, suppose that we know $\theta_1$). It is very likely that it would be better to use Monte Carlo methods only to estimate $\theta_2$ and take as our estimate of $\theta$ the sum of the known $\theta_1$ and the estimated value of $\theta_2$. This seems intuitively obvious, and it is generally true unless there is some relationship between $f(x_1)$ and $f(x_2)$, where $x_1$ is in $D_1$ and $x_2$ is in $D_2$. If there is some known relationship, however, it may be possible to improve the estimate $\widehat{\theta}_2$ of $\theta_2$ by using a transformation of the same random numbers used for $\widehat{\theta}_1$ to estimate $\theta_1$. For example, if $\widehat{\theta}_1$ is larger than the known value of $\theta_1$, the proportionality of the overestimate, $(\widehat{\theta}_1 - \theta_1)/\theta_1$, may be used to adjust $\widehat{\theta}_2$. This is the same principle as ratio or regression estimation in ordinary sampling theory.

Now, consider a different representation of the integral, in which $f$ is expressed as $g + h$, where $g$ and $h$ have the same signs. We have

$$\theta = \int_D (g(x) + h(x)) \, dx$$
$$= \int_D g(x)\,dx + \int_D h(x)\,dx$$
$$= \theta_3 + \theta_4,$$

and suppose that a part of this decomposition, say $\theta_3$, is known. In this case, the use of the known value of $\int_D g(x)\,dx$ is likely to help only if $g(x)$ tends to vary similarly with $f(x)$. In this case, it would be better to use Monte Carlo methods only to estimate $\theta_4$ and take as our estimate of $\theta$ the sum of the known $\theta_3$ and the estimated value of $\theta_4$. This is because $|h(x)|$ is less rough than $|f(x)|$. Also, as in the case above, if there is some known relationship between $g(x)$ and

$h(x)$, such as one tends to decrease as the other increases, it may be possible to use the negative correlation of the individual estimates to reduce the variance of the overall estimate.

## 7.5.2 Stratified Sampling and Importance Sampling

In *stratified sampling*, certain proportions of the total sample are taken from specified regions (or "strata") of the sample space. The objective in stratified sampling may be to ensure that all regions are covered. Another objective is to reduce the overall variance of the estimator by sampling more heavily where the function is rough (that is, where the values $f(x_i)$ are likely to exhibit a lot of variability) so as

Stratified sampling is usually performed by forming distinct subregions with different importance functions in each. This is the same idea as in analytic reduction except that Monte Carlo sampling is used in each region.

Stratified sampling is based on exactly the same principle in sampling methods in which the allocation is proportional to the variance (see Särndal, Swensson, and Wretman, 1992). In some of the literature on Monte Carlo methods, stratified sampling is called "geometric splitting".

In *importance sampling*, just as may be the case in stratified sampling, regions corresponding to large values of the integrand are sampled more heavily. In importance sampling, however, instead of a finite number of regions, we allow the relative sampling density to change continuously. This is accomplished by careful choice of $w$ in the decomposition implied by equation (7.4) on page 232. We have

$$\theta = \int_D f(x)\,dx$$
$$= \int_D \frac{f(x)}{p(x)} p(x)\,dx, \tag{7.10}$$

where $p(x)$ is a probability density over $D$. The density $p(x)$ is called the *importance function*. Stratified sampling can be thought of as importance sampling in which the importance function is composed of a mixture of densities. In some of the literature on Monte Carlo methods, stratified sampling and importance sampling are said to use "weight windows".

From a sample of size $m$ from the distribution with density $p$, we have the estimator,

$$\widehat{\theta} = \frac{1}{m} \sum \frac{f(x_i)}{p(x_i)}. \tag{7.11}$$

Generating the random variates from the distribution with density $p$ weights the sampling into regions of higher probability with respect to $p$. By judicious choice of $p$, we can reduce the variance of the estimator.

The variance of the estimator is

$$V(\widehat{\theta}) = \frac{1}{m} V\left(\frac{f(X)}{p(X)}\right),$$

where the variance is taken with respect to the distribution of the random variable $X$ with density $p(x)$. Now,

$$V\left(\frac{f(X)}{p(X)}\right) = E\left(\frac{f^2(X)}{p^2(X)}\right) - \left(E\left(\frac{f(X)}{p(X)}\right)\right)^2.$$

The objective in importance sampling is to choose $p$ so that this variance is minimized. Because

$$\left(E\left(\frac{f(X)}{p(X)}\right)\right)^2 = \left(\int_D f(x)\,dx\right)^2,$$

the choice involves only the first term in the expression for the variance. By Jensen's inequality, we have a lower bound on that term:

$$E\left(\frac{f^2(X)}{p^2(X)}\right) \geq \left(E\left(\frac{|f(X)|}{p(X)}\right)\right)^2$$

$$= \left(\int_D |f(x)|\,dx\right)^2.$$

That bound is obviously achieved when

$$p(x) = \frac{|f(x)|}{\int_D |f(x)|\,dx}.$$

Of course, if we knew $\int_D |f(x)|\,dx$, we would probably know $\int_D f(x)\,dx$ and would not even be considering the Monte Carlo procedure to estimate the integral. In practice, for importance sampling we would seek a probability density $p$ that is nearly proportional to $|f|$; that is, such that $|f(x)|/p(x)$ is nearly constant.

The problem of choosing an importance function is very similar to the problem of choosing a majorizing function for the acceptance/rejection method, as we discussed in Sections 4.5 and 4.6. Selection of an importance function involves the principles of function approximation with the added constraint that the approximating function be a probability density from which it is easy to generate random variates.

Let us now consider another way of developing the estimator (7.11). Let $h(x) = f(x)/p(x)$ (where $p(x)$ is positive; otherwise, let $h(x) = 0$) and generate $y_1,\ldots,y_m$ from a density $g(y)$ with support $D$. Compute *importance weights*,

$$w_i = p(y_i)/g(y_i), \tag{7.12}$$

and form the estimate of the integral as

$$\hat{\theta} = \frac{1}{m}\frac{\sum w_i h(y_i)}{\sum w_i}. \tag{7.13}$$

In this form of the estimator, $g(y)$ is a trial density, just as in the acceptance/rejection methods of Section 4.5. This form of the estimator has similarities to the weighted resampling method discussed in Section 4.12 on page 149.

By the same reasoning as above, we see that the trial density should be "close" to $f$; that is, optimally, $g(x) = c|f(x)|$ for some constant $c$.

Although the variance of the estimator in equations (7.11) and (7.13) may appear rather simple, the term $E((f(X)/p(X))^2)$ could be quite large if $p$ (or $g$) becomes small at some point where $f$ is large. Of course, the objective in importance sampling is precisely to prevent that, but if the functions are not well-understood, it may happen. An element of the Monte Carlo sample at a point where $p$ is small and $f$ is large has an unduly large influence on the overall estimate. Because of this kind of possibility, importance sampling must be used with some care.

We should note that the weights $w_i$ do not sum to 1. Hesterberg (1995) points out a number of situations in which normalized weights would improve the estimation procedure. (If the weights do not sum to 1, the estimator is not equivariant with respect to addition of a constant.) He describes various ways of normalizing the weights by assuming that the weights themselves are covariates in a sampling design. One method is the regression estimate adjustment, in which the weights in equation (7.12) are adjusted as

$$\tilde{w}_i = w_i(1 + b(w_i - \bar{w})/m),$$

where $\bar{w}$ is the sample mean of the $w_i$ and $b$ is the least squares slope adjustment, $m(1 - \bar{w})/(\sum(w_i - \bar{w})^2)$. Although this adjustment results in a biased estimator (because of the random variables in the denominators), Hesterberg reports favorable results for the regression-based weights except in cases involving very small probabilities.

If the integrand $f$ is multimodal, it is very likely that the best importance function would be a mixture of densities. Oh and Berger (1993) suggested use of a mixture of Student's $t$ distributions and described an adaptive method for selecting the mixture that works well in certain cases.

Importance sampling is similar to hit-or-miss Monte Carlo sampling (see Exercise 7.2, page 271), and the relationship is particularly apparent when the weighting function is sampled by acceptance/rejection. (See Caflisch and Moskowitz, 1995, for some variations of the acceptance/rejection method for this application. Their variations allow a weighting of the acceptance/rejection decision, called smoothed acceptance/rejection, that appears to be particularly useful if the sampling is done from a quasirandom sequence as discussed in Chapter 3.)

In higher dimensions, as we have seen in Section 4.14 and in Exercise 4.4f, because the content of geometric regions far from the origin (which would be in the rejection area) is large relative to the content near the origin, acceptance/rejection and hit-or-miss Monte Carlo methods become inefficient. Importance sampling can help to alleviate these problems by decreasing the content of the outer regions. On the other hand, in higher dimensions, the problems of very large ratios mentioned above are exacerbated, and importance sampling must be used with extreme caution.

## Sequential Importance Sampling

Direct application of importance sampling in sequential Monte Carlo methods (Section 7.2 on page 233) would involve identification of a density $g_t(x_{0:t})$ and formation of importance weights of the form

$$w_t(x_{0:t}) = \frac{p_t(x_{0:t})}{g_t(x_{0:t})}, \tag{7.14}$$

where we are using a simpler notation for the posterior density of interest, $p_{X_{0:t}}(x_{0:t})$.

For the filtering problem, we would similarly identify a density $g(x_{0:t}|y_{1:t})$ and form importance weights

$$w_t(x_{0:t}) = \frac{p(x_{0:t}|y_{1:t})}{g(x_{0:t}|y_{1:t})}.$$

If the density $g_t(x_{0:t})$ is close to the density $p_t(x_{0:t})$, we may be able to represent the weights in equation (7.14) recursively as

$$w_t(x_{0:t}) = w_{t-1}(x_{0:t-1}) \frac{p_t(x_{0:t-1})p_{t|t-1}(x_t|x_{0:t-1})}{p_{t-1}(x_{0:t-1})g_t(x_t|x_{0:t-1})}.$$

In computations, at the $i^{\text{th}}$ trajectory in a simulation of $m$ trajectories, it may be more efficient to work with normalized weights,

$$\tilde{w}_t^{(i)} = \frac{w^{(i)}(x_{0:t})}{\sum_{j=1}^{m} w^{(j)}(x_{0:t})}.$$

In the filtering problem, these normalized weights can be computed recursively in the trajectory, up to constants of proportionality, by

$$\tilde{w}_t^{(i)} \propto \tilde{w}_{t-1}^{(i)} \frac{p(y_t|x_t^{(i)})p(x_t^{(i)}|x_{t-1}^{(i)})}{g(x_t^{(i)}|x_{0:t-1}^{(i)}, y_{1:t})}.$$

Unfortunately, these importance weights become unstable in sequential Monte Carlo sampling. The process has an increasing variance. Two approaches to dealing with the problem are the resampling method of Gordon, Salmond, and Smith (1993) and the rejection control method of Liu, Chen, and Wong (1998). We refer the reader to those papers for the details of the methods.

## 7.5.3 Use of Covariates

Another way of reducing the variance, just as in ordinary sampling, is to use covariates. Any variable that is correlated with the variable of interest has potential value in reducing the variance of the estimator. Such a variable is useful if it is easy to generate and if it has properties that are known or that can be computed easily. In the general case in Monte Carlo sampling, covariates are called control variates. Two special cases are called antithetic variates and common variates. We first describe the general case, and then the two special cases. We then relate the use of covariates to the statistical method sometimes called "Rao-Blackwellization".

### Control Variates

Suppose that $Y$ is a random variable, and the Monte Carlo method involves estimation of $E(Y)$. Suppose that $X$ is a random variable with known expectation, $E(X)$, and consider the random variable

$$\widetilde{Y} = Y - b(X - E(X)). \tag{7.15}$$

The expectation of $\widetilde{Y}$ is the same as that of $Y$, and its variance is

$$V(\widetilde{Y}) = V(Y) - 2b\text{Cov}(Y, X) + b^2 V(X).$$

For reducing the variance, the optimal value of $b$ is $\text{Cov}(Y, X)/V(X)$. With this choice $V(\widetilde{Y}) < V(Y)$ as long as $\text{Cov}(Y, X) \neq 0$. Even if $\text{Cov}(Y, X)$ is not known, there is a $b$ that depends only on the sign of $\text{Cov}(Y, X)$ for which the variance of $\widetilde{Y}$ is less than the variance of $Y$.

The variable $X$ is called a *control variate*. This method has long been used in survey sampling, where $\widetilde{Y}$ in equation (7.15) is called a regression estimator.

Use of these facts in Monte Carlo methods requires identification of a control variable $X$ that can be simulated simultaneously with $Y$. If the properties of $X$ are not known but can be estimated (by Monte Carlo methods), the use of $X$ as a control variate can still reduce the variance of the estimator.

These ideas can obviously be extended to more than one control variate:

$$\widetilde{Y} = Y - b_1(X_1 - E(X_1)) - \cdots - b_k(X_k - E(X_k)).$$

The optimal values of the $b_i$s depend on the full variance-covariance matrix. The usual regression estimates for the coefficients can be used if the variance-covariance matrix is not known.

Identification of appropriate control variates often requires some ingenuity. In some special cases, there may be techniques that are almost always applicable. In the case of Monte Carlo estimation of significance levels or quantiles, for example, Hesterberg and Nelson (1998) describe some methods that are easy to apply.

**Antithetic Variates**

Again consider the problem of estimating the integral

$$\theta = \int_a^b f(x)\,dx$$

by Monte Carlo methods. The standard crude Monte Carlo estimator, equation (7.2), is $(b-a)\sum f(x_i)/n$, where $x_i$ is uniform over $(a,b)$. It would seem intuitively plausible that our estimate would be subject to less sampling variability if, for each $x_i$, we used its "mirror"

$$\tilde{x}_i = a + (b - x_i).$$

This mirror value is called an antithetic variate, and use of antithetic variates can be effective in reducing the variance of the Monte Carlo estimate, especially if the integral is nearly uniform. For a sample of size $n$, the estimator is

$$\frac{b-a}{n}\sum_{i=1}^{\frac{n}{2}}(f(x_i) + f(\tilde{x}_i)).$$

The variance of the sum is the sum of the variances plus twice the covariance. Antithetic variates have negative covariances, thus reducing the variance of the sum.

Antithetic variates from distributions other than the uniform can also be formed. The linear transformation that works for uniform antithetic variates cannot be used, however. A simple way of obtaining negatively correlated variates from other distributions is just to use antithetic uniforms in the inverse CDF. Schmeiser and Kachitvichyanukul (1990) discuss other ways of doing this. If the variates are generated using acceptance/rejection, for example, antithetic variates can be used in the majorizing distribution.

**Common Variates**

Often, in Monte Carlo simulation, the objective is to estimate the differences in parameters of two random processes. The two parameters are likely to be positively correlated. If that is the case, then the variance in the individual differences is likely to be smaller than the variance of the difference of the overall estimates.

Suppose, for example, that we have two statistics, $T$ and $S$, that are unbiased estimators of some parameter of a given distribution. We would like to know the difference in the variances of these estimators,

$$V(T) - V(S)$$

(because the one with the smaller variance is better). We assume that each statistic is a function of a random sample: $\{x_1, \ldots, x_n\}$. A Monte Carlo estimate of the variance of the statistic $T$ for a sample of size $n$ is obtained by generating $m$ samples of size $n$ from the given distribution, computing $T_i$ for the $i^{\text{th}}$ sample, and then computing

$$\widehat{V}(T) = \frac{\sum_{i=1}^{m}(T_i - \bar{T})^2}{m-1}.$$

Rather than doing this for $T$ and $S$ separately, using the unbiasedness, we could first observe

$$
\begin{aligned}
V(T) - V(U) &= E(T^2) - E(S^2) \\
&= E(T^2 - S^2)
\end{aligned}
\tag{7.16}
$$

and hence estimate the latter quantity. Because the estimators are likely to be positively correlated, the variance of the Monte Carlo estimator $\widehat{E}(T^2 - S^2)$ is likely to be smaller than the variance of $\widehat{V}(T) - \widehat{V}(S)$. If we compute $T^2 - S^2$ from each sample (that is, if we use *common variates*), we are likely to have a more precise estimate of the difference in the variances of the two estimators, $T$ and $S$.

**Rao-Blackwellization**

As in the discussion of control variates above, suppose that we have two random variables $Y$ and $X$ and we want to estimate $E(f(Y, X))$ with an estimator of the form $T = \sum f(Y_i, X_i)/m$. Now suppose that we can evaluate $E(f(Y, X)|X = x)$. (This is similar to what is done in using equation (7.15) above.) Now, $E(E(f(Y, X)|X = x)) = E(f(Y, X))$, so the estimator

$$\widetilde{T} = \sum E(f(Y_i, X)|X = x_i)/m$$

has the same expectation as $T$. However, we have

$$V(f(Y, X)) = V(E(f(Y, X)|X = x)) + E(V(f(Y, X)|X = x));$$

that is,

$$V(f(Y, X)) \geq V(E(f(Y, X)|X = x)).$$

Therefore, $\widetilde{T}$ is preferable to $T$ because it has the same expectation but no larger variance. (The function $f$ may depend on $Y$ only. In that case, if $Y$ and $X$ are independent we can gain nothing.)

The principle of minimum variance unbiased estimation leads us to consider statistics such as $\widetilde{T}$ conditioned on other statistics. The Rao-Blackwell Theorem (see any text on mathematical statistics) tells us that if a sufficient statistic exists, the greatest improvement in variance while still requiring unbiasedness occurs when the conditioning is done with respect to a sufficient statistic. This process of conditioning a given estimator on another statistic is called Rao-Blackwellization. (This name is often used even if the conditioning statistics is not sufficient.)

## 7.5.4  Constrained Sampling

Sometimes, in Monte Carlo methods it is desirable that certain sample statistics match the population parameters exactly; for example, the sample mean may be adjusted by transforming each observation in the Monte Carlo sample by

$$\tilde{x}_i = x_i + \mu - \bar{x},$$

where $\mu$ is the mean of the target population, and $\bar{x}$ is the mean of the original Monte Carlo sample.

This idea can be extended to more than one sample statistic but requires more algebra to match up several statistics with the corresponding parameters. Pullin (1979) describes the obvious Helmert transformations for univariate normal deviates so that the sample has a specified mean and variance, and Cheng (1985) gives an extension for the multivariate normal distribution so that the samples generated have a specified mean and variance-covariance matrix. Cheng (1984) gives a method for generating a sample of inverse Gaussian variates with specified mean and variance and discusses applications of this kind of constrained sampling in variance reduction.

Variance estimators that result from constrained samples must be used with care. This is because the constraints change the sampling variability, usually by reducing it. In using a finite mixture distribution, for example, the variance of estimators of first-order parameters, such as means, can be reduced by generating data with the exact proportions of the mixture. If the purpose is to estimate the variance or the mean squared error of the estimators, however, sampling under such constraints biases the estimators of the variance.

## 7.5.5  Stratification in Higher Dimensions: Latin Hypercube Sampling

The techniques of sampling theory are generally designed for reducing sampling variance for single variables, often using one or just a few covariates. The statistical developments in the design of experiments provide a more powerful set of tools for reducing the variance in cases where several factors are to be investigated simultaneously. Such techniques as balanced or partially balanced fractional factorial designs allow the study of a large number of factors while keeping the total experiment size manageably small. Some processes are so complex that even with efficient statistical designs, experiments to study the process would involve a prohibitively large number of factors and levels. For some processes, it may not be possible to apply the treatments whose effects are to be studied, and data are available only from observational studies. The various processes determining weather are examples of phenomena that are not amenable to traditional experimental study. Such processes can often be modeled and studied by computer experiments.

There are some special concerns in experimentation using the computer (see Sacks et al., 1989), but the issues of statistical efficiency remain. Rather than a model involving ordinary experimental units, a computer experimental model receives a fixed input and produces a deterministic output. An objective in computer experimentation (just as in any experimentation) is to provide a set of inputs that effectively (or randomly) spans a space of interest. McKay, Conover, and Beckman (1979) introduce a technique called Latin hypercube sampling (as a generalization of the ideas of a Latin square design) for providing input to a computer experiment.

If each of $k$ factors in an experiment is associated with a random input that is initially a U(0, 1) variate, a sample of size $n$ that efficiently covers the factor space can be formed by selecting the $i^{\text{th}}$ realization of the $j^{\text{th}}$ factor as

$$v_j = \frac{\pi_j(i) - 1}{n} + \frac{u_j}{n},$$

where

- $\pi_1(\cdot), \ldots, \pi_k(\cdot)$ are permutations of the integers $1, \ldots, n$, sampled randomly, independently, and with replacement from the set of $n!$ possible permutations; and $\pi_j(i)$ is the $i^{\text{th}}$ element of the $j^{\text{th}}$ permutation.

- The $u_j$ are sampled independently from U(0, 1).

It is easy to see that the realizations of $v_j$ for the $j^{\text{th}}$ factor are also independent realizations from a U(0, 1) distribution. We can see heuristically that such numbers tend to be "spread out" over the space. This is because there is a separation of at least $1/n$ in the terms $(\pi_j(i) - 1)/n$. The $i^{\text{th}}$ realization of all factors is constrained in a manner reminiscent of quasirandom sequences, but with a much more regular pattern.

Expressing the Monte Carlo problem as a multidimensional integral and considering residuals from lower-dimensional integrals, Stein (1987) shows that the variance of Monte Carlo estimators using Latin hypercube sampling is asymptotically smaller than the variance of Monte Carlo estimators using unrestricted random sampling. Stein's approach was to decompose the variance of the estimator of the integral into a mean, main effects of the integrated residuals of the lower-dimensional integrals, and an additional residual.

Beckman and McKay (1987) and Tang (1993) provide empirical evidence that its performance is superior to simple random sampling. Owen (1992a, 1992b, 1994a), Tang (1993), and Avramidis and Wilson (1995) derive various properties of estimators based on Latin hypercube samples.

Using Stein's decomposition of the variance, Owen (1994b) showed how the variance could be reduced by controlling correlations in Latin hypercube samples. Use of Latin hypercube sampling is particularly useful in higher dimensions (Owen, 1998).

# 7.6 The Distribution of a Simulated Statistic

In some Monte Carlo applications, the objective is to estimate some parameter of a distribution. The parameter may be a simple scalar, such as a mean or

a variance, or it may be a vector or matrix parameter, perhaps corresponding to moments of the underlying random variables. These applications are all
essentially equivalent to estimating an integral, as in Section 7.1. A more complicated problem is the estimation of the distribution of some random variable,
which may be some statistic that is a function of the simulated input random
variables. The estimate for the distribution is a function. If the random variable of interest is discrete, the estimate is essentially a histogram corresponding
to the probabilities of the points in its support. If the random variable of interest has a continuous distribution, the objective is to estimate the probability
density function.

There are various ways of estimating a continuous probability density. In
one type of approach, we assume some parametric family and estimate the parameters that characterize the distribution. Other approaches are nonparametric.
They may be based on histograms or variations of histograms, kernel functions,
or orthogonal series expansions (see Gentle, 2002, Chapters 8 and 9).

Graphs give very useful pictures of the location, spread, and shape of the
distribution. To return to the simple example mentioned at the beginning
of this chapter, suppose that we are interested in the midrange of the gamma
distribution. We know, of course, that the distribution of the midrange depends
on the parameters of the gamma distribution and on the sample size. We can
represent the density of the midrange analytically, but it depends on some rather
complicated integrals. Alternatively, for a given combination of parameters and
sample size, we can generate a sample of the midrange computed from each of
a number of samples of the given size from the gamma distribution. In Monte
Carlo applications such as this example, we often speak of two samples: the
sample from the underlying distribution of interest (in this case, the gamma)
and the sample of the statistic of interest, called the Monte Carlo sample (in
this case, the sample of midranges).

Figure 7.1, for example, shows the results of a small Monte Carlo study of
the distribution of the midrange of samples from a gamma(2,2) distribution.
From the histograms, we see how the distribution of a linear combination of
central order statistics is somewhat skewed for a small sample but is more like
a normal distribution for a larger sample.

## 7.7  Computational Statistics

The field of computational statistics includes computationally intensive statistical methods and the supporting theory. Many of these methods involve the use
of Monte Carlo methods, in some cases methods that simulate a hypothesis and
in other cases methods that involve resampling from a given sample or multiple
partitioning of a given sample.

Some of the methods for dealing with missing data, for example, are computationally intensive and may involve Monte Carlo resampling. An extensive
discussion of resampling methods for dealing with missing data is given by

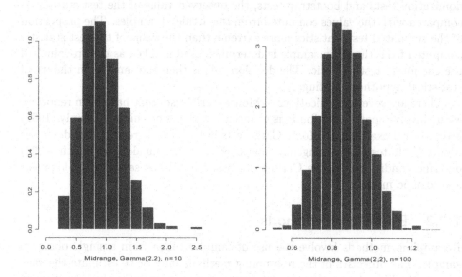

Figure 7.1: Histograms of Simulated Midranges from Gamma Distributions

Schafer (1997). The books by Robert and Casella (1999) and by Gentle (2002) describe various uses of Monte Carlo methods in statistical data analysis.

## 7.7.1   Monte Carlo Methods for Inference

Barnard (1963), in a discussion of a paper read before the Royal Statistical Society, suggested that a test statistic be evaluated for simulated random samples to determine the significance level of the test statistic computed from a given dataset. This kind of procedure is called a Monte Carlo test. In Barnard's Monte Carlo test, the observed value of the test statistic is compared with the values of the test statistic computed in each of $m$ simulated samples of the same size as the given sample. For a fixed significance level, $\alpha$, Barnard suggested fixing $r$ so that $r/(m+1) \approx \alpha$ and then basing the test decision on whether fewer than $r$ of the simulated values of the test statistics exceeded the observed value of the test statistic. Hope (1968) and Marriott (1979) studied the power of the test and found that the power of Monte Carlo tests can be quite good even for $r$ as small as 5, which would require only a small number of simulated samples, often of the order of 100 (which would correspond to a test of level approximately 0.05 if $r = 5$).

Monte Carlo tests are used when the distribution of the test statistic is not known. To use a Monte Carlo test, however, the distribution of the random component in an assumed model (usually the "error" term in the model) must be known. In a Monte Carlo test, new, artificial samples are generated using the distribution of the random component in the model. For each Monte Carlo sample thus generated, the value of the test statistic is computed. As in randomization tests and bootstrap tests, the observed value of the test statistic is compared with the values computed from the artificial samples. The proportion of the simulated test statistics more extreme than the value of the test statistic computed from the given sample is determined and is taken as the "p-value" of the computed test statistic. The decision rule is then the same as in the usual statistical hypothesis testing.

There are several applications of Monte Carlo tests that have been reported. Many involve spatial distributions of species of plants or animals. Manly (1997) gives several examples of Monte Carlo tests in biology. Agresti (1992) describes Monte Carlo tests in contingency tables. Forster, McDonald, and Smith (1996) describe conditional Monte Carlo tests based on Gibbs sampling in loglinear and logistic models.

## 7.7.2   Bootstrap Methods

Resampling methods involve the use of samples taken from a single observed sample. The objective of the resampling methods may be to estimate the variance or the bias of an estimator. Of course, if we can estimate the bias, we may be able to correct for it to give us an unbiased estimator. Resampling can be used to estimate the significance level of a test statistic or to form confidence intervals for a parameter. The methods can be used when very little is known about the underlying distribution.

One form of resampling, the "jackknife" and its generalizations, involves forming subsamples without replacement from the observed sample. Although the jackknife has more general applications, one motivation for its use is that in certain situations it can reduce the bias of an estimator.

### The Bootstrap Rationale

Another form of resampling, the nonparametric "bootstrap", involves forming samples, with replacement, from the observed sample. The basic step in the nonparametric bootstrap is, because the observed sample contains all of the available information about the underlying population, to consider the *observed sample* to be the mass points of a *discrete population* that represents the population from which the original sample came. The distribution of any relevant test statistic can be simulated by taking random samples from the "population" consisting of the original sample.

Suppose that a sample $S = \{x_1, \ldots, x_n\}$ is to be used to estimate a population parameter, $\theta$. We form a statistic $T$ that estimates $\theta$. We wish to know the sampling distribution of $T$ to set confidence intervals for our estimate of $\theta$. The sampling distribution of $T$ is often intractable in applications of interest.

The basic idea of the bootstrap is that the true population can be approximated by an infinite population in which each of the $n$ sample points are equally likely. Using this approximation of the true population, we can approximate distributions of statistics formed from the sample. The basic tool is the empirical distribution function, which is the distribution function of the finite population that is to be used as an approximation of the underlying population of interest. We denote the empirical cumulative distribution function based on a sample of size $n$ as $P_n(\cdot)$. It is defined as

$$P_n(x) = \frac{1}{n} \sum_{i=1}^{n} I_{(-\infty, x]}(x_i),$$

where the indicator function $I_S(x)$ is defined by

$$
\begin{aligned}
I_S(x) &= 1 \quad \text{if } x \in S; \\
&= 0 \quad \text{otherwise.}
\end{aligned}
$$

The parameter is a functional of a population distribution function:

$$\theta = \int g(x) \, dP(x).$$

The estimator is often the same functional of the empirical distribution function:

$$T = \int g(x) \, dP_n(x).$$

Various properties of the distribution of $T$ can be estimated by use of "bootstrap samples", each of the form $S^* = \{x_1^*, \ldots, x_n^*\}$, where the $x_i^*$s are chosen from the original $x_i$s in $S$ with replacement.

**Nonparametric Bootstrap Bias Estimation**

The problem in its broadest setting is to estimate some function $h(\cdot)$ of the expected value of some function $g(\cdot)$ of the random variable; that is, to estimate $h(E(g(X)))$. The approach is to find a functional $f_T$ (from some class of functionals) that allows us to relate the distribution function of the sample $P_n$ to the population distribution function $P$; that is, such that

$$E(f_T(P, P_n) \mid P) = 0.$$

We then wish to estimate $h(\int g(x) \, dP(x))$.

For example, suppose that we wish to estimate

$$\theta = \left( \int x \, dP(x) \right)^r.$$

The estimator

$$T = \left( \int x \, dP_n(x) \right)^r = \bar{x}^r$$

is biased. Determining the bias is equivalent to finding $b$ that solves the equation

$$f_T(P, P_n) = T(P_n) - \theta(P) + b$$

so that $f_T$ has zero expectation with respect to $P$.

The *bootstrap principle* suggests repeating this whole process. We now take a sample from the "population" with distribution function $P_n$. We look for $f_T^{(1)}$ so that

$$\mathrm{E}\left(f_T^{(1)}(P_n, P_n^{(1)}) \mid P_n\right) = 0,$$

where $P_n^{(1)}$ is the empirical distribution function for a sample from the discrete distribution formed from the original sample.

The difference between this and the original problem is that we know more about this equation because we know more about $P_n$. Our knowledge of $P_n$ comes either from parametric assumptions about the underlying distribution or from just working with the discrete distribution in which the original sample is given equal probabilities at each mass point.

Using the sample, we have

$$\mathrm{E}(T(P_n^{(1)}) - T(P_n) + b_1 \mid P_n) = 0$$

or

$$b_1 = T(P_n) - \mathrm{E}(T(P_n^{(1)}) \mid P_n).$$

An estimate with less bias is therefore

$$\tilde{T}_1 = 2T(P_n) - \mathrm{E}(T(P_n^{(1)}) \mid P_n).$$

We may be able to compute $\mathrm{E}(T(P_n^{(1)}) \mid P_n)$, but generally we must resort to Monte Carlo methods to estimate it. The Monte Carlo estimate is based on $m$ random samples, each of size $n$, taken with replacement from the original sample.

The basic nonparametric bootstrap procedure is to take $m$ random samples $S_j^*$ each of size $n$ and *with replacement* from the given set of data (that is, the original sample $S = \{x_1, \ldots, x_n\}$) and for each sample compute an estimate $T_j^*$ of the same functional form as the original estimator $T$. The distribution of the $T_j^*$s is related to the distribution of $T$. The variability of $T$ about $\theta$ can be assessed by the variability of $T_j^*$ about $T$. The bias of $T$ can be assessed by the mean of $T_j^* - T$.

## Parametric Bootstrap

In the parametric bootstrap, the distribution function of the population of interest, $F$, is assumed known up to a finite set of unknown parameters, $\lambda$. The estimate of $P$, $\widehat{P}$, instead of $P_n$, is $P$ with $\lambda$ replaced by its sample estimates (of some kind). Likewise, $\widehat{P}^{(1)}$ is formed from $\widehat{P}$ by using estimates that are the same function of a sample from a population with distribution function $\widehat{P}$.

This population is more like the original assumed population, and the sample is not just drawn with replacement from the original sample as is done in the nonparametric bootstrap.

In a parametric bootstrap procedure, the first step is to obtain estimates of the parameters that characterize the distribution within the assumed family. After this, the procedure is very similar to that described above: generate $m$ random samples, each of size $n$, from the estimated distribution, and for each sample, compute an estimate $T_j$ of the same functional form as the original estimator $T$. The distribution of the $T_j$s is used to make inferences about the distribution of $T$.

Extensive general discussions of the bootstrap are available in Efron and Tibshirani (1993), Shao and Tu (1995), Davison and Hinkley (1997), and Chernick (1999).

## 7.7.3 Evaluating a Posterior Distribution

In the Bayesian approach to data analysis, the parameter in the probability model for the data is taken to be a random variable. In this approach, the model is a prior distribution of the parameter together with a conditional distribution of the data, given the parameter. Instead of statistical inference about the parameters being based on the estimation of the distribution of a statistic, Bayesian inference is based on the estimation of the conditional distribution of the parameter, given the data. This conditional distribution of the parameter is called the posterior. Rather than the data being reduced to a statistic whose distribution is of interest, the data are used to form the conditional distribution of the parameter.

The prior distribution of the parameter, of course, has a parameter, so the Bayesian analysis must take into account this "hyperparameter". (When we speak of a "parameter", we refer to an object that may be a scalar, a vector, or even some infinite-dimensional object such as a function. Generally, however, "parameter" refers to a real scalar or vector.) The hyperparameter for the parameter leads to a hierarchical model with the phenomenon of interest, modeled by the "data" random variable, having a probability distribution conditional on a parameter, which has a probability distribution conditional on another parameter, and so on. With an arbitrary starting point for the model, we may have the following random variables:

hyperprior parameter: $\Psi$;

prior parameter: $\Phi$;

data model parameter: $\Theta$;

data: $Y$.

The probability distributions of these random variables (starting with $\Phi$) are:

prior parameter, $\phi$, with hyperprior distribution $p_\Phi(\cdot)$;

data model parameter, $\theta$, with prior distribution $p_{\Theta|\phi}(\cdot)$;

data, $y$, with distribution $p_{Y|\theta,\phi}(\cdot)$.

The distributions of interest are the conditional distributions, given the data: the posterior distributions. These densities are obtained by forming the joint distributions using the hierarchical conditionals and marginals above and forming the marginals by integration. Except for specially chosen prior distributions, however, this integration is generally difficult. Moreover, in many interesting models, the integrals are multidimensional; hence, Monte Carlo methods are the tools of choice. The Markov chain Monte Carlo methods discussed in Sections 4.10 and 4.14 are used to generate the variates.

Instead of directly performing an integration, the analysis is performed by generating a sample of the parameter of interest, given the data. This sample is used for inference about the posterior distribution (see Gelfand and Smith, 1990, who used a Gibbs sampler).

Geweke (1991b) discusses general Monte Carlo methods for evaluating the integrals that arise in Bayesian analysis. Carlin and Louis (1996) and Gelman et al. (1995) provide extensive discussions of the methods of the data analysis and the associated computations. Also, Smith and Roberts (1993) give an overview of the applications of Gibbs sampling and other MCMC methods in Bayesian computations.

The computer program BUGS ("**B**ayesian inference **U**sing **G**ibbs **S**ampling") allows the user to specify a hierarchical model and then evaluates the posterior by use of Gibbs sampling; see Gilks, Thomas, and Spiegelhalter (1992, 1994) or Thomas, Spiegelhalter, and Gilks (1992). The program is available from the authors at the Medical Research Council Biostatistics Unit at the University of Cambridge. Information can be obtained at the URL

$$\texttt{http://www.mrc-bsu.cam.ac.uk/bugs/}$$

Chen, Shao, and Ibrahim (2000) provide extensive discussion and examples of the use of Monte Carlo methods in Bayesian computations.

## 7.8   Computer Experiments

Some of the most important questions in science and industry involve the relationship of an entity of interest to other entities that can either be controlled or more easily measured than the quantity of interest. We envision a relationship expressed by a model

$$y \approx f(x).$$

The quantity of interest $y$, usually called a "response" (although it may not be a response to any of the other entities), may be the growth of a crystal, the growth of a tumor, the growth of corn, the price of a stock one month hence, etc. The other variables $x$, called "factors", "regressors", or just "input variables", may be temperature, pressure, amount of a drug, amount of a type of fertilizer, interest rates, etc. Both $y$ and $x$ may be vectors. An objective is to determine a suitable form of $f$ and the nature of the approximation. The simplest type of approximation is one in which an additive deviation can be identified with a random variable:

$$Y = f(x) + E.$$

The most important objective, whatever the nature of the approximation, usually is to determine values of $x$ that are associated with optimal realizations of $Y$. The association may or may not be one of causation.

One of the major contributions of the science of statistics to the scientific method is the experimental methods that efficiently help to determine $f$, the nature of an unexplainable deviation $E$, and the values of $x$ that yield optimal values of $y$. Design and analysis of experiments is a fairly mature subdiscipline of statistics.

In computer experiments, the function $f$ is a computer program, $x$ is the input, and $y$ is the output. The program implements known or supposed relationships among the phenomena of interest. In cases of practical interest, the function is very complicated, the number of input variables may be in the hundreds, and the output may consist of many elements. The objective is to find a tractable function, $\widehat{f}$, that approximates the true behavior, at least over ranges of interest, and to find the values of the input, say $\widehat{x}_0$, such that $\widehat{f}(\widehat{x}_0)$ is optimal. How useful $\widehat{x}_0$ is depends on how close $\widehat{f}(\widehat{x}_0)$ is to $f(x_0)$, where $x_0$ yields the optimal value of $f$.

What makes this an unusual statistical problem is that the relationships are deterministic. The statistical approach to computer experiments introduces randomness into the problem. The estimate $\widehat{f}(\widehat{x}_0)$ can then be described in terms of probabilities or variances.

In a Bayesian approach, randomness is introduced by considering the function $f$ to be a realization of a random function, $F$. The prior on $F$ may be specified only at certain points, say $F(x_0)$. A set of input vectors $x_1, \ldots, x_n$ is chosen, and the output $y_i = f(x_i)$ is used to estimate a posterior distribution for $F(x)$ or at least for $F(x_0)$. See Sacks et al. (1989), Currin et al. (1991), or Koehler and Owen (1996) for descriptions of this approach. The Bayesian approach generally involves extensive computations.

In a frequentist approach, randomness is introduced by taking random values of the input, $x_1, \ldots, x_n$. This randomness in the input yields randomness in the output $y_i = f(x_i)$, which is used to obtain the estimates $\widehat{x}_0$ and $\widehat{f}(\widehat{x}_0)$ and estimates of the variances of the estimators. See Koehler and Owen (1996) for further discussion of this approach.

# 7.9 Computational Physics

Many models of physics that describe the behavior of an ensemble of particles are stochastic. Generally, the individual particles obey simple laws that govern their motion or state.

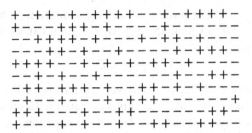

Figure 7.2: A Lattice in 2-D

One of the most widely studied models is the Ising model, introduced by Ernst Ising in the 1920s. The model can be used effectively to study phase transitions in ferromagnetism, which was the original use by Ising, to study state transitions, and to model binary amalgamations. In the Ising model, a lattice is used to locate positions of the entities of interest. In applications in physics, of course, the lattice is generally of three dimensions, but often a two-dimensional lattice is useful.

In the applications in magnetism, we think of the lattice as representing locations of atoms that have binary magnetic moments, either $+$ or $-$. If a linear ordering is imposed on the lattice, say by a systematic traversal of rows, then columns, then planes, and so on, the system can be described by a configuration vector $\sigma = (\sigma_1, \ldots, \sigma_n)$. In a two-dimensional lattice, the state of the system shown in Figure 7.2, for example, can be represented by $\sigma = (+1, -1, +1, +1, \ldots, -1, -1)$, in which the ordering is rowwise, beginning with the top row.

Each particle interacts with all of the others in the system. The total energy in the system is given by the Hamiltonian

$$H(\sigma) = -\sum_{i<j} J_{ij}\sigma_i\sigma_j - h\sum_i \sigma_i, \qquad (7.17)$$

where $J_{ij}$ represents the strength of attraction between particles $i$ and $j$, and $h$ is the strength of an external magnetic field. In the Ising model, only interactions between particles at adjacent points on the lattice are considered, and the strength of attraction is assumed to be equal for all adjacent pairs. Hence, $J_{ij} = 0$ for nonadjacent pairs, and $J_{ij} = J$ for adjacent pairs.

The system is subject to "thermal agitation"; that is, random changes in state. The probability model for states is a *Boltzmann distribution* or a *Gibbs distribution*. In this model, the probability of a state $\sigma$ is given by

$$\Pr(\sigma) \propto e^{-\beta H(\sigma)}, \qquad (7.18)$$

where $\beta$ includes units; often $\beta = 1/(kT)$, where $k$ is Boltzmann's constant and $T$ is the temperature in absolute degrees.

The normalizing constant for this probability distribution is

$$Z(\beta, J, h, n) = \sum_{\text{all configurations}} e^{-\beta H(\sigma)},$$

which in physics is called the *partition function*. The probability function then becomes

$$p(\sigma) = \frac{1}{Z} e^{-\beta H(\sigma)}. \qquad (7.19)$$

In a continuous medium, the free energy density is

$$F(\beta, J, h) = \lim_{n \to \infty} \frac{1}{n} Z(\beta, J, h, n).$$

The system changes randomly but generally in such a way that the energy decreases. The changes can be modeled as a Markov process, with transition probabilities favoring an energy decrease. It may be possible to determine a steady-state distribution analytically in lower dimensions, but in higher dimensions (currently three or greater), it is necessary to resort to simulation.

We can simulate the system changes by randomly generating a change and then using the probability (7.18) in the Metropolis acceptance criterion (4.15), page 140. (The Metropolis algorithm is also sometimes called "M(RT)$^2$" after the names of all five authors of the original 1953 paper, Metropolis et al., 1953.) This method moves the system generally in a direction of lower energy. It allows the system to move to higher energy states, however, so the system does not get stuck in local minimum energy states. If the temperature $T$ is decreased, the probability of going to a higher energy state decreases, similar to the annealing of iron during cooling. This method of "simulated annealing" has general applications in optimization problems, especially ones with local optima (see Exercise 7.16 on page 277, and Kirkpatrick, Gelatt, and Vecchi, 1983).

The Metropolis algorithm method can be applied to one site at a time (that is, to a single spin), but more efficient algorithms consider clusters of points. One of the earliest cluster algorithms is the Swendsen–Wang algorithm, which can be viewed from a statistical perspective as a data augmentation algorithm.

Currently, the most widely used method for simulating the Ising model is the Wolff algorithm.

### Algorithm 7.1 Wolff's Algorithm for the Ising Model for Fixed $\beta$

1. Choose a lattice point $x$ at random uniformly.

2. Form a cluster about the chosen point:

   (a) check each neighbor point, and if the point has the same polarity as $x$, add it to the cluster with probability $1 - e^{-2\beta J}$;

   (b) for each point added to the cluster, check each of its unchecked neighbor points, and if the point has the same polarity as $x$, add it to the cluster with probability $1 - e^{-2\beta J}$;

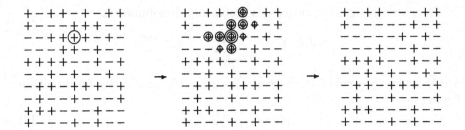

Figure 7.3: One Step of Wolff's Algorithm

    (c) repeat step 2b until there are no points in the cluster with any
    unchecked neighbors.

3. Flip the cluster. ∎

Figure 7.3 shows one step in Wolff's algorithm on a portion of the system
represented by the lattice in Figure 7.2. The lattice on the left-hand side represents the original configuration with the randomly chosen point shown by a
circle. In the lattice in the middle, the point shown with three concentric circles
is the one initially chosen, and the points with a single small circle represent
ones that were inspected during the process. Those with a second larger concentric circle were accepted into the cluster. The lattice on the right-hand side
represents the state of the system following that step of Wolff's algorithm.

Newman and Barkema (1999) discuss programming issues for Wolff's algorithm and give a C function implementing the algorithm. Shchur and Blöte (1997)
give an equivalent formulation of Wolff's algorithm as a random walk in one
dimension.

Cipra (1987) provides a very readable account of the development of the
Ising model and some of the interesting mathematics underlying the model.
Newman and Barkema (1999) discuss many approaches and details of simulations of the Ising model. As mentioned in Chapter 2, Vattulainen, Ala-Nissila,
and Kankaala (1994, 1995) have developed tests of random number generators
based on the Ising model.

In the Ising model, there are only two possible states at each lattice point.
The Potts model is a generalization to allow any finite number of discrete states.
The XY model and Heisenberg model allow continuous states; see Newman and
Barkema (1999).

In most methods of simulated annealing, there are several possible variations
that affect the efficiency of the algorithm, and it is somewhat difficult to know
a priori what variations work best in a given problem. A general rule is that
more variation should be allowed early in the simulation and then less variation
as the system reaches the optimal state. Knowing whether a state is only a
local optimum or the global optimum is not easy. Consistent with the rule of
decreasing the amount of variation as the simulation proceeds, we generally find

that the Metropolis single-flip algorithm is more efficient far from the critical point (that is, presumably, early on in the simulation), and a cluster algorithm is more efficient near the critical point.

In general, simulated annealing methods work well on optimization problems of this type. When there are local optima separated by strongly suboptimal regions, it may be useful to allow the temperature $T$ to increase occasionally. This is called "simulated tempering", and there are several ways that it can be done, as described by Marinari and Parisi (1992). One way is just to run multiple streams in parallel and occasionally swap states in two streams with a Metropolis acceptance criterion using the Boltzmann distribution (7.18).

Any of these types of methods can be programmed in parallel (see Newman and Barkema, 1999).

## 7.10 Computational Finance

Important questions in finance concern the fair price for various assets. Although there are many fundamental issues that must be considered in order to determine fair prices for some types of assets, there are other assets, called "derivatives", for which prices depend primarily on the prices and price movement characteristics of other assets (the underlying assets or the "underlying") but are conditionally independent of other fundamental economic conditions. Derivatives are financial instruments having values dependent on constraints on their trading (their "exercise") and on the price of other assets (the underlying) or on some measure of the state of the economy or nature. A derivative is an agreement between two sides: a *long position* and a *short position*.

Many pricing problems in finance cannot be solved analytically. The traditional approach has been to develop overly simplified models that approximate what is believed to be a more realistic description of market behavior. Pricing of various derivative instruments is an area in finance in which Monte Carlo methods can be used to analyze more realistic models.

### Pricing Forward Contracts

One of the simplest kinds of derivative is a *forward contract*, which is an agreement to buy or sell an asset at a specified time at a specified price. The agreement to buy is a long position, and the agreement to sell is a short position. The agreed upon price is the *delivery price*. The consummation of the agreement is an *execution*.

Forward contracts are relatively simple to price, and their analysis helps in developing pricing methods for other derivatives. Let $k$ be the delivery price or "strike price" at the settlement time $t_s$, and let $X_t$ be the value of the underlying. (We consider all times to be measured in years.) A graph of the profit or the payoff at the settlement date is shown in Figure 7.4 for both long and short positions.

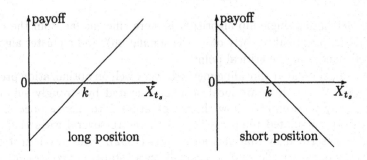

Figure 7.4: Payoff of a Forward Contract

**Principles and Procedures for Pricing Derivatives**

A basic assumption in pricing financial assets is that there exists a fixed rate of
return on some available asset that is constant and "guaranteed"; that is, there
exists an asset that can be purchased, and the value of that asset changes at a
fixed, or "riskless", rate. The concept of "risk-free rate of return" is a financial
abstraction based on the assumption that there is some absolute controller of
funds that will pay a fixed rate of interest over an indefinite length of time.
In current world markets, the rate of interest on a certain financial instrument
issued by the United States Treasury is used as this value.

Given a riskless positive rate of return, another key assumption in economic
analyses is that there are no opportunities for arbitrage. An *arbitrage* is a
trading strategy with a guaranteed rate of return that exceeds the riskless rate
of return. In financial analysis, we assume that arbitrages do not exist. This is
the "no-arbitrage principle". This means, basically, that all markets have two
liquid sides and that there are market participants who can establish positions
on either side of the market.

The assumption of a riskless rate of return leads to the development of
a *replicating portfolio*, having fluctuations in total valuation that can match
the expected rate of return of any asset. The replication approach to pricing
derivatives is to determine a portfolio and an associated trading strategy that
will provide a payout that is identical to that of the underlying. This portfolio
and trading strategy *replicates* the derivative.

Consider a forward contract that obligates one to pay $k$ at $t_s$ for the underlying. The value of the contract at expiry is $X_{t_s} - k$, but of course we do
not know $X_{t_s}$. If we have a riskless (annual) rate of return $r$, we can use the
no-arbitrage principle to determine the correct price of the contract.

To apply the no-arbitrage principle, consider the following strategy:

- take a long position in the forward contract;

- take a short position of the same amount in the underlying (sell the underlying short).

With this strategy, the investor immediately receives $X_t$ for the short sale of the underlying. At the settlement time $t_s$, this amount can be guaranteed to be

$$X_t e^{r(t_s - t)} \qquad (7.20)$$

using the risk-free rate of return. Now, if the settlement price $k$ is such that

$$k < X_t e^{r(t_s - t)}, \qquad (7.21)$$

a long position in the forward contract and a short position in the underlying is an arbitrage, so by the no-arbitrage principle, this is not possible. Conversely, if

$$k > X_t e^{r(t_s - t)}, \qquad (7.22)$$

a short position in the forward contract and a long position in the underlying is an arbitrage, and again, by the no-arbitrage principle, this is not possible. Therefore, under the no-arbitrage assumption, the correct value of the forward contract, or its "fair price", at time $t$ is $X_t e^{r(t_s - t)}$.

There are several modifications to the basic forward contract that involve different types of underlying, differences in when the agreements can be executed, and differences in the nature of the agreement: whether it conveys a right (that is, a "contingent claim") or an obligation.

The common types of derivatives include stock options, index options, commodity futures, and rate futures. Stock options are used by individual investors and by investment companies for leverage, hedging, and income. Index options are used by individual investors and by investment companies for hedging and speculative income. Commodity futures are used by individual investors for speculative income, by investment companies for income, and by producers and traders for hedging. Rate futures are used by individual investors for speculative income, by investment companies for income and hedging, and by traders for hedging.

**Stock Options**

A single call option on a stock conveys to the owner the right to buy (usually) 100 shares of the underlying stock at a fixed price, the strike price, anytime before the expiration date. A single put option conveys to the owner the right to sell (usually) 100 shares of the underlying stock at a fixed price, the strike price, anytime before the expiration date. Most real-world stock options can be exercised at any time (during trading hours) prior to expiration. Such options that can be exercised at any time are called "American options". We will consider a modification, the "European option", which can only be exercised at a specified time. There are some European options that are actually traded, but they are generally for large amounts, and they are rarely traded by individuals. European options are studied because the analysis of their fair price is easier.

A major difference between stock options and forward contracts is that stock options depend on the fluctuating (and unpredictable) prices of the underlying.

Another important difference between stock options and forward contracts is that stock options are *rights*, not obligations. The payoff therefore cannot be negative. Because the payoff cannot be negative, there must be a cost to obtain a stock option. The profit is the difference between the payoff and the price paid.

## Pricing of Stock Options

It is difficult to determine the appropriate price of stock options because stocks are *risky assets*; that is, they are assets whose prices vary randomly (or at least unpredictably).

Pricing formulas for stock options can be developed from a simple model of the market that assumes two types of assets: the risky asset (that is, the stock) with price at time $t$ of $X_t$ and a *riskless asset* with price at time $t$ of $\beta_t$. The price of a stock option can be analyzed based on trading strategies involving these two assets, as we briefly outline below. (See Hull, 2000, for a much more extensive discussion.)

The price of the riskless asset, like the price of a forward contract, follows the deterministic ordinary differential equation

$$\mathrm{d}\beta_t = r\beta_t\mathrm{d}t, \tag{7.23}$$

where $r$ is the instantaneous risk-free interest rate.

A useful model for the price of the risky asset is the stochastic differential equation

$$\mathrm{d}X_t = \mu X_t\mathrm{d}t + \sigma X_t\mathrm{d}B_t, \tag{7.24}$$

where $\mu$ is a constant mean rate of return, $\sigma$ is a positive constant, called the "volatility", and $B_t$ is a Brownian motion; that is,

1. the change $\Delta B_t$ during the small time interval $\Delta t$ is

$$\Delta B_t = Z\sqrt{\Delta t},$$

where $Z$ is a random variable with distribution $N(0,1)$;

2. $\Delta B_{t_1}$ and $\Delta B_{t_2}$ are independent for $t_1 \neq t_2$.

Notice, therefore, that for $0 < t_1 < t_2$, $B_{t_2} - B_{t_1}$ has a $N(0, t_2 - t_1)$ distribution. Notice also that the change in the time interval $\Delta t$ is randomly proportional to $\sqrt{\Delta t}$. By convention, we set $B_0 = 0$, so $B_{t_2}$ has a $N(0, t_2)$ distribution. A process following equation (7.24) is a special case of an *Ornstein-Uhlenbeck process*.

Given a starting stock price, $X_0$, the differential equation (7.24) specifies a random *path* of stock prices. Any realization of $X_0$, $x(0)$, and any realization of $B_t$ at $t \in [0, t_1]$, $b(t)$, yields a realized path, $x(t)$.

We should note three simplifying assumptions in this model:

- $\mu$ is constant;

- $\sigma$ is constant;

- the stochastic component is a Brownian motion (that is, i.i.d. normal).

The instantaneous rate of return from equation (7.24) is

$$\frac{\mathrm{d}X_t}{X_t} = \mu\mathrm{d}t + \sigma\mathrm{d}B_t, \tag{7.25}$$

so under the assumptions of the model (7.24), the price of the stock itself follows a lognormal distribution.

A discrete-time version of the change in the stock price that corresponds to the stochastic differential equation (7.24) is the stochastic difference equation

$$\begin{aligned} \Delta X_t &= \mu X_t \Delta t + \sigma X_t \Delta B_t \\ &= \mu X_t \Delta t + \sigma X_t Z \sqrt{\Delta t}, \end{aligned} \tag{7.26}$$

and a discrete-time version of the rate of return is

$$R_t(\Delta t) = \mu\Delta t + \sigma Z \sqrt{\Delta t},$$

where, as before, $Z$ is a random variable with distribution $N(0,1)$, so $R_t(\Delta t)$ is $N(\mu\Delta t, \sigma^2\Delta t)$. The quantity $\mu\Delta t$ is the expected value of the return in the time period $\Delta t$ and by assumption is constant. The quantity $\sigma Z\sqrt{\Delta t}$ is the stochastic component of the return, where, by assumption, $\sigma$, the "volatility" or the "risk", is constant.

Figure 7.5 shows 100 simulated paths of the price of a stock for one year using the model (7.26) with $x(0) = 20$, $\Delta t = 0.01$, $\mu = 0.1$, and $\sigma = 0.2$.

We first consider the fair price of the option at the time of its expiration. The *payoff*, $h$, of the option at time $t_s$ is either 0 or the excess of the price of the underlying $X_{t_s}$ over the strike price $k$. Once the parameters $k$ and $t_s$ are fixed, $h$ is a function of the realized value of the random variable $X_{t_s}$:

$$h(x_{t_s}) = \begin{cases} x_{t_s} - k & \text{if } x_{t_s} > k, \\ 0 & \text{otherwise.} \end{cases}$$

The price of the option at any time $t$ prior to expiration $t_s$ is a function of $t$ and the price of the underlying $x$. We denote the price as $P(t,x)$. With full generality, we can set the time of interest as $t = 0$.

The price of the European call option should be the expected value of the payoff of the option at expiration discounted back to $t = 0$,

$$P(0,x) = \mathrm{e}^{-qt_s}\mathrm{E}(h(X_{t_s})), \tag{7.27}$$

where $q$ is the rate of growth of an asset. Likewise, for an American option, we could maximize the expected value over all stopping times, $0 < \tau < t_s$:

$$P(0,x) = \sup_{\tau \le t_s} \mathrm{e}^{-q\tau}\mathrm{E}(h(X_\tau)). \tag{7.28}$$

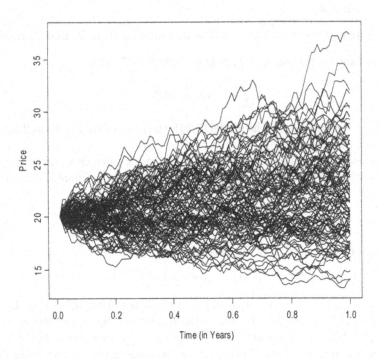

Figure 7.5: 100 Simulated Paths of the Price of a Stock with $\mu = 0.1$ and $\sigma = 0.2$

The problem with expressions (7.27) and (7.28), however, is the choice of the rate of growth $q$. If the rate of growth $r$ of the riskless asset in equation (7.23) is different from the mean rate of growth $\mu$ in equation (7.24), then there is an opportunity for arbitrage. We must therefore consider a completely different approach. Consideration of a "replicating strategy" leads us to a fair price for options under the assumptions of the model in equation (7.24) and consistent with a no-arbitrage assumption.

### Replicating Strategies

A replicating strategy involves both long and short positions that together match the price fluctuations in the underlying, and thus in the fair price of the derivative. We will generally assume that every derivative can be replicated by positions in the underlying and a risk-free asset. (In that case, the economy or market is said to be *complete*.) We assume a finite universe of assets, all priced consistently with some pricing unit. (In general, we call the price, or the pricing unit, a *numeraire*. A more careful development of this concept rests on the idea of a *pricing kernel*.) The set of positions, both long and short, is called a *portfolio*.

The value of a derivative changes in time and as a function of the value of the underlying. Therefore, a replicating portfolio must be changing in time or "dynamic". In analyses with replicating portfolios, transaction costs are ignored. Also, the replicating portfolio must be self-financing; that is, once the portfolio is initiated, no further capital is required. Every purchase is financed by a sale.

Now, using our simple market model, with a *riskless asset* with price at time $t$ of $\beta_t$ and a *risky asset* with price at time $t$ of $X_t$ (with the usual assumptions on the prices of these assets), we can construct a portfolio with a value that will be the payoff of a European call option on the risky asset at time $T$.

At time $t$, the portfolio consists of $a_t$ units of the risky asset and $b_t$ units of the riskless asset. Therefore, the value of the portfolio is $a_t X_t + b_t \beta_t$. If we scale $\beta_t$ so that $\beta_0 = 1$ and adjust $b_t$ accordingly, the expression simplifies so that $\beta_t = \mathrm{e}^{rt}$.

The portfolio replicates the value of the option at time $t_s$ if almost surely

$$a_{t_s} X_{t_s} + b_{t_s} \mathrm{e}^{rt_s} = h(X_{t_s}). \qquad (7.29)$$

The portfolio is self-financing if at any time $t$

$$\mathrm{d}(a_t X_t + b_t \mathrm{e}^{rt}) = a_t \mathrm{d}X_t + r b_t \mathrm{e}^{rt} \mathrm{d}t. \qquad (7.30)$$

## The Black–Scholes Differential Equation

Consider the price $P$ of a European call option at time $t < T$. At any time, this is a function of both $t$ and the price of the underlying $X_t$. We would like to construct a dynamic, self-financing portfolio $(a_t, b_t)$ that will replicate the derivative at maturity. If we can, then the no-arbitrage principle requires that

$$a_t X_t + b_t \mathrm{e}^{rt} = P(t, X_t) \qquad (7.31)$$

for $t < t_s$.

Now, differentiate both sides of this equation. If $a_t$ is constant, the differential of the left-hand side is the left-hand side of equation (7.30), which must be satisfied by a self-financing portfolio. (*The assumption that $a_t$ is constant is not correct, but the approximation does not seem to have serious consequences.*)

The derivative of $P(t, X_t)$ is rather complicated because of its dependence on $X_t$ and the fact that $\mathrm{d}X_t$ has components of both $\mathrm{d}t$ and $\mathrm{d}B_t$ in the stochastic differential equation (7.24). If $P(t, X_t)$ is continuously twice-differentiable, we can use Itô's formula (see Øksendal, 1998, for example) to develop the expression for the differential of the right-hand side of equation (7.31),

$$\mathrm{d}P(t, X_t) = \left( \frac{\partial P}{\partial t} + \frac{\partial P}{\partial X_t} \mu X_t + \frac{1}{2} \frac{\partial^2 P}{\partial X_t^2} \sigma^2 X_t^2 \right) \mathrm{d}t + \frac{\partial P}{\partial X_t} (\sigma X_t) \mathrm{d}B_t. \qquad (7.32)$$

By inserting the market model (7.24) for $dX_t$ into the differential of the left-hand side, we have

$$(a_t \mu X_t + r b_t e^{rt}) dt + a_t \sigma X_t dB_t.$$

Now, equating the coefficients of $dB_t$, we have

$$a_t = \frac{\partial P}{\partial X_t}.$$

From our equation for the replicating portfolio, we have

$$b_t = (P(t, X_t) - a_t X_t) e^{-rt}.$$

Finally, equating coefficients of $dt$ and substituting for $a_t$ and $b_t$, we have the Black–Scholes differential equation,

$$r\left(P - X_t \frac{\partial P}{\partial X_t}\right) = \frac{\partial P}{\partial t} + \frac{1}{2} \sigma^2 X_t^2 \frac{\partial^2 P}{\partial X_t^2}. \tag{7.33}$$

Instead of European calls, we can consider European puts and proceed in the same way. We arrive at the same Black–Scholes differential equation (written in a slightly different but equivalent form from the equation above):

$$\frac{\partial P}{\partial t} + r X_t \frac{\partial P}{\partial X_t} + \frac{1}{2} \sigma^2 X_t^2 \frac{\partial^2 P}{\partial X_t^2} = rP. \tag{7.34}$$

It is interesting to note that $\mu$ is not in the equations, but $r$ has effectively replaced it.

A similar development could be used for American options. Other approaches can be used to develop these equations. One method is called "delta hedging" (see Hull, 2000).

## The Black–Scholes Formula

The solution of the differential equations depends on the boundary conditions. These come from the price of the option at expiration. In the case of European options, these are simple. For calls, they are

$$P_c(t_s, X_{t_s}) = \max(X_{t_s} - k, 0),$$

and, for puts, they are

$$P_p(t_s, X_{t_s}) = \max(k - X_{t_s}, 0),$$

where $k$ is the strike price in either case. With these boundary conditions, the Black–Scholes differential equations have closed-form solutions. For the call, the solution is the "Black–Scholes formula",

$$P_c(t, X_t) = X_t \Phi(d_1) - k e^{-r(t_s - t)} \Phi(d_2), \tag{7.35}$$

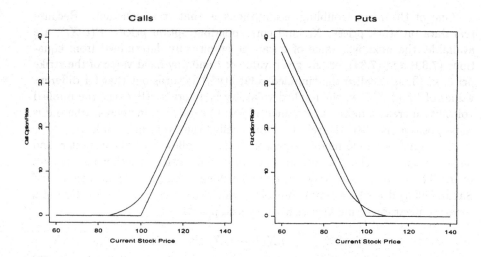

Figure 7.6: The Black–Scholes Call Pricing Function

and, for the put, it is

$$P_p(t, X_t) = k e^{-r(t_s - t)} \Phi(-d_2) - X_t \Phi(-d_1), \tag{7.36}$$

where $\Phi(\cdot)$ is the standard normal CDF,

$$d_1 = \frac{\log(X_t/k) + (r + \frac{1}{2}\sigma^2)(t_s - t)}{\sigma \sqrt{t_s - t}},$$

and

$$d_2 = d_1 - \sigma \sqrt{t_s - t}.$$

The Black–Scholes formulas are widely used in determining a fair price for options on risky assets. In practice, of course, $\sigma^2$ is not known. The standard procedure is to use price data over some fixed time interval, perhaps a year, compute the sample variance of the rates of return during some fixed-length subintervals, perhaps subintervals of length one day, and use the sample variance as an estimate of $\sigma^2$. A data analyst who has looked at such data will see the effects of the rather arbitrary choice of the fixed times.

The prices have systematic relationships to the prices of the underlying, as shown in Figure 7.6.

## More Realistic Models

As we have seen, several simplifying assumptions were made in the development of the Black–Scholes formulas. As in most financial analyses, we have assumed throughout that there are no costs for making transactions. In trades involving derivatives, the transaction costs (commissions) can be quite high.

One of the most troubling assumptions is that $\sigma$ is constant. Because real data on stock prices, $X_t$, and corresponding option prices, $P(t, X_t)$, are available, the unknown value of $\sigma$ can be empirically determined from equations (7.33) and (7.34) for any given value of $t$ and any fixed value of the strike price, $k$. (This is called the "implied volatility".) It turns out that for different values of $t$ and $X_t - k$, the implied volatility is different. (Because the implied volatility increases more or less smoothly as $|X_t + d - k|$ increases, where $d$ is some positive number, the relationship is called the "volatility smile".)

One approach is to modify equation (7.24) to allow for nonconstant $\sigma$ and augment the model by a second stochastic differential equation for changes in $\sigma$. There are various ways this can be done. Fouque, Papanicolaou, and Sircar (2000) describe a simple model in which the volatility is a function of a separate mean-reverting Ornstein-Uhlenbeck process:

$$
\begin{aligned}
dX_t &= \mu X_t dt + \sigma_t X_t dB_t, \\
d\sigma_t^2 &= f(Y_t), \\
dY_t &= \alpha(\mu_Y - Y_t)dt + \beta d\tilde{B}_t,
\end{aligned}
\tag{7.37}
$$

where $\alpha$ and $\beta$ are constants and $\tilde{B}_t$ is a linear combination of $B_t$ and an independent Brownian motion. The function $f$ can incorporate various degrees of complexity, including the simple identity function. These coupled equations provide a better match for observed stock and option prices. Economists refer to the condition of nonconstant volatility as "stochastic volatility". Pricing in the presence of stochastic volatility is discussed extensively by Fouque, Papanicolaou, and Sircar (2000). The fact that the implied volatility is not constant for a given stock does not mean that a model with an assumption of constant volatility cannot be useful. It only implies that there are some aspects of the model (7.24) that do not correspond to observational reality.

Another very questionable assumption in the model given by equation (7.24) is that the changes in stock prices follow an i.i.d. normal distribution.

There are several other simplifications, such as the restriction to European options, the assumption that the stocks do not pay dividends, the assumption that the derivative of the left-hand side of equation (7.31) can be done as if $a_t$ were constant, and so on. All of these assumptions allow the derivation of a closed-form solution.

More realistic models can be studied by Monte Carlo methods, and this is currently a fruitful area of research. Paths of prices of the underlying can be simulated using a model similar to equation (7.26), as we did to produce Figure 7.5, but with different distributions on $Z$. A very realistic modification of the model is to assume that $Z$ has a superimposed jump or shock on its $N(0, 1)$ distribution. The simulated paths of the price of the underlying provide a basis for determining a fair price for the options. This price is just the break-even value discounted back in time by the risk-free rate $r$. Thompson (2000), Barndorff-Nielsen and Shephard (2001), and Jäckel (2002) all discuss use of Monte Carlo simulation in pricing options under various models. Thompson uses a model in which $\Delta X_t$ is subjected to a fixed relative decrease as a Poisson event. Such "bear jumps", of course, would decrease the fair price of a call option from the Black–Scholes price and would increase the fair price of a put option. Other modifications to the underlying distribution of $\Delta X_t$ result in other differences in the fair price of options. Barndorff-Nielsen and Shephard use a nonnormal process in an Ornstein-Uhlenbeck type of model.

Another type of approach to the problem of a nonconstant $\sigma$ in the pricing model (7.24) is to use a GARCH model (equation (6.2) on page 226). This model is easy to simulate with various distributions.

# Exercises

7.1. Use Monte Carlo methods to estimate the expected value of the fifth order statistic from a sample of size 25 from a N(0, 1) distribution. As with any estimate, you should also compute an estimate of the variance of your estimate. Compare your estimate of the expected value with the true expected value of 0.90501. Is your estimate reasonable? (Your answer must take into account your estimate of the variance.)

7.2. The "hit-or-miss" method is another Monte Carlo method to evaluate an integral. To simplify the exposition, let us assume that $f(x) \geq 0$ on $[a, b]$, and we wish to evaluate the integral, $I = \int_a^b f(x)\,dx$. First, determine $c$ such that $c \geq f(x)$ on $[a, b]$. Generate a random sample of $m$ pairs of uniform deviates $(x_i, y_i)$, in which $x_i$ is from a uniform distribution over $[a, b]$, $y_i$ is from a uniform distribution over $[0, c]$, and $x_i$ and $y_i$ are independent. Let $m_1$ be the number of pairs such that $y_i \leq f(x_i)$. Estimate $I$ by $c(b - a)m_1/m$. (Sketch a picture, and you can see why it is called hit-or-miss. Notice also the similarity of the hit-or-miss method to acceptance/rejection. It can be generalized by allowing the $y$s to arise from a more general distribution (that is, any distribution with a majorizing density). Another way to think of the hit-or-miss method is as importance sampling in which the sampling of the weighting function is accomplished by acceptance/rejection.)

(a) Is this a better method than the crude method described in Section 7.1? (What does this question mean? Think "bias" and "variance". To simplify the computations for answering this question, consider the special case in which $[a, b]$ is $[0, 1]$ and $c = 1$. For a further discussion, see Hammersley and Handscomb, 1964.)

(b) Suppose that $f$ is a probability density and the hit-or-miss method is used. Consider the set of the $m$ $x_i$s for which $y_i \leq f(x_i)$. What can you say about this set with respect to the probability density $f$?

Because a hit-or-miss estimate is a rational fraction, the methods are subject to granularity. See the discussion following the Buffon's needle problem in Exercise 7.10 below.

7.3. Describe a Monte Carlo method for evaluating each of the integrals

(a)

$$\int_{-\infty}^{\infty}\int_0^2\int_0^{\infty} y\cos(\pi(x+y+z))e^{-x^2}e^{-z}\,dz\,dy\,dx.$$

(b)

$$\int_{-\infty}^{\infty}\int_0^2\int_0^{\infty} y\cos(\pi y)e^{-x^2 y}e^{-yz}\,dz\,dy\,dx.$$

7.4. Use Monte Carlo methods to tabulate the cumulative distribution function of the Anderson–Darling $A^2$ statistic (equation (2.9), page 76) for the uniform null distribution and for $n = 10, 100$, and $1000$. Use Monte Carlo sample sizes $m = 500$, and compute sample quantiles as $(i - .5)/m$ (so that your tabulated values correspond to $0.001, 0.003, \ldots 0.999$).

Could you use these values in testing your uniform generator?

7.5. Obtain Monte Carlo estimates of the base of the natural logarithm, e.

(a) For Monte Carlo sample sizes $n = 8, 64, 512, 4096$, compute estimates and the errors of the estimates using the known value of e. Plot the errors versus the sample sizes on log-log axes. What is the order of the error?

(b) For your estimate with a sample of size 512, compute 95% confidence bounds.

7.6. Consider the following Monte Carlo method to evaluate the integral:

$$I = \int_a^b f(x)\,dx.$$

Generate a random sample of uniform order statistics $x_{(1)}, x_{(2)}, \ldots, x_{(n)}$ on the interval $(a, b)$, and define $x_{(0)} = a$ and $x_{(n+1)} = b$. Estimate $I$ by $\widehat{I}$:

$$\frac{1}{2}\left(\sum_{i=1}^n (x_{(i+1)} - x_{(i-1)})f(x_{(i)}) + (x_{(2)} - a)f(x_{(1)}) + (b - x_{(n-1)})f(x_{(n)})\right).$$

This method is similar to approximation of the integral by Riemann sums except that in this case the intervals are random. Determine the variance of $\widehat{I}$. What is the order of the variance in terms of the sample size? How would this method compare in efficiency with the crude Monte Carlo method?

7.7. Obtain a simplified expression for the variance of the Monte Carlo estimator (7.5) on page 232.

7.8. Random walks on a two-dimensional uniform grid can be used as models in a variety of scientific applications. The walk usually progresses from point $(i, j)$ at time $t$ to one of the four points $(i \pm 1, j)$ and $(i, j \pm 1)$ at time $t + 1$. Sometimes, the points $(i, j)$ (no movement) and $(i \pm 1, j \pm 1)$ are also allowed. The probability of moving to any one of the allowed points is usually chosen to be equal, although by assigning different probabilities, a trend can be induced in the direction of the walk.

   (a) Write a program to generate a two-dimensional random walk that goes to each of the four adjacent points with equal probabilities. Use Monte Carlo to develop a relationship between the expected distance from the starting point to the point at time $t$, $D_t$, and $t$; that is, experimentally determine the function $f$ in

$$\mathrm{E}(D_t) = f(t).$$

   (b) A self-avoiding random walk is one that does not revisit any point. (It is not a Markov process.) This process has been used to simulate the growth of large molecules (polymers) from smaller sets of atoms (monomers). A self-avoiding random walk will usually stop because it reaches a point from which no unvisited point is reachable. Experimentally determine the average time at which this stoppage occurs in a two-dimensional random walk that goes to each of the four adjacent points with equal probabilities. (Modifications of this process are necessary if the model is to be used for studying the growth of polymers. See Liu, Chen, and Logvinenko, 2001, for discussions of the problem.)

   (c) There are various ways of constraining random walks. One type of constrained random walk is one that begins at point $(x_0, y_0)$ and at time $t$ is at the specified point $(x_t, y_t)$. Develop an algorithm that generates a realization of a two-dimensional random walk that goes to each of the four adjacent points with equal probabilities that meets these conditions. Note, of course, that $(x_t, y_t)$ must be close enough to $(x_0, y_0)$ that it can be reached in $t$ steps. (This problem is harder than it may appear at first.)

7.9. Consider the integral

$$\int_0^\infty x^2 \sin(\pi x) e^{-\frac{x}{2}} \, dx.$$

   (a) Use the crude Monte Carlo method with an exponential weight to estimate the integral.

   (b) Use antithetic exponential variates to estimate the integral.

(c) Compare the variance of the estimator in Exercise 7.9a with that of the estimator in Exercise 7.9b by performing the estimation 100 times in each case with sample sizes of 10,000.

7.10. The French naturalist Comté de Buffon showed that the probability that a needle of length $l$ thrown randomly onto a grid of parallel lines with distance $d$ ($\geq l$) apart intersects a line is

$$\frac{2l}{\pi d}.$$

(a) Write a program to simulate the throwing of a needle onto a grid of parallel lines, and obtain an estimate of $\pi$. Remember that all estimates are to be accompanied by an estimate of the variability (that is, a standard deviation, a confidence interval, etc.).

(b) Derive the variance of your estimator in Exercise 7.10a, and determine the optimum ratio for $l$ and $d$.

(c) Now, consider a variation. Instead of a grid of just parallel lines, use a square grid, as in Figure 7.7. For simplicity, let us estimate $\theta = 1/\pi$. If $n_v$ is the number of crossings of vertical lines and $n_h$ is the number of crossings of horizontal lines, then for a total of $n$ tosses both

$$\frac{dn_v}{2ln} \quad \text{and} \quad \frac{dn_h}{2ln}$$

are unbiased estimates of $\theta$. Show that the estimator

$$\frac{dn_v + dn_h}{4ln}$$

is more than twice as efficient (has variance less than half) as either of the separate estimators. This is an application of antithetic variates.

(d) Perlman and Wichura (1975) consider this problem and show that a complete sufficient statistic for $\theta$ is $n_1 + n_2$, where $n_1$ is the number of

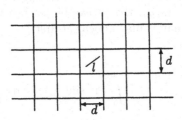

Figure 7.7: Buffon's Needle on a Square Grid

times that the needle crosses exactly one line and $n_2$ is the number of times that the needle crosses two lines. Derive an unbiased estimator of $\theta$ based on this sufficient statistic, and determine its variance. Determine the optimum ratio for $l$ and $d$.

(e) Now, consider another variation. Use a triangular grid, as in Figure 7.8. Write a program to estimate $\pi$ using the square grid and

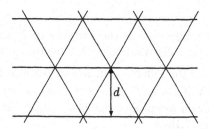

Figure 7.8: Buffon's Needle on a Triangular Grid

using the triangular grid. Using 10,000 trials, obtain estimates of $\pi$ using each type of grid. Obtain an estimate of the variance of each of your estimators.

See Robertson and Wood (1998) for further discussion of the efficiency of Buffon experiments using different types of grids.

C. R. Rao (in *Statistics and Truth*, Council of Scientific & Industrial Research, New Delhi, 1989) relates some interesting facts about reports of Buffon needle experiments. An Italian mathematician, Lazzarini, reported in 1901 that he had obtained an estimate of 3.1415929 for $\pi$ using the simple Buffon method of one set of parallel lines. This estimate differs from the true value only in the seventh decimal place. This estimate came from 1808 successes in 3408 trials. Why 3408, instead of 3400 or 3500? The answer may be related to the fact that the best rational approximation of $\pi$ for integers less than 15,000 is $\frac{355}{113}$, and that is the value Lazzarini reported, to seven places. The next rational approximation that is as close is $\frac{52163}{16604}$. In Lazzarini's experiment, $\frac{l}{d} = \frac{5}{6}$, so he could have obtained his accurate approximation only if the number of trials he used was 213, 426, ..., 3408, 3621, ..., or else much larger numbers. The choice of 3408 does appear suspect. One wonders if exactly 1808 successes occurred the first time he tried 3408 trials.)

7.11. Generate a pseudorandom sample of size 100 from a $N(0,1)$ distribution that has a sample mean of 0 and sample variance of 1.

7.12. Consider two estimators for the mean of a $t$ distribution with 2 degrees of freedom. (The mean obviously is 0, but we may think of this problem as

arising in a model that has an unknown mean plus a stochastic component that has a heavy-tailed distribution much like a $t$ with 2 degrees of
freedom.) Two unbiased estimators are $T$, the $\alpha$-trimmed mean (which
is the mean of the remaining data after the most extreme $\alpha$ fraction of
the data has been discarded), and $U$, the $\alpha$-Winsorized mean (which is
the mean of the dataset formed by shrinking the most extreme $\alpha$ fraction
of the data to the extreme order statistics of the central $1 - \alpha$ fraction of
the data).

(a) Estimate the difference in the variance of $T$ and $U$ for samples of
size 100 from a $t$ distribution with 2 degrees of freedom.

(b) Now, estimate the difference in the variance of $\widehat{E}(T^2 - U^2)$ and the
variance of $\widehat{V}(T) - \widehat{V}(U)$, where the estimators $\widehat{E}$ and $\widehat{V}$ are based
on Monte Carlo samples of size 500. (Read this problem carefully!
See equation (7.16).)

7.13. Consider a common application in statistics: three different treatments
are to be compared by applying them to randomly selected experimental
units. This, of course, usually leads us to "analysis of variance" using a
model such as $y_{ij} = \mu + \alpha_i + e_{ij}$, with the standard meanings of these
symbols and the usual assumptions about the random component $e_{ij}$ in
the model. Suppose that, instead of the usual assumptions, we assume
that the $e_{ij}$ have independent and identical beta distributions centered on
zero (the density is proportional to $(\frac{c+x}{2c})^p(\frac{c-x}{2c})^p$ over $(-c, c)$). Describe
how you would perform a Monte Carlo test instead of the usual AOV test.
Be clear in stating the alternative hypothesis.

7.14. Let $S = \{x_1, x_2, \ldots, x_n\}$ be a random sample from a population with
mean $\mu$, variance $\sigma^2$, and distribution function $P$. Let $\widehat{P}$ be the empirical distribution function. Let $\bar{x}$ be the sample mean for $S$. Let
$S^* = \{x_1^*, x_2^*, \ldots, x_n^*\}$ be a random sample taken with replacement from
$S$. Let $\bar{x}^*$ be the sample mean for $S^*$.

(a) Show that
$$E_{\widehat{P}}(\bar{x}^*) = \bar{x}.$$

(b) Show that
$$E_P(\bar{x}^*) = \mu.$$

(c) Note that, in the questions above, there was no replication of the
bootstrap sampling. Now, suppose that we take $m$ samples $S_j^*$, compute $\bar{x}_j^*$ for each, and compute
$$V = \frac{1}{m-1}\sum_j \left(\bar{x}_j^* - \overline{\bar{x}_j^*}\right)^2.$$

Derive $E_{\widehat{P}}(V)$.

(d) Derive $E_P(V)$.

7.15. Monte Carlo methods are often useful to ensure that our thinking is reasonable. A good example of this kind of use is to investigate a simple problem that generated much attention several years ago and for which many mathematicians obtained an incorrect solution. The problem was the analysis of the optimal strategy in a television game show popular at the time. The show was *Let's Make a Deal* with host Monty Hall. At some point in the show, a contestant was given a choice of selecting one of three possible items, each concealed behind one of three closed doors. The items varied considerably in value. After the contestant made a choice but before the chosen door was opened, the host, who knew where the most valuable item was, would open one of the doors not selected and reveal a worthless item. The host would then offer to let the contestant select a different door from what was originally selected. The question, of course, is should the contestant switch? Much interest in this problem was generated when it was analyzed by a popular magazine writer, Marilyn vos Savant, who concluded that the optimal strategy is to switch. This strategy is counterintuitive to many mathematicians, who would say that there is nothing to be gained by switching; that is, that the probability of improving the selection is 0.5. Study this problem by Monte Carlo methods. What is the probability of improving the selection by switching? Be careful to understand all of the assumptions, and then work the problem analytically also. (A Monte Carlo study is no substitute for analytic study.)

7.16. *Simulated annealing* is a method that simulates a thermodynamic process in which a metal is heated to its melting temperature and then is allowed to cool slowly so that its structure is frozen at the crystal configuration of lowest energy. In this process, the atoms go through continuous rearrangements, moving toward a lower energy level as they gradually lose mobility due to the cooling. The rearrangements do not result in a monotonic decrease in energy, however. If the energy function has local minima, going uphill occasionally is desirable.

Metropolis et al. (1953) developed a stochastic relaxation technique that simulates the behavior of a system of particles approaching thermal equilibrium. (This is the same paper that described the Metropolis sampling algorithm.) The energy associated with a given configuration of particles is compared to the energy of a different configuration. In simulated annealing, the system moves through a sequence of states $s^{(1)}, s^{(2)}, \ldots$, and the energy at each state, $f(s^{(1)}), f(s^{(2)}), \ldots$, generally decreases. The algorithm proceeds from state $s^{(j)}$ by choosing a trial state $r$ and comparing the energy at $r$, $f(r)$, with the energy at the current state, $f(s^{(j)})$. If the energy of the new configuration is lower than that of the previous one (that is, if $f(r) < f(s^{(j)})$), the new configuration is immediately accepted. If the new configuration has a larger energy, the new state is accepted with some nonzero probability that depends on the amount of the increase in energy, $f(r) - f(s^{(j)})$. In simulated annealing, the probability of moving to a state with larger energy also depends on a "temperature" parameter $T$. When the temperature is high, the probability of acceptance of any given point is high. When the temperature is low, however, the probability of accepting any given point is low.

Although the technique is heuristically related to the cooling of a metal, as in the original application, it can be used in other kinds of optimization problems. The objective function of the general optimization problem is used in place of the energy function of the original application. Simulated annealing is particularly useful in optimization problems that involve configurations of a discrete set, such as a set of particles whose configuration can continuously change, or a set of cities in which the interest is an ordering for shortest distance of traversal. Kirkpatrick, Gelatt, and Vecchi (1983) discussed various applications, and the method has become widely used since the publication of that article.

To use simulated annealing, a "cooling schedule" must be chosen; that is, we must decide the initial temperature and how to decrease the temperature as a function of time. If $T(0)$ is the initial temperature, a simple schedule is $T(k+1) = qT(k)$, for some fixed $q$, such that $0 < q < 1$. A good cooling schedule depends on the problem being solved, and often a few passes through the algorithm are used to choose a schedule. Another choice that must be made is how to update the state of the system; that is, at any point, how to choose the next trial point.

The probability of accepting a higher energy state must also be chosen. Although this probability could be chosen in various ways, it is usually taken as

$$e^{-(f(r)-f(s^{(j)}))/T},$$

which is proportional to the probability of an energy change of $f(r) - f(s^{(j)})$ at temperature $T$ and comes from the original application of Metropolis et al. (1953). In addition to the cooling schedule, the method of generating trial states, and the probability of accepting higher energy states, there are a number of tuning parameters to choose in order to apply the simulated annealing algorithm. These include the number of trial states to consider before adjusting the temperature. It is also good to adjust the temperature if a certain number of state changes have been made at the given temperature even if the limit on the number of trial states to consider has not been reached. Finally, there must be some kind of overall stopping criterion.

The simulated annealing method consists of the following steps:

0. Set $k = 1$ and initialize state $s$.

1. Compute $T(k)$ based on a cooling schedule.

2. Set $i = 0$ and $j = 0$.

3. Generate state $r$, and compute $\delta f = f(r) - f(s)$.

4. Based on $\delta f$, decide whether to move from state $s$ to state $r$.
   If $\delta f \leq 0$,
   > accept;

   otherwise,
   > accept with a probability $P(\delta f, T(k))$.

   If state $r$ is accepted, set $i = i + 1$.

5. If $i$ is equal to the limit for the number of successes at a given temperature, $i_{max}$, go to step 1.

6. Set $j = j + 1$. If $j$ is less than the limit for the number of iterations at the current temperature, $j_{max}$, go to step 3.

7. If $i = 0$,
   > deliver $s$ as the optimum; otherwise,
   > if $k < k_{max}$,
   >> set $k = k + 1$ and go to step 1;
   >
   > otherwise,
   >> issue message that
   >> "algorithm did not converge in $k_{max}$ iterations".                ∎

The traveling salesperson problem can serve as a prototype of the problems in which the simulated annealing method has had good success. In this problem, a state is an ordered list of points ("cities"), and the objective function is the total distance between all points in the order given plus the return distance from the last point to the first point. One simple rearrangement of the list is the reversal of a sublist; that is, for example,

$$(1, \underline{2, 3, 4, 5, 6}, 7, 8, 9) \rightarrow (1, \underline{6, 5, 4, 3, 2}, 7, 8, 9).$$

Another simple rearrangement is the movement of a sublist to some other chosen point in the list; for example,

$$(1, \underline{2, 3, 4, 5, 6}, 7, 8,_\uparrow 9) \rightarrow (1, 7, 8, \underline{2, 3, 4, 5, 6}, 9).$$

Both of these rearrangements are called "2-changes" because in the graph defining the salesperson's circuit, exactly two edges are replaced by two others.

(a) Write a program to use simulated annealing to solve the traveling salesperson problem for $d$ cities. Your progrm should take as input a $d \times d$ matrix of distances between the cities. Use the simple cooling schedule, $T(k + 1) = qT(k)$, and use random 2-changes as described above. The program should accept the beginning temperature $T(0)$ and the temperature reduction factor $q$. Finally, the program should also accept the control parameters $i_{max}$, $j_{max}$, and $k_{max}$ from the algorithm description above.

Although the program might seem rather complicated, it is not too difficult if it is built in separate modules. One special task, for example, is the implementation of the 2-change rule. You should write a module to perform these random changes and thoroughly test it before incorporating it in the full program.

(b) Now, use your program to determine the optimal order in which to visit the cities in the mileage chart below. Assume that you return to the starting city (it does not matter which one it is).

| Alexandria | ↓ | | | | | | | | | |
|---|---|---|---|---|---|---|---|---|---|---|
| Blacksburg | 263 | ↓ | | | | | | | | |
| Charlottesville | 117 | 151 | ↓ | | | | | | | |
| Culpeper | 70 | 193 | 47 | ↓ | | | | | | |
| Fairfax | 15 | 249 | 102 | 55 | ↓ | | | | | |
| Front Royal | 71 | 203 | 124 | 44 | 57 | ↓ | | | | |
| Lynchburg | 178 | 94 | 66 | 108 | 163 | 157 | ↓ | | | |
| Manassas | 28 | 238 | 91 | 44 | 23 | 45 | 157 | ↓ | | |
| Richmond | 104 | 220 | 71 | 89 | 106 | 133 | 110 | 96 | ↓ | |
| Roanoke | 233 | 41 | 120 | 164 | 218 | 174 | 52 | 207 | 189 | ↓ |
| Williamsburg | 148 | 257 | 120 | 133 | 150 | 177 | 165 | 140 | 51 | 215 |

Distances Between Cities in Virginia

7.17. Simulation of stock prices and Monte Carlo pricing of options.

In simulation applications, the values of the parameters can be set to any desired values. This is one characteristic that makes Monte Carlo simulation so useful. In any applications, however, we need some general feel for the magnitude of the parameters for which values are to be chosen. Before doing the simulations in this exercise, you may want to look at some real price data. Daily historical price data are available at

http://chart.yahoo.com/d

These data can easily be downloaded in a spreadsheet format and entered into an analysis program. The distribution of the rates of return, of course, is different depending on the length of the time period. The data from Yahoo can be used for periods of one day or longer.

Historical option prices can be obtained from some brokerage web sites. E*Trade, for example, provides actual daily option prices and the corresponding prices yielded by the Black–Scholes formula.

(a) Write a program to simulate the path of a stock price that follows the stochastic difference equation (7.26) on page 265.

(b) For fixed values of $\mu$, $\sigma$, and $X_0$, simulate 1000 price paths from $t = 0$ to $t = 1$, and, for each, determine the value at $t = 1$ of call options with three different strike prices: $X_0 - \mu X_0$, $X_0$, and $X_0 + \mu X_0$. Using your average values of the call, discount them back to $t = 0$ for a fixed value of the riskless rate $r$. (Most options do not have a life of one year, as in this exercise.) How do your three estimated fair prices at $t = 0$ compare to the price given by the Black–Scholes formula (7.35)?

(c) Now, repeat the previous exercise with only one modification. Assume that an event occurs with a Poisson frequency having mean 1 (using the same time units as $t$) in which the change $\Delta X_t = \gamma_1 X_t$, where $\gamma_1 = -.2$ and an event occurs with a Poisson frequency with mean $1/2$ in which the change $\Delta X_t = \gamma_2 X_t$, where $\gamma_2 = .1$. How do your three estimated fair prices at $t = 0$ compare to the price given by the estimates in the previous exercise and with the Black–Scholes formula (7.35)?

7.18. Use of GARCH models for stock prices.

In applications of either the stochastic differential equations or a GARCH model, the first problem is to get estimates for the parameters in the model. The parameters in the differential equation (7.24) or the stochastic difference equation (7.26) are generally estimated from the moments of an observed price series. The parameters in the GARCH model (6.2) are usually estimated by maximum likelihood, and programs for estimating them are available in a number of analysis packages.

(a) Use your program as in Exercise 7.17b to generate one price path using your fixed values of $\mu$, $\sigma$, and $X_0$, and then obtain the maximum likelihood estimates of the parameters in a GARCH(1,1) model (assuming a normal distribution). Now, using your GARCH model, simulate a price path, and estimate $\mu$ and $\sigma$ for the stochastic difference equation (7.26). How do they compare?

(b) Now, study the variance of the maximum likelihood estimates of the parameters in a GARCH(1,1) model by simulating 1000 price paths as in Exercise 7.17b and computing the estimates for each.

# Chapter 8

# Software for Random Number Generation

Random number generators are widely available in a variety of software packages. Some programming languages such as C/C++, Fortran 95, and Ada 95 provide built-in uniform random number generators, but the standards for these languages do not specify the algorithm, and it is often difficult to determine from the documentation what algorithm is implemented or whether the algorithm is correctly implemented. Except for small studies, these built-in generators should be avoided.

A good software package for random number generation provides a variety of basic uniform generators with long periods and that have undergone stringent testing. The documentation for the package should include reports of results of tests using such standard suites as DIEHARD, the NIST Test Suite, and TestU01 (see page 79). The software package should provide the capability of skipping ahead in the sequence or of interrupting the sequence and then beginning again at the same point in the sequence. The software package also should provide generators for some of the standard distributions, such as the normal distribution. The package should allow the user to specify which uniform generator is to be used for generating deviates from other distributions. The package should also allow incorporation of the user's own basic uniform generator and allow the user to specify it as the one to be used for generating deviates from other distributions. These last two things are not often found in scientific software packages.

Because of the impossibility of fully testing a random number generator, we advocate the use of ad hoc tests of streams that are used in simulation. The software package should include a basic test suite, so the user can easily perform empirical tests on the generators. This is an capability not often found in scientific software packages.

Test suites are often designed to test the bit stream produced by a given algorithm. The user should be able not only to test unformatted bit streams but also the actual numbers represented by formatted bits, either in floating point or in (signed) fixed point. It is also useful to compare results from simulations using different basic uniform generators.

### Efficiency

Monte Carlo simulations often involve many hours of computer time, so computational efficiency is very important in software for random number generation.

Implementing one of the simple methods to convert a uniform deviate to that of another distribution may not be as efficient as a special method for the target distribution, and, as we have indicated, those special methods may be somewhat complicated. The IMSL Libraries and S-Plus and R have a number of modules that use efficient methods to generate variates from several of the more common distributions. Matlab has a basic uniform generator, rand, and a standard normal generator, randn. The Matlab Statistics Toolbox also contains generators for several other distributions.

In programming random number generators, the standard principles of efficiency apply: precompute and store constants, remove any constants within a loop from the loop, rearrange computations to their simplest form (in terms of computer operations), and so on.

It should be noted that the algorithms described in Chapters 4 and 5 are written with an emphasis on clarity rather than on computational efficiency; therefore, in some cases, the code should not correspond directly to the algorithm description.

### Choice of Software for Monte Carlo Studies

Monte Carlo studies typically require many repetitive computations, which are usually implemented through looping program control structures. Some higher-level languages do not provide efficient looping structures. For this reason, it is usually desirable to conduct moderate- to large-scale Monte Carlo studies using a lower-level language such as C or Fortran together with a high-quality library callable with the language.

Higher-level languages are often used in Monte Carlo studies because of the analysis capabilities that they provide. (After all, a Monte Carlo study involves more than just generating random numbers!) Some higher-level languages provide the capability to produce compiled code, which will execute faster. If Monte Carlo studies are to be conducted using an interpretive language, and if the production of compiled code is an option, that option should be chosen for the Monte Carlo work. The higher-level language may also allow the user to replace its basic uniform generator with one supplied by the user. The user should view this option as a major advantage of the software package, even if there is no immediate intent to incorporate a different generator in the program.

A number of Fortran or C/C++ programs are available in collections published by *Applied Statistics* and by *ACM Transactions on Mathematical Software*. These collections are available online at statlib and netlib, respectively. See the bibliography for more information on `statlib` and `netlib`. In Section 8.3 below, we discuss other sources for Fortran and C/C++.

The *Guide to Available Mathematical Software*, or GAMS, (see the bibliography) can be used to locate special software for various distributions.

## 8.1  The User Interface for Random Number Generators

Software for random number generation must provide a certain amount of control by the user, including the ability to

- set or retrieve the seed;

- select seeds that yield separate streams;

- possibly select the method from a limited number of choices.

Whenever the user invokes a random number generator for the first time in a program or a session, the software should not require the specification of a seed, but it should allow the user to set it if desired. If the user does not specify the seed, the software should use some mechanism, such as accessing the system clock, to form a "random" seed. On a subsequent invocation of the random number generator, unless the user specifies a seed, the software should use the value of the seed at the end of the previous invocation. This means that the routine for generating random numbers must produce a "side effect"; that is, it changes something other than the main result. It is a basic tenet of software engineering that careful note must be taken of side effects. At one time, side effects were generally to be avoided. In object-oriented programming, however, objects may encapsulate many entities, and as the object is acted upon, any of the components may change. Therefore, in object-oriented software, side effects are to be expected. In object-oriented software for random number generation, the state of the generator is an object.

Another issue to consider in the design of a user interface for a random number generator is whether the output is a single value (and an updated seed) or an array of values. Although a function that produces a single value such as the C/C++ function `rand()` is convenient to use, it can carry quite a penalty in execution time because of the multiple invocations required to generate an array of random numbers. It is generally better to provide both single- and multivalued procedures for random number generation, especially for the basic uniform generator.

## 8.2   Controlling the Seeds in Monte Carlo Studies

There are four reasons why the user must be able to control the seeds in Monte Carlo studies: for testing of the program, for use of blocks in Monte Carlo experiments, for combining results of Monte Carlo studies, and for strict reproducibility of the research results.

In the early phases of programming for a Monte Carlo study, it is very important to be able to test the output of the program. To do this, it is necessary to use the same seed from one run of the program to another.

As discussed in Chapter 7, there are many situations in which the Monte Carlo study involves different settings for major factors in the study. The experiments in the separate classes may be performed in parallel or in other unsynchronized ways. Separate streams begun by separate seeds are useful in such cases.

When a Monte Carlo study is conducted as a set of separate computer runs, it is necessary that the separate runs not use overlapping sequences of random numbers. The software must provide the capability of interrupting the sequence and then beginning again at the same point in the sequence.

Reproducibility is one of the standard requirements of scientific experimentation. Use of pseudorandom number generators allows Monte Carlo experimentation to be *strictly* reproducible if the software is preserved and if the seeds used are available.

Controlling seeds in a parallel random number generator is much more complicated than in a serial generator. Performing Monte Carlo computations in parallel requires some way of ensuring the independence of the parallel streams (see Section 1.10, page 51).

## 8.3   Random Number Generation in Programming Languages

The built-in generator in C/C++ is the function rand() in stdlib.h. This function returns an integer in the range 0 through RAND_MAX, so the result must be normalized to the range $(0, 1)$. (The scaling should be done with care. Recall that it is desirable to have uniforms in $(0, 1)$ rather than $[0, 1]$; see Exercise 1.13, page 59.) The seed for the C/C++ built-in random number generator is set in srand().

In Fortran 95, the built-in generator is the subroutine random_number, which returns $U(0, 1)$ numbers. (The user must be careful, however; the generator may yield either a 0 or a 1.) The seed can be set in the subroutine random_seed. The design of the Fortran 95 module as a subroutine yields a major advantage over the C function in terms of efficiency. (Of course, because Fortran 95 has the basic advantage of arrays, the module could have been designed as an array function and would still have had an advantage over the C function.)

A basic problem with the built-in generator of C, Fortran 95, and Ada 95 is the problem of portability discussed in Section 1.11, page 54. The bindings are portable, but none of the generators will necessarily generate the same sequence on different platforms. (The "bindings" are the application programming interfaces.)

The freely distributed GNU Scientific Library (GSL) contains several C functions for random number generation. There are several different basic uniform generators in the library, including the Mersenne twister (MT19937), three versions of RANLUX, a combined multiple recursive generator of L'Ecuyer (1996, 1999) (equation 1.46) on page 48), a fifth-order multiple recursive generator of L'Ecuyer, Blouin, and Couture (1993) (equation (1.29) on page 33), a four-tap generalized feedback shift register generator of Ziff (1998), and the R250 generator of Kirkpatrick and Stoll (1981). The library also includes a number of basic uniform generators that yield output sequences that correspond (or almost correspond) to legacy generators provided by various system developers, such as the IBM RANDU and generators associated with various Unix distributions.

GSL also includes programs for generating deviates from a few of the most common nonuniform distributions. The type of the basic uniform generator to be used is determined by an environmental variable, which can be specified by a utility function but also can be set outside of the program (so that different generators can be selected without recompiling the program). A simple seed is also determined by an environmental variable, but utility functions in the library must be used to save or set more complicated states of the generators. Information about the GNU Scientific Library, including links to sites from which source code can be obtained, is available at

```
http://sources.redhat.com/gsl/
```

In addition to the random number generators in Fortran, C, and C++ available at statlib and netlib or through the GNU Scientific Library, *Numerical Recipes* by Press et al. provides random number generators in Fortran (Press et al., 1992) and in C++ (Press et al., 2002). Those generators, especially ran2, which implements a combined shuffled generator of L'Ecuyer (1988), are generally of high quality. *Numerical Recipes* also includes programs for generating deviates from a few of the most common nonuniform distributions.

Given a uniform random number generator, it is usually not too difficult to generate variates from other distributions using the techniques discussed in Chapters 4 and 5. For example, in Fortran 95, the inverse CDF technique for generating a random deviate from a Bernoulli distribution with parameter $\pi$ shown in Algorithm 4.1, page 105, can be implemented by the code in Figure 8.1.

```
integer, parameter      :: n = 100  ! INITIALIZE THIS
real, parameter (pi)    :: pi = .5   ! INITIALIZE THIS
real, dimension (n)     :: uniform
real, dimension (n)     :: bernoulli
call random_number (uniform)
where (uniform .le. pi)
        bernoulli = 1.0
elsewhere
        bernoulli = 0.0
endwhere
```

Figure 8.1: A Fortran 95 Code Fragment to Generate $n$ Bernoulli Random Deviates with Parameter $\pi$

## 8.4  Random Number Generation in IMSL Libraries

For doing Monte Carlo studies, it is usually better to use a software system with a compilable programming language, such as Fortran or C. Not only do such systems provide more flexibility and control, but the programs built in the compiler languages execute faster. To do much work in such a system, however, a library or routines both to perform the numerical computations in the inner loop of the Monte Carlo study and to generate the random numbers driving the study are needed.

The IMSL Libraries contain a large number of routines for random number generation. The libraries are available in both Fortran and C, each providing the same capabilities and with essentially the same interface within the two languages. In Fortran, the basic uniform generator is provided in both function and subroutine forms.

The uniform generator allows the user to choose among seven different algorithms: a linear congruential generator with modulus of $2^{31} - 1$ and with three choices of multiplier, each with or without shuffling, and the generalized feedback shift generator described by Fushimi (1990), which has a period of $2^{521} - 1$. The multipliers that the user can choose are the "minimal standard" one of Park and Miller (1988), which goes back to Lewis, Goodman, and Miller (1969) (see page 20), and two of the "best" multipliers found by Fishman and Moore (1982, 1986), one of which is used in Exercise 1.10 of Chapter 1.

The user chooses which of the basic uniform generators to use by means of the Fortran routine rnopt or the C function imsls_random_option. For whatever choice is in effect, that form of the uniform generator will be used for whatever types of pseudorandom events are to be generated. The states of the generators are maintained in a common block (for the simple congruential generators, the state is a single seed; for the shuffled generators and the GFSR generator, the state is maintained in a table). There are utility routines for setting and saving states of the generators and a utility routine for obtaining a seed to skip ahead a fixed amount.

There are routines to generate deviates from most of the common distributions. Most of the routines are subroutines, but some are functions. The algorithms used often depend on the values of the parameters to achieve greater efficiency. The routines are available in both single and double precisions. (Because the basic underlying sequence is the same, double precision is more for the purpose of convenience for the user than it is for increasing accuracy of the algorithm.)

A single-precision IMSL Fortran subroutine for generating from a specific distribution has the form

$$\text{rn}\textit{name} \ (\textit{number, parameter\_1, parameter\_2, ..., output\_array})$$

where "*name*" is an identifier for the distribution, "*number*" is the number of random deviates to be generated, "*parameter\_i*" are parameters of the distribution, and "*output\_array*" is the output argument with the generated deviates. The Fortran subroutines generate variates from standard distributions, so location and scale parameters are not included in the argument list. The subroutine and formal arguments to generate gamma random deviates, for example, are

$$\text{rngam (nr, a, r)}$$

where a is the shape parameter ($\alpha$) of the gamma distribution. The other parameter in the common two-parameter gamma distribution (usually called $\beta$) is a scale parameter. The deviates produced by the routine rngam have a scale parameter of 1; hence, for a scale parameter of b, the user would follow the call above with a call to a BLAS routine:

$$\text{sscal (nr, b, r, 1)}$$

Identifiers of distributions include those shown in Table 8.1.

For general distributions, the IMSL Libraries provide routines for an alias method and for table lookup for either discrete or continuous distributions. The user specifies a discrete distribution by providing a vector of the probabilities at the mass points and specifies a continuous distribution by giving the values of the cumulative distribution function at a chosen set of points. In the case of a discrete distribution, the generation can be done either by an alias method or by an efficient table-lookup method. For a continuous distribution, a cubic spline is first fit to the given values of the cumulative distribution function, and then an inverse CDF method is used to generate the random numbers from the target distribution. Another routine uses the Thompson–Taylor data-based scheme (Taylor and Thompson, 1986, and Algorithm 5.11, page 212) to generate deviates from an unknown population from which only a sample is available.

Other routines in the IMSL Libraries generate various kinds of time series, random permutations, and random samples. The routine rnuno, which generates order statistics from a uniform distribution, can be used to generate order statistics from other distributions.

Table 8.1: Root Names for IMSL Random Number Generators

| Continuous Distributions | | Discrete Distributions | |
|---|---|---|---|
| un or unf | uniform | bin | binomial |
| nor, noa, or nof | normal | nbn | negative binomial |
| mvn | multivariate normal | poi | Poisson |
| chi | chi-squared | geo | geometric |
| stt | Student's $t$ | hyp | hypergeometric |
| tri | triangular | lgr | logarithmic |
| lnl | lognormal | und | discrete uniform |
| exp | exponential | mtn | multinomial |
| gam | gamma | tab | two-way tables |
| wib | Weibull | | |
| chy | Cauchy | | |
| beta | beta | | |
| vms | von Mises | | |
| stab | stable | | |
| ext | exponential mixture | | |
| cor | correlation matrices | | |
| sph | points on a circle or sphere | | |
| nos | order statistics from a normal | | |
| uno | order statistics from a uniform | | |
| arm | ARMA process | | |
| npp | nonhomogeneous Poisson process | | |

All of the IMSL routines for random number generation are available in both Fortran and C. The C functions have more descriptive names, such as random_normal. Also, the C functions may allow specification of additional arguments, such as location and scale parameters. For example, random_normal has optional arguments IMSLS_MEAN and IMSLS_VARIANCE.

### Controlling the State of the Generators

Figure 8.2 illustrates the way to save the state of an IMSL generator and then restart it. The functions to save and to set the seed are rnget and rnset, respectively.

In a collection of numerical routines such as the IMSL Libraries, it is likely that some of the routines will use random numbers in regular deterministic computations, such as an optimization routine generating random starting points. In a well-designed system, before a routine in the system uses a random number generator in the system, it will retrieve the current value of the seed if one has been set, use the generator, and then reset the seed to the former value. IMSL subprograms are designed in this way. This allows the user to control the seeds in the routines called directly.

```
call rnget (iseed)        ! save it
call rnun (nr, y)         ! get sample, analyze, etc.
...
call rnset (iseed)        ! restore seed
call rnun (nr, yagain)    ! will be the same as y
```

Figure 8.2: Fortran Code Fragment to Save and Restart a Random Sequence Using the IMSL Library

# 8.5 Random Number Generation in S-Plus and R

The software system called S was developed at Bell Laboratories in the mid-1970s. Work on S has continued at Bell Labs, and the system has evolved considerably since the early versions (see Becker, Chambers, and Wilks, 1988, and Chambers, 1997). S is both a data analysis system and an object-oriented programming language.

S-Plus is an enhancement of S developed by StatSci, Inc. (now a part of Insightful Corporation). The enhancements include graphical interfaces, more statistical analysis functionality, and support.

There is a freely available package, called R, that provides generally the same functionality in the same language as S (see Gentleman and Ihaka, 1997). The R programming system is available at

http://www.r-project.org/

S-Plus and R do not use the same random number generators. *Monte Carlo studies conducted using built-in random number generators in one system cannot reliably be reproduced exactly in the other system with its built-in generators.*

Random number generators in S-Plus are all based upon a single uniform random number generator that is a combination of a linear congruential generator and a Tausworthe generator. The original generator, called "Super-Duper", was developed by George Marsaglia in the 1970s. It is described in Learmonth and Lewis (1973). McCullough (1999) reports results of the DIEHARD tests on the S-Plus generator. The tests raise some questions about the quality of the generator.

Several choices for the basic uniform generator are available in R. The function RNGkind can be used to choose the generator. One of the choices is Super-Duper, but the implementation is slightly different from the implementation in S-Plus. The user can also specify a user-defined and programmed generator. The chosen (or default) basic uniform generator is used in the generation of nonuniform variates.

In S-Plus and R, there are some basic functions with the form

*rname* (*number* [, *parameters*])

where "*name*" is an identifier for the distribution, "*number*" is the number of random deviates to be generated (which can be specified by an array argument, in which case the number is the number of elements in the array), and "*parameters*" are parameters of the distribution, which may or may not be required.

For distributions with standard forms, such as the normal, the parameters may be optional, in which case they take on default values if they are not specified. For other distributions, such as the gamma or the $t$, there are required parameters. Optional parameters are both positional and keyword.

For example, the normal variate generation function is

```
rnorm (n, mean=0, sd=1)
```

so

| | |
|---|---|
| `rnorm (n)` | yields $n$ normal (0,1) variates, |
| `rnorm (n, 100, 10)` | yields $n$ normal (100,100) variates, |
| `rnorm (n, 100)` | yields $n$ normal (100,1) variates, |
| `rnorm (n, sd=10)` | yields $n$ normal (0,100) variates. |

(Note that S-Plus and R consider one of the natural parameters of the normal distribution to be the standard deviation or the scale rather than the variance, as is more common.)

For the gamma distribution, at least one parameter (the shape parameter, see page 178) is required. The function reference

```
rgamma (100,5)
```

generates 100 random numbers from a gamma distribution with a shape parameter of 5 and a scale parameter of 1 (a standard gamma distribution).

Identifiers of distributions include those shown in Table 8.2.

The `sample` function generates a random sample with or without replacement. Sampling with replacement is equivalent to generating random numbers from a (finite) discrete distribution. The mass points and probabilities can be specified in optional arguments:

```
xx <- sample(massp, n, replace=T, probs)
```

Order statistics in S-Plus and R can be generated using the beta distribution and the inverse distribution function. For example, ten maximum order statistics from normal samples of size 30 can be generated by

```
x <- qnorm(rbeta(10,30,1))
```

### Controlling the State of the Generators

Both S-Plus and R use an object called `.Random.seed` to maintain the state of the random number generators. In R, `.Random.seed` also maintains an indicator of which of the basic uniform random number generators is the current choice. Whenever random number generation is performed, if `.Random.seed`

Table 8.2: Root Names for S-Plus and R Functions for Distributions

| Continuous Distributions | | Discrete Distributions | |
|---|---|---|---|
| `unif` | uniform | `binom` | binomial |
| `norm` | normal | `nbinom` | negative binomial |
| `mvnorm` | multivariate normal | `pois` | Poisson |
| `chisq` | chi-squared | `geom` | geometric |
| `t` | $t$ | `hyper` | hypergeometric |
| `f` | $F$ | `wilcox` | Wilcoxon rank sum statistic |
| `lnorm` | lognormal | | |
| `exp` | exponential | | |
| `gamma` | gamma | | |
| `weibull` | Weibull | | |
| `cauchy` | Cauchy | | |
| `beta` | beta | | |
| `logis` | logistic | | |
| `stab` | stable | | |
| `arima.sim` | ARIMA process | | |

does not exist in the user's working directory, it is created. If it exists, it is used to initiate the pseudorandom sequence and then is updated after the sequence is generated. Setting a different working directory will change the state of the random number generator.

The function `set.seed(i)` provides a convenient way of setting the value of the `.Random.seed` object in the working directory to one of a fixed number of values. The argument `i` is an integer between 0 and 1023, and each value represents a state of the generator, which is "far away" from the other states that can be set in `set.seed`.

To save the state of the generator, just copy `.Random.seed` into a named object, and to restore, just copy the named object back into `.Random.seed`, as in Figure 8.3.

```
oldseed <- .Random.seed  # save it
y <- runif(1000)          # get sample, analyze, etc.
...
.Random.seed <- oldseed  # restore seed
yagain <- rnorm(1000)     # will be the same as y
```

Figure 8.3: Code Fragment to Save and Restart a Random Sequence Using S-Plus or R

A common situation is one in which computations for a Monte Carlo study are performed intermittently and are interspersed with other computations, perhaps broken over multiple sessions. In such a case, we may begin by setting the seed using the function set.seed(i), save the state after each set of computations in the study, and then restore it prior to resuming the computations, similar to the code shown in Figure 8.4.

```
set.seed(10)                # set seed at beginning of study
... # perform some computations for the Monte Carlo study
MC1seed <- .Random.seed  # save the generator state
... # do other computations
.Random.seed <- MC1seed  # restore seed
... # perform some computations for the Monte Carlo study
MC1seed <- .Random.seed  # save the generator state
```

Figure 8.4: Starting and Restarting Monte Carlo Studies in S-Plus or R

The built-in functions in S-Plus that use the random number generators have the side effect of changing the state of the generators, so the user must be careful in Monte Carlo studies where the computational nuclei, such as ltsreg for robust regression, for example, invoke an S-Plus random number generator. In this case, the user must retrieve the state of the generator prior to calling the function and then reset the state prior to the next invocation of a random number generator.

In order to avoid the side effect of changing the state of the generator, when writing a function in S-Plus or R, the user can preserve the state upon entry to the function and restore it prior to exit. The assignment

```
.Random.seed <- oldseed
```

in Figure 8.3, however, does not work if it occurs within a user-written function in S-Plus or R. Within a function, the assignment must be performed by the <<- operator. A well-designed S-Plus or R function that invokes a random number generator would have code similar to that in Figure 8.5.

```
oldseed <- .Random.seed   # save seed on entry
...
.Random.seed <<- oldseed  # restore seed on exit
return(...)
```

Figure 8.5: Saving and Restoring the State of the Generator within an S-Plus or R Function

## Monte Carlo in S-Plus and R

Explicit loops in S-Plus or R execute very slowly. For that reason, it is best to use array arguments for functions rather than to loop over scalar values of the arguments. Consider, for example, the problem of evaluating the integral

$$\int_0^2 \log(x+1)x^2(2-x)^3\, dx.$$

This could be estimated in a loop as follows:

```
# First, initialize n.
uu <- runif(n, 0, 2)
eu <- 0
for (i in 1:n) eu <- eu + log(uu[i]+1)*uu[i]^2*(2-uu[i])^3
eu <- 2*eu/n
```

A much more efficient way, without the for loop but still using the uniform, is

```
uu <- runif(n, 0, 2)
eu <- 2*sum(log(uu+1)*uu^2*(2-uu)^3)/n
```

Alternatively, using the beta density as a weight function, we have

```
eb <- (16/15)*sum(log(2*rbeta(n,3,4)+1))/n
```

(Of course, if we recognize the relationship of the integral to the beta distribution, we would not use Monte Carlo as the method of integration.)

For large-scale Monte Carlo studies, an interpretive language such as S-Plus or R may require an inordinate amount of running time. These systems are very useful for prototyping Monte Carlo studies, but it is often better to do the actual computations in a compiled language such as Fortran or C.

# Exercises

8.1. Identify as many random number generators as you can that are available on your computer system. Try to determine what method each uses. Do the generators appear to be of high quality?

8.2. Consider the problem of evaluating the integral

$$\int_{-\pi}^{\pi}\int_0^4\int_0^{\infty} x^2 y^3 \sin(z)(\pi+z)^2(\pi-z)^3 e^{-\frac{x}{2}}\, dx\, dy\, dz.$$

Note the gamma and beta weighting functions.

(a) Write a Fortran or C program to use the IMSL Libraries to evaluate this integral by Monte Carlo methods. Use a sample of size 1000, and save the state of the generator, so you can restart it. Now, use a sample of size 10,000, starting where you left off in the first 1000. Combine your two estimates.

(b) Now, do the same thing in S-Plus.

(c) Now, do the same thing in Fortran 90 using its built-in random number functions. You may use other software to evaluate special functions if you wish.

8.3. Obtain the programs for Algorithm 738 for generating quasirandom numbers (Bratley, Fox, and Niederreiter, 1994) from the *Collected Algorithms of the ACM*. The programs are in Fortran and may require a small number of system-dependent modifications, which are described in the documentation embedded in the source code.

Devise some tests for Monte Carlo evaluation of multidimensional integrals, and compare the performance of Algorithm 738 with that of a pseudorandom number generator. (Just use any convenient pseudorandom generator available to you.) The subroutine TESTF accompanying Algorithm 738 can be used for this purpose.

Can you notice any difference in the performance?

8.4. Obtain the code for SPRNG, the scalable parallel random number generators. The source code is available at

http://sprng.cs.fsu.edu

Get the code running, preferably on a parallel system. (It will run on a serial machine also.) Choose some simple statistical tests, and apply them to sample output from single streams and also to the output of separate streams. (In the latter case, the primary interest is in correlations across the streams.)

# Chapter 9

# Monte Carlo Studies in Statistics

In statistical inference, certain properties of the test statistic or estimator must be assumed to be known. In simple cases, under rigorous assumptions, we have complete knowledge of the statistic. In testing a mean of a normal distribution, for example, we use a $t$ statistic, and we know its exact distribution. In other cases, however, we may have a perfectly reasonable test statistic but know very little about its distribution. For example, suppose that a statistic $T$, computed from a differenced time series, could be used to test the hypothesis that the order of differencing is sufficient to yield a series with a zero mean. If enough information about the distribution of $T$ is known under the null hypothesis, that value may be used to construct a test that the differencing is adequate. This, in fact, was what Erastus Lyman de Forest studied in the 1870s in one of the earliest documented Monte Carlo studies of a statistical procedure. De Forest studied ways of smoothing a time series by simulating the data using cards drawn from a box. A description of De Forest's Monte Carlo study is given in Stigler (1978). Stigler (1991) also describes other Monte Carlo simulation by nineteenth-century scientists and suggests that "simulation, in the modern sense of that term, may be the oldest of the stochastic arts".

Another early use of Monte Carlo was the sampling experiment (using biometric data recorded on pieces of cardboard) that led W. S. Gosset to the discovery of the distribution of the $t$-statistic and the correlation coefficient. (See Student, 1908a, 1908b. Of course, it was Ronald Fisher who later worked out the distributions.)

Major advances in Monte Carlo techniques were made during World War II and afterward by mathematicians and scientists working on problems in atomic physics. (In fact, it was the group led by John von Neumann and S. M. Ulam who coined the term "Monte Carlo" to refer to these methods.) The use of Monte Carlo techniques by statisticians gradually increased from the time of

De Forest, but after the widespread availability of digital computers, the usage greatly expanded.

In the mathematical sciences, including statistics, simulation has become an important tool in the development of theory and methods. For example, if the properties of an estimator are very difficult to work out analytically, a Monte Carlo study may be conducted to estimate those properties.

Often, the Monte Carlo study is an informal investigation whose main purpose is to indicate promising research directions. If a "quick and dirty" Monte Carlo study indicates that some method of inference has good properties, it may be worth the time of the research worker in developing the method and perhaps doing the difficult analysis to confirm the results of the Monte Carlo study.

In addition to quick Monte Carlo studies that are mere precursors to analytic work, Monte Carlo studies often provide a significant amount of the available knowledge of the properties of statistical techniques, especially under various alternative models. A large proportion of the articles in the statistical literature include Monte Carlo studies. In recent issues of the *Journal of the American Statistical Association*, for example, almost half of the articles report on Monte Carlo studies that supported the research.

One common use of Monte Carlo studies is to compare statistical methods. For example, we may wish to compare a procedure based on maximum likelihood with a procedure using least squares. The comparison of methods is often carried out for different distributions for the random component of the model used in the study. It is especially interesting to study how standard statistical methods perform when the distribution of the random component has heavy tails or when the distribution is contaminated by outliers. Monte Carlo methods are widely used in these kinds of studies of the robustness of statistical methods.

## 9.1    Simulation as an Experiment

A simulation study that incorporates a random component is an experiment. The principles of statistical design and analysis apply just as much to a Monte Carlo study as they do to any other scientific experiment. The Monte Carlo study should adhere to the same high standards of any scientific experimentation:

- control;

- reproducibility;

- efficiency;

- careful and complete documentation.

In simulation, *control*, among other things, relates to the fidelity of a *nonrandom* process to a *random* process. The experimental units are only simulated.

Questions about the computer model must be addressed (tests of the random number generators and so on).

Likewise, *reproducibility* is predicated on good random number generators (or else on equally bad ones!). Portability of the random number generators enhances reproducibility and in fact can allow *strict* reproducibility. Reproducible research also requires preservation and documentation of the computer programs that produced the results (see Buckheit and Donoho, 1995).

The principles of good statistical design can improve the efficiency. Use of good designs (fractional factorials, etc.) can allow efficient simultaneous exploration of several factors. Also, there are often many opportunities to reduce the variance (improve the efficiency). Hammersley and Hanscomb (1964, page 8) note

> ... statisticians were insistent that other experimentalists should design experiments to be as little subject to unwanted error as possible, and had indeed given important and useful help to the experimentalist in this way; but in their own experiments they were singularly inefficient, nay negligent in this respect.

Many properties of statistical methods of inference are analytically intractable. Asymptotic results, which are often easy to work out, may imply excellent performance, such as consistency with a good rate of convergence, but the finite sample properties are ultimately what must be considered. Monte Carlo studies are a common tool for investigating the properties of a statistical method, as noted above. In the literature, the Monte Carlo studies are sometimes called "numerical results". Some numerical results are illustrated by just one randomly generated dataset; others are studied by averaging over thousands of randomly generated sets.

In a Monte Carlo study, there are usually several different things ("treatments" or "factors") that we want to investigate. As in other kinds of experiments, a factorial design is usually more efficient. Each factor occurs at different "levels", and the set of all levels of all factors that are used in the study constitutes the "design space". The measured responses are properties of the statistical methods, such as their sample means and variances.

The factors commonly studied in Monte Carlo experiments in statistics include the following.

- statistical method (estimator, test procedure, etc.)

- sample size

- the problem for which the statistical method is being applied (that is, the "true" model, which may be different from the one for which the method was developed). Factors relating to the type of problem may be:

  - distribution of the random component in the model (normality?)

  - correlation among observations (independence?)

  – homogeneity of the observations (outliers?, mixtures?)

  – structure of associated variables (leverage?)

The factor whose effect is of primary interest is the statistical method. The other factors are generally just blocking factors. There is, however, usually an interaction between the statistical method and these other factors.

As in physical experimentation, observational units are selected for each point in the design space and measured. The measurements, or "responses" made at the same design point, are used to assess the amount of random variation, or variation that is not accounted for by the factors being studied. A comparison of the variation among observational units at the same levels of all factors with the variation among observational units at different levels is the basis for a decision as to whether there are real (or "significant") differences at the different levels of the factors. This comparison is called analysis of variance. The same basic rationale for identifying differences is used in simulation experiments.

A fundamental (and difficult) question in experimental design is how many experimental units to observe at the various design points. Because the experimental units in Monte Carlo studies are generated on the computer, they are usually rather inexpensive. The subsequent processing (the application of the factors, in the terminology of an experiment) may be very extensive, however, so there is a need to design an efficient experiment.

## 9.2   Reporting Simulation Experiments

The reporting of a simulation experiment should receive the same care and consideration that would be accorded the reporting of other scientific experiments. Hoaglin and Andrews (1975) outline the items that should be included in a report of a simulation study. In addition to a careful general description of the experiment, the report should include mention of the random number generator used, any variance-reducing methods employed, and a justification of the simulation sample size. The *Journal of the American Statistical Association* includes these reporting standards in its style guide for authors.

Closely related to the choice of the sample size is the standard deviation of the estimates that result from the study. The sample standard deviations actually achieved should be included as part of the report. Standard deviations are often reported in parentheses beside the estimates with which they are associated. A formal analysis, of course, would use the sample variance of each estimate to assess the significance of the differences observed between points in the design space; that is, a formal analysis of the simulation experiment would be a standard analysis of variance.

The most common method of reporting the results is by means of tables, but a better understanding of the results can often be conveyed by graphs.

# 9.3 An Example

One area of statistics in which Monte Carlo studies have been used extensively is robust statistics. This is because the finite sampling distributions of many robust statistics are very difficult to work out, especially for the kinds of underlying distributions for which the statistics are to be studied. A well-known use of Monte Carlo methods is in the important study of robust statistics described by Andrews et al. (1972), who introduced and examined many alternative estimators of location for samples from univariate distributions. This study, which involved many Monte Carlo experiments, employed innovative methods of variance reduction and was very influential in subsequent Monte Carlo studies reported in the statistical literature.

As an example of a Monte Carlo study, we will now describe a simple experiment to assess the robustness of a statistical test in linear regression analysis. The purpose of this example is to illustrate some of the issues in designing a Monte Carlo experiment. The results of this small study are not of interest here. There are many important issues about the robustness of the procedures that we do not address in this example.

### The Problem

Consider the simple linear regression model

$$Y = \beta_0 + \beta_1 x + E,$$

where a response or "dependent variable", $Y$, is modeled as a linear function of a single regressor or "independent variable", $x$, plus a random variable, $E$, called the "error". Because $E$ is a random variable, $Y$ is also a random variable. The statistical problem is to make inferences about the unknown, constant parameters $\beta_0$ and $\beta_1$ and about distributional parameters of the random variable, $E$. The inferences are made based on a sample of $n$ pairs, $(y_i, x_i)$, with which are associated unobservable realizations of the random error, $\epsilon_i$, and are assumed to have the relationship

$$y_i = \beta_0 + \beta_1 x_i + \epsilon_i. \tag{9.1}$$

We also generally assume that the realizations of the random error are independent and are unrelated to the value of $x$.

For this example, let us consider just the specific problem of testing the hypothesis

$$H_0\colon \beta_1 = 0 \tag{9.2}$$

versus the universal alternative. If the distribution of $E$ is normal and we make the additional assumptions above about the sample, the optimal test for the hypothesis (using the common definitions of optimality) is based on a least squares procedure that yields the statistic

$$t = \frac{\widehat{\beta}_1 \sqrt{(n-2)\sum(x_i - \bar{x})^2}}{\sqrt{\sum r_i^2}}, \tag{9.3}$$

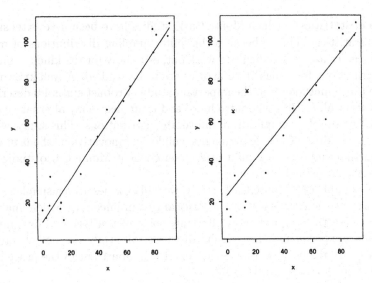

Figure 9.1: Least Squares Fit Using Two Datasets that are the Same Except for Two Outliers

where $\bar{x}$ is the mean of the $x$s, $\widehat{\beta}_1$ together with $\widehat{\beta}_0$ minimizes the function

$$L_2(b_0, b_1) = \sum_{i=1}^{n}(y_i - b_0 - b_1 x_i)^2,$$

and

$$r_i = y_i - (\widehat{\beta}_0 + \widehat{\beta}_1 x_i).$$

If the null hypothesis is true, then $t$ is a realization of a Student's $t$ distribution with $n - 2$ degrees of freedom. The test is performed by comparing the $p$-value from the Student's $t$ distribution with a preassigned significance level, $\alpha$, or by comparing the observed value of $t$ with a critical value. The test of the hypothesis depends on the estimates of $\beta_0$ and $\beta_1$ used in the test statistic $t$.

Often, a dataset contains outliers (that is, observations that have a realized error that is very large in absolute value) or observations for which the model is not appropriate. In such cases, the least squares procedure may not perform so well. We can see the effect of some outliers on the least squares estimates of $\beta_0$ and $\beta_1$ in Figure 9.1. For well-behaved data, as in the plot on the left, the least squares estimates seem to fit the data fairly well. For data with two outlying points, as in the plot on the right in Figure 9.1, the least squares estimates are affected so much by the two points in the upper left part of the graph that the estimates do not provide a good fit for the bulk of the data.

Another method of fitting the linear regression line that is robust to outliers in $E$ is to minimize the absolute values of the deviations. The least absolute values procedure chooses estimates of $\beta_0$ and $\beta_1$ to minimize the function

$$L_1(b_0, b_1) = \sum_{i=1}^{n} |y_i - b_0 - b_1 x_i|.$$

Figure 9.2 shows the same two datasets as before with the least squares (LS) fit and the least absolute values (LAV) fit plotted on both graphs. We see that the least absolute values fit does not change because of the outliers.

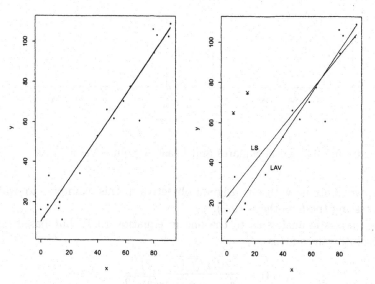

Figure 9.2: Least Squares Fits and Least Absolute Values Fits

Another concern in regression analysis is the unduly large influence that some individual observations exert on the aggregate statistics because the values of $x$ in those observations lie at a large distance from the mean of all of the $x_i$s (that is, those observations whose values of the independent variables are outliers). The influence of an individual observation is called *leverage*. Figure 9.3 shows two datasets together with the least squares and the least absolute values fits for both. In both datasets, there is one value of $x$ that lies far outside the range of the other values of $x$. All of the data in the plot on the left in Figure 9.3 lie relatively close to a line, and both fits are very similar. In the plot on the right, the observation with an extreme value of $x$ also happens to have an outlying value of $E$. The effect on the least squares fit is marked, while the least absolute values fit is not affected as much. (Despite this example, least absolute values fits are generally not very robust to outliers at high leverage points, especially if there are multiple such outliers. There are other methods of fitting that are more robust to outliers at high leverage points. We refer the interested reader to Rousseeuw and Leroy, 1987, for discussion of these issues.)

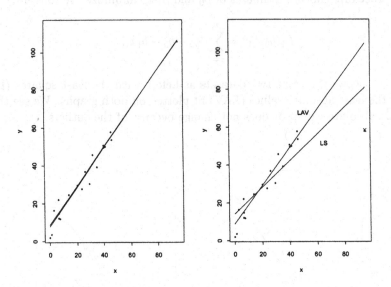

Figure 9.3: Least Squares and Least Absolute Values Fits

Now, we continue with our original objective in this example: to evaluate ways of testing the hypothesis (9.2).

A test statistic analogous to the one in equation (9.3), but based on the least absolute values fit, is

$$t_1 = \frac{2\tilde{\beta}_1 \sqrt{\sum(x_i - \bar{x})^2}}{(e_{(k_2)} - e_{(k_1)})\sqrt{n-2}}, \qquad (9.4)$$

where $\tilde{\beta}_1$ together with $\tilde{\beta}_0$ minimizes the function

$$L_1(b_0, b_1) = \sum_{i=1}^{n} |y_i - b_0 - b_1 x_i|,$$

$e_{(k)}$ is the $k^{\text{th}}$ order statistic from

$$e_i = y_i - (\tilde{\beta}_0 + \tilde{\beta}_1 x_i),$$

$k_1$ is the integer closest to $(n-1)/2 - \sqrt{n-2}$, and $k_2$ is the integer closest to $(n-1)/2 + \sqrt{n-2}$. This statistic has an approximate Student's $t$ distribution with $n-2$ degrees of freedom (see Birkes and Dodge, 1993, for example).

If the distribution of the random error is normal, inference based on minimizing the sum of the absolute values is not nearly as efficient as inference based on least squares. This alternative to least squares should therefore be used with some discretion. Furthermore, there are other procedures that may warrant consideration. It is not our purpose here to explore these important issues in robust statistics, however.

## The Design of the Experiment

At this point, we should have a clear picture of the problem: we wish to compare two ways of testing the hypothesis (9.2) under various scenarios. The data may have outliers, and there may be observations with large leverage. We expect that the optimal test procedure will depend on the presence of outliers or, more generally, on the distribution of the random error and on the pattern of the values of the independent variable. The possibilities of interest for the distribution of the random error include

- the family of the distribution (that is, normal, double exponential, Cauchy, and so on);

- whether the distribution is a mixture of more than one basic distribution, and, if so, the proportions in the mixture;

- the values of the parameters of the distribution (that is, the variance, the skewness, or any other parameters that may affect the power of the test).

In textbooks on the design of experiments, a simple objective of an experiment is to perform a $t$ test or an $F$ test of whether different levels of response are associated with different treatments. Our objective in the Monte Carlo experiment that we are designing is to investigate and characterize the dependence of the performance of the hypothesis test on these factors. The principles of design are similar to those of other experiments, however.

It is possible that the optimal test of the hypothesis will depend on the sample size or on the true values of the coefficients in the regression model, so some additional issues that are relevant to the performance of a statistical test of this hypothesis are the sample size and the true values of $\beta_0$ and $\beta_1$.

In the terminology of statistical models, the factors in our Monte Carlo experiment are the estimation method and the associated test, the distribution of the random error, the pattern of the independent variable, the sample size, and the true value of $\beta_0$ and $\beta_1$. The estimation method together with the associated test is the "treatment" of interest. The "effect" of interest (that is, the measured response) is the proportion of times that the null hypothesis is rejected using the two treatments.

We now can see our objective more clearly: for each setting of the distribution, pattern, and size factors, we wish to measure the power of the two tests. These factors are similar to blocking factors except that there is likely to be an interaction between the treatment and these factors. Of course, the power depends on the nominal level of the test, $\alpha$. It may be the case that the nominal level of the test affects the relative powers of the two tests.

We can think of the problem in the context of a binary response model,

$$\mathrm{E}(P_{ijklqsr}) = f(\tau_i, \delta_j, \phi_k, \nu_l, \alpha_q, \beta_{1s}), \qquad (9.5)$$

where the parameters represent levels of the factors listed above ($\beta_{1s}$ is the $s^{\text{th}}$ level of $\beta_1$), and $P_{ijklqsr}$ is a binary variable representing whether the test rejects the null hypothesis on the $r^{\text{th}}$ trial at the $(ijklqs)^{\text{th}}$ setting of the design factors. It is useful to write down a model like this to remind ourselves of the issues in designing an experiment.

At this point, it is necessary to pay careful attention to our terminology. We are planning to use a statistical procedure (a Monte Carlo experiment) to evaluate a statistical procedure (a statistical test in a linear model). For the statistical procedure that we will use, we have written a model (9.5) for the observations that we will make. Those observations are indexed by $r$ in that model. Let $m$ be the sample size for each combination of factor settings. This is the Monte Carlo sample size. It is not to be confused with the data sample size, $n$, which is one of the factors in our study.

We now choose the levels of the factors in the Monte Carlo experiment.

- For the estimation method, we have decided on two methods: least squares and least absolute values. Its differential effect in the binary response model (9.5) is denoted by $\tau_i$ for $i = 1, 2$.

- For the distribution of the random error, we choose three general ones:

   1. normal $(0, 1)$;

   2. normal $(0, 1)$ with $c\%$ outliers from normal $(0, d^2)$;

   3. standard Cauchy.

  We choose different values of $c$ and $d$ as appropriate. For this example, let us choose $c = 5$ and 20 and $d = 2$ and 5. Thus, in the binary response model (9.5), $j = 1, 2, 3, 4, 5, 6$.

- For the pattern of the independent variable, we choose three different arrangements:

   1. uniform over the range;

   2. a group of extreme outliers;

   3. two groups of outliers.

  In the binary response model (9.5), $k = 1, 2, 3$. We use fixed values of the independent variable.

- For the sample size, we choose three values: 20, 200, and 2000. In the binary response model (9.5), $l = 1, 2, 3$.

- For the nominal level of the test, we choose two values: 0.01 and 0.05. In the binary response model (9.5), $q = 1, 2$.

- The true value of $\beta_0$ is probably not relevant, so we just choose $\beta_0 = 1$. We are interested in the power of the tests at different values of $\beta_1$. We expect the power function to be symmetric about $\beta_1 = 0$ and to approach 1 as $|\beta_1|$ increases.

The estimation method is the "treatment" of interest.

Restating our objective in terms of the notation introduced above, for each of two tests, we wish to estimate the power curve,

$$\Pr(\text{reject } H_0) = g(\beta_1 \mid \tau_i, \delta_j, \phi_k, \nu_l, \alpha_q),$$

for any combination $(\tau_i, \delta_j, \phi_k, \nu_l, \alpha_q)$. For either test, this curve should have the general appearance of the curve shown in Figure 9.4.

The minimum of the power curve should occur at $\beta_1 = 0$ and should be $\alpha$. The curve should approach 1 symmetrically as $|\beta_1|$.

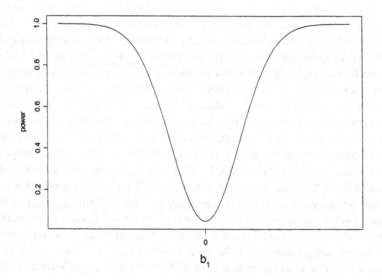

Figure 9.4: Power Curve for Testing $\beta_1 = 0$

To estimate the curve, we use a discrete set of points, and because of symmetry, all values chosen for $\beta_1$ can be nonnegative. The first question is at what point does the curve flatten out just below 1. We might arbitrarily define the region of interest to be that in which the power is less than 0.99 approximately. The abscissa of this point is the maximum $\beta_1$ of interest. This point, say $\beta_1^*$, varies, depending on all of the factors in the study. We could work this out in the least squares case for uncontaminated normal errors using the noncentral Student's $t$ distribution, but, for other cases, it is analytically intractable. Hence, we compute some preliminary Monte Carlo estimates to determine the maximum $\beta_1$ for each factor combination in the study.

To do a careful job of fitting a curve using a relatively small number of points, we would choose points where the second derivative is changing rapidly and especially near points of inflection where the second derivative changes sign. Because the problem of determining these points for each combination of

$(i, j, k, l, q)$ is not analytically tractable (otherwise, we would not be doing the study!), we may conveniently choose a set of points equally spaced between 0 and $\beta_1^*$. Let us decide on five such points for this example. It is not important that the $\beta_1^*$s be chosen with a great deal of care. The objective is that we be able to calculate two power curves between 0 and $\beta_1^*$ that are meaningful for comparisons.

### The Experiment

The observational units in the experiment are the values of the test statistics (9.3) and (9.4). The measurements are the binary variables corresponding to rejection of the hypothesis (9.2). At each point in the factor space, there will be $m$ such observations. If $z$ is the number of rejections observed, then the estimate of the power is $z/m$, and the variance of the estimator is $\pi(1 - \pi)/m$, where $\pi$ is the true power at that point. ($z$ is a realization of a binomial random variable with parameters $m$ and $\pi$.) This leads us to a choice of the value of $m$. The coefficient of variation at any point is $\sqrt{(1 - \pi)/(m\pi)}$, which increases as $\pi$ decreases. At $\pi = 0.50$, a 5% coefficient of variation can be achieved with a sample of size 400. This yields a standard deviation of 0.025. There may be some motivation to choose a slightly larger value of $m$ because we can assume that the minimum of $\pi$ will be approximately the minimum of $\alpha$. To achieve a 5% coefficient of variation at the point at which $\alpha_1 = 0.05$ would require a sample of size approximately 160,000. That would correspond to a standard deviation of 0.0005, which is probably much smaller than we need. A sample size of 400 would yield a standard deviation of 0.005. Although that is large in a relative sense, it may be adequate for our purposes. Because this particular point (where $\beta_1 = 0$) corresponds to the null hypothesis, however, we may choose a larger sample size, say 4000, at that special point. A reasonable choice therefore is a Monte Carlo sample size of 4000 at the null hypothesis and 400 at all other points. We will, however, conduct the experiment in such a way that we can combine the results of this experiment with independent results from a subsequent experiment.

The experiment is conducted by running a computer program. The main computation in the program is to determine the values of the test statistics and to compare them with their critical values to decide on the hypothesis. These computations need to be performed at each setting of the factors and for any given realization of the random sample.

We design a program that allows us to loop through the settings of the factors and, at each factor setting, to use a random sample. The result is a nest of loops. The program may be stopped and restarted, so we need to be able to control the seeds (see Section 8.2, page 286).

Recalling that the purpose of our experiment is to obtain estimates, we may now consider any appropriate methods of reducing the variance of those estimates. There is not much opportunity to apply methods of variance reduction discussed in Section 7.5, but at least we might consider at what points to use common realizations of the pseudorandom variables. Because the things that we want to compare most directly are the powers of the tests, we perform the tests on the same pseudorandom datasets. Also, because we are interested in the shape of the power curves, we may want to use the same pseudorandom data-sets at each value of $\beta_1$; that is, to use the same set of errors in the model (9.1). Finally, following similar reasoning, we may use the same pseudorandom data-sets at each setting of the pattern of the independent variable. This implies that our program of nested loops has the structure shown in Figure 9.5.

Initialize a table of counts.
>  Fix the data sample size. (Loop over the sample sizes $n = 20$, $n = 200$, and $n = 2000$.)
>> Generate a set of residuals for the linear regression model (9.1). (This is the loop of $m$ Monte Carlo replications.)
>>> Fix the pattern of the independent variable. (Loop over patterns $P_1$, $P_2$, and $P_3$.)
>>>> Choose the distribution of the error term. (Loop over the distributions $D_1$, $D_2$, $D_3$, $D_4$, $D_5$, and $D_6$.)
>>>>> For each value of $\beta_1$, generate a set of observations (the $y$ values) for the linear regression model (9.1), and perform the tests using both procedures and at both levels of significance. Record results.
>>>> End distributions loop.
>>> End patterns loop.
>> End Monte Carlo loop.
> End sample size loop.
Perform computations of summary statistics.

Figure 9.5: Program Structure for the Monte Carlo Experiment

After writing a computer program with this structure, the first thing is to test the program on a small set of problems and determine appropriate values of $\beta_1^*$. We should compare the results with known values at a few points. (As mentioned earlier, the only points that we can work out correspond to the normal case with the ordinary $t$ statistic. One of these points, at $\beta_1 = 0$, is easily checked.) We can also check the internal consistency of the results. For example, does the power curve increase? We must be careful, of course, in applying such consistency checks because we do not know the behavior of the tests in most cases.

## Reporting the Results

The report of this Monte Carlo study should address as completely as possible the results of interest. The relative values of the power are the main points of interest. The estimated power at $\beta_1 = 0$ is of interest. This is the actual significance level of the test, and how it compares to the nominal level $\alpha$ is of particular interest.

The presentation should be in a form easily assimilated by the reader. This may mean graphs similar to Figure 9.4, except only the nonnegative half, and with the tick marks on the horizontal axis. Two graphs, for the two test procedures, should be shown on the same set of axes. It is probably counterproductive to show a graph for each factor setting. (There are 108 combinations of factor settings.)

In addition to the graphs, tables may allow presentation of a large amount of information in a compact format.

The Monte Carlo study should be described so carefully that the study could be replicated exactly. This means specifying the factor settings, the loop nesting, the software and computer used, the seed used, and the Monte Carlo sample size. There should also be at least a simple statement explaining the choice of the Monte Carlo sample size.

As mentioned earlier, the statistical literature is replete with reports of Monte Carlo studies. Some of these reports (and, likely, the studies themselves) are woefully deficient. An example of a careful Monte Carlo study and a good report of the study are given by Kleijnen (1977). He designed, performed, and reported on a Monte Carlo study to investigate the robustness of a multiple ranking procedure. In addition to reporting on the study of the question at hand, another purpose of the paper was to illustrate the methods of a Monte Carlo study.

## Exercises

9.1. Write a computer program to implement the Monte Carlo experiment described in Section 9.3. The S-Plus functions lsfit and l1fit or the IMSL Fortran subroutines rline and rlav can be used to calculate the fits. See Chapter 8 for discussions of other software that you may use in the program.

9.2. Choose a recent issue of the *Journal of the American Statistical Association* and identify five articles that report on Monte Carlo studies of statistical methods. In each case, describe the Monte Carlo experiment.

   (a) What are the factors in the experiment?

   (b) What is the measured response?

   (c) What is the design space (that is, the set of factor settings)?

   (d) What random number generators were used?

   (e) Critique the report in each article. Did the author(s) justify the sample size? Did the author(s) report variances or confidence intervals? Did the author(s) attempt to reduce the experimental variance?

9.3. Select an article that you identified in Exercise 9.2 that concerns a statistical method that you understand and that interests you. Choose a design space that is not a subspace of that used in the article but has a nonnull intersection with it, and perform a similar experiment. Compare your results with those reported in the article.

# Appendix A

# Notation and Definitions

All notation used in this work is "standard", and in most cases it conforms to the ISO conventions. (The notable exception is the notation for vectors.) I have opted for simple notation, which, of course, results in a one-to-many map of notation to object classes. Within a given context, however, the overloaded notation is generally unambiguous. I have endeavored to use notation consistently.

This appendix is not intended to be a comprehensive listing of definitions. The subject index, beginning on page 377, is a more reliable set of pointers to definitions, except for symbols that are not words.

## General Notation

Uppercase italic Latin and Greek letters, $A$, $B$, $E$, $\Lambda$, and so on, are generally used to represent either matrices or random variables. Random variables are usually denoted by letters nearer the end of the Latin alphabet, $X$, $Y$, $Z$, and by the Greek letter $E$. Parameters in models (that is, unobservables in the models), whether or not they are considered to be random variables, are generally represented by lowercase Greek letters. Uppercase Latin and Greek letters, especially $P$, in general, and $\Phi$, for the normal distribution, are also used to represent cumulative distribution functions. Also, uppercase Latin letters are used to denote sets.

Lowercase Latin and Greek letters are used to represent ordinary scalar or vector variables and functions. *No distinction in the notation is made between scalars and vectors*; thus, $\beta$ may represent a vector, and $\beta_i$ may represent the $i^{\text{th}}$ element of the vector $\beta$. In another context, however, $\beta$ may represent a scalar. All vectors are considered to be column vectors, although we may write a vector as $x = (x_1, x_2, \ldots, x_n)$. Transposition of a vector or a matrix is denoted by a superscript $^{\text{T}}$.

Uppercase calligraphic Latin letters, $\mathcal{F}$, $\mathcal{V}$, $\mathcal{W}$, and so on, are generally used to represent either vector spaces or transforms.

313

Subscripts generally represent indexes to a larger structure; for example, $x_{ij}$ may represent the $(i,j)^{\text{th}}$ element of a matrix, $X$. A subscript in parentheses represents an order statistic. A superscript in parentheses represents an iteration, for example, $x_i^{(k)}$ may represent the value of $x_i$ at the $k^{\text{th}}$ step of an iterative process. The following are some examples:

$x_i$          The $i^{\text{th}}$ element of a structure (including a sample, which is a multiset).

$x_{(i)}$          The $i^{\text{th}}$ order statistic.

$x^{(i)}$          The value of $x$ at the $i^{\text{th}}$ iteration.

Realizations of random variables and placeholders in functions associated with random variables are usually represented by lowercase letters corresponding to the uppercase letters; thus, $\epsilon$ may represent a realization of the random variable $E$.

A single symbol in an italic font is used to represent a single variable. A Roman font or a special font is often used to represent a standard operator or a standard mathematical structure. Sometimes, a string of symbols in a Roman font is used to represent an operator (or a standard function); for example, exp represents the exponential function, but a string of symbols in an italic font on the same baseline should be interpreted as representing a composition (probably by multiplication) of separate objects; for example, $exp$ represents the product of $e$, $x$, and $p$.

A fixed-width font is used to represent computer input or output; for example,

```
a = bx + sin(c).
```

In computer text, a string of letters or numerals with no intervening spaces or other characters, such as bx above, represents a single object, and there is no distinction in the font to indicate the type of object.

Some important mathematical structures and other objects are:

$\mathbb{R}$          The field of reals or the set over which that field is defined.

$\mathbb{R}^d$          The usual $d$-dimensional vector space over the reals or the set of all $d$-tuples with elements in $\mathbb{R}$.

$\mathbb{R}_+^d$          The set of all $d$-tuples with positive real elements.

| | |
|---|---|
| $\mathbb{C}$ | The field of complex numbers or the set over which that field is defined. |
| $\mathbb{Z}$ | The ring of integers or the set over which that ring is defined. |
| $\mathbb{G}(n)$ | A Galois field defined on a set with $n$ elements. |
| $C^0, C^1, C^2, \ldots$ | The set of continuous functions, the set of functions with continuous first derivatives, and so forth. |
| i | The imaginary unit $\sqrt{-1}$. |

## Computer Number Systems

Computer number systems are used to simulate the more commonly used number systems. It is important to realize that they have different properties, however. Some notation for computer number systems follows.

| | |
|---|---|
| $\mathbb{F}$ | The set of floating-point numbers with a given precision, on a given computer system, or this set together with the four operators +, -, *, and /. $\mathbb{F}$ is similar to $\mathbb{R}$ in some useful ways; it is not, however, closed under the two basic operations, and not all reciprocals of the elements exclusive of the additive identity exist, so it is clearly not a field. |
| $\mathbb{I}$ | The set of fixed-point numbers with a given length, on a given computer system, or this set together with the four operators +, -, *, and /. $\mathbb{I}$ is similar to $\mathbb{Z}$ in some useful ways; it is not, however, closed under the two basic operations, so it is clearly not a ring. |
| $e_{min}$ and $e_{max}$ | The minimum and maximum values of the exponent in the set of floating-point numbers with a given length. |
| $\epsilon_{min}$ and $\epsilon_{max}$ | The minimum and maximum spacings around 1 in the set of floating-point numbers with a given length. |
| $\epsilon$ or $\epsilon_{mach}$ | The machine epsilon, the same as $\epsilon_{min}$. |
| $[\cdot]_c$ | The computer version of the object $\cdot$. |
| NaN | Not-a-Number. |

## Notation Relating to Random Variables

A common function used with continuous random variables is a *density function*, and a common function used with discrete random variables is a *probability function*. The more fundamental function for either type of random variable is the *cumulative distribution function*, or CDF. The CDF of a random variable $X$, denoted by $P_X(x)$ or just by $P(x)$, is defined by

$$P(x) = \Pr(X \le x),$$

where "Pr", or "probability", can be taken here as a primitive (it is defined in terms of a measure). For vectors (of the same length), "$X \le x$" means that each element of $X$ is less than or equal to the corresponding element of $x$. Both the CDF and the density or probability function for a $d$-dimensional random variable are defined over $\mathbb{R}^d$. (It is unfortunately necessary to state that "$P(x)$" means the "function $P$ evaluated at $x$", and likewise "$P(y)$" means the *same* "function $P$ evaluated at $y$" unless $P$ has been redefined. Using a different expression as the argument *does not redefine* the function despite the sloppy convention adopted by some statisticians—including myself sometimes!)

The density for a continuous random variable is just the derivative of the CDF (if it exists). The CDF is therefore the integral. To keep the notation simple, we likewise consider the probability function for a discrete random variable to be a type of derivative (a Radon–Nikodym derivative) of the CDF. Instead of expressing the CDF of a discrete random variable as a sum over a countable set, we often also express it as an integral. (In this case, however, the integral is over a set whose ordinary Lebesgue measure is 0.)

A useful analog of the CDF for a random sample is the *empirical cumulative distribution function*, or ECDF. For a sample of size $n$, the ECDF is

$$P_n(x) = \frac{1}{n} \sum_{i=1}^{n} \mathrm{I}_{(-\infty,x]}(x_i)$$

for the indicator function $\mathrm{I}_{(-\infty,x]}(\cdot)$.

Functions and operators such as Cov and E that are commonly associated with Latin letters or groups of Latin letters are generally represented by that letter in a Roman font.

| | |
|---|---|
| $\Pr(A)$ | The probability of the event $A$. |
| $p_X(\cdot)$ or $P_X(\cdot)$ | The probability density function (or probability function), or the cumulative probability function, of the random variable $X$. |
| $p_{XY}(\cdot)$ or $P_{XY}(\cdot)$ | The joint probability density function (or probability function), or the joint cumulative probability function, of the random variables $X$ and $Y$. |

$p_{X|Y}(\cdot)$
or $P_{X|Y}(\cdot)$ — The conditional probability density function (or probability function), or the conditional cumulative probability function, of the random variable $X$ given the random variable $Y$ (these functions are random variables).

$p_{X|y}(\cdot)$
or $P_{X|y}(\cdot)$ — The conditional probability density function (or probability function), or the conditional cumulative probability function, of the random variable $X$ given the realization $y$.

Sometimes, the notation above is replaced by a similar notation in which the arguments indicate the nature of the distribution; for example, $p(x, y)$ or $p(x|y)$.

$p_\theta(\cdot)$
or $P_\theta(\cdot)$ — The probability density function (or probability function), or the cumulative probability function, of the distribution characterized by the parameter $\theta$.

$Y \sim D_X(\theta)$ — The random variable $Y$ is distributed as $D_X(\theta)$, where $X$ is the name of a random variable associated with the distribution, and $\theta$ is a parameter of the distribution. The subscript may take forms similar to those used in the density and distribution functions, such as $X|y$, or it may be omitted. Alternatively, in place of $D_X$, a symbol denoting a specific distribution may be used. An example is $Z \sim N(0, 1)$, which means that $Z$ has a normal distribution with mean 0 and variance 1.

CDF — A cumulative distribution function.

ECDF — An empirical cumulative distribution function.

i.i.d. — Independent and identically distributed.

$X^{(i)} \xrightarrow{d} X$
or $X_i \xrightarrow{d} X$ — The sequence of random variables $X^{(i)}$ or $X_i$ converges in distribution to the random variable $X$. (The difference in the notation $X^{(i)}$ and $X_i$ is generally unimportant. The former notation is often used to emphasize the iterative nature of a process.)

$E(g(X))$ — The expected value of the function $g$ of the random variable $X$. The notation $E_P(\cdot)$, where $P$ is a cumulative distribution function or some other identifier of a probability distribution, is sometimes used to indicate explicitly the distribution with respect to which the expectation is evaluated.

$V(g(X))$ — The variance of the function $g$ of the random variable $X$. The notation $V_P(\cdot)$ is also often used.

$\mathrm{Cov}(X, Y)$     The covariance of the random variables $X$ and $Y$. The notation $\mathrm{Cov}_P(\cdot, \cdot)$ is also often used.

$\mathrm{Cov}(X)$     The variance-covariance matrix of the vector random variable $X$.

$\mathrm{Corr}(X, Y)$     The correlation of the random variables $X$ and $Y$. The notation $\mathrm{Corr}_P(\cdot, \cdot)$ is also often used.

$\mathrm{Corr}(X)$     The correlation matrix of the vector random variable $X$.

$\mathrm{Bias}(T, \theta)$ or $\mathrm{Bias}(T)$     The bias of the estimator $T$ (as an estimator of $\theta$); that is,

$$\mathrm{Bias}(T, \theta) = \mathrm{E}(T) - \theta.$$

$\mathrm{MSE}(T, \theta)$ or $\mathrm{MSE}(T)$     The mean squared error of the estimator $T$ (as an estimator of $\theta$); that is,

$$\mathrm{MSE}(T, \theta) = \big(\mathrm{Bias}(T, \theta)\big)^2 + \mathrm{V}(T).$$

## General Mathematical Functions and Operators

Functions such as sin, max, span, and so on that are commonly associated with groups of Latin letters are generally represented by those letters in a Roman font.

Generally, the argument of a function is enclosed in parentheses: $\sin(x)$. Often, for the very common functions, the parentheses are omitted: $\sin x$. In expressions involving functions, parentheses are generally used for clarity, for example, $(\mathrm{E}(X))^2$ instead of $\mathrm{E}^2(X)$.

Operators such as d (the differential operator) that are commonly associated with a Latin letter are generally represented by that letter in a Roman font.

$|x|$     The modulus of the real or complex number $x$; if $x$ is real, $|x|$ is the absolute value of $x$.

$\lceil x \rceil$     The ceiling function evaluated at the real number $x$: $\lceil x \rceil$ is the smallest integer greater than or equal to $x$.

$\lfloor x \rfloor$     The floor function evaluated at the real number $x$: $\lfloor x \rfloor$ is the largest integer less than or equal to $x$.

$\#S$     The cardinality of the set $S$.

$I_S(\cdot)$        The indicator function:

$$I_S(x) \;=\; 1 \text{ if } x \in S;$$
$$=\; 0 \quad \text{otherwise.}$$

If $x$ is a scalar, the set $S$ is often taken as the interval $(-\infty, y]$, and, in this case, the indicator function is the Heaviside function, H, evaluated at the difference of the argument and the upper bound on the interval:

$$I_{(-\infty,y]}(x) = H(y - x).$$

(An alternative definition of the Heaviside function is the same as this, except that $H(0) = \frac{1}{2}$.) In higher dimensions, the set $S$ is often taken as the product set,

$$A^d \;=\; (-\infty, y_1] \times (-\infty, y_2] \times \cdots \times (-\infty, y_d]$$
$$=\; A_1 \times A_2 \times \cdots \times A_d,$$

and, in this case,

$$I_{A^d}(x) = I_{A_1}(x_1) I_{A_2}(x_2) \cdots I_{A_d}(x_d),$$

where $x = (x_1, x_2, \ldots, x_d)$. The derivative of the indicator function is the Dirac delta function, $\delta(\cdot)$.

$\delta(\cdot)$        The Dirac delta "function", defined by

$$\delta(x) = 0 \quad \text{for } x \neq 0,$$

and

$$\int_{-\infty}^{\infty} \delta(t)\, dt = 1.$$

The Dirac delta function is not a function in the usual sense. For any continuous function $f$, we have the useful fact

$$\int_{-\infty}^{\infty} f(y)\, dI_{(-\infty,y]}(x) \;=\; \int_{-\infty}^{\infty} f(y)\, \delta(y - x)\, dy$$
$$=\; f(x).$$

$\min f(\cdot)$   The minimum value of the real scalar-valued function $f$, or the
or $\min(S)$   smallest element in the countable set of real numbers $S$.

$\operatorname{argmin} f(\cdot)$   The value of the argument of the real scalar-valued function $f$ that yields its minimum value.

$\oplus$  Bitwise binary exclusive-or (see page 39).

$O(f(n))$  Big O; $g(n) = O(f(n))$ means that there exists a positive constant $M$ such that $|g(n)| \leq M|f(n)|$ as $n \to \infty$. $g(n) = O(1)$ means that $g(n)$ is bounded from above.

d  The differential operator. The derivative with respect to the variable $x$ is denoted by $\frac{d}{dx}$.

$f', f'', \ldots, f^{k'}$  For the scalar-valued function $f$ of a scalar variable, differentiation (with respect to an implied variable) taken on the function once, twice, ..., $k$ times.

$\bar{x}$  The mean of a sample of objects generically denoted by $x$.

$x_\bullet$  The sum of the elements in the object $x$. More generally, $x_{i \bullet k} = \sum_j x_{ijk}$.

$x^-$  The multiplicative inverse of $x$ with respect to some modulus (see page 36).

## Special Functions

$\log x$  The natural logarithm evaluated at $x$.

$\sin x$  The sine evaluated at $x$ (in radians) and similarly for other trigonometric functions.

$x!$  The factorial of $x$. If $x$ is a positive integer, $x! = x(x-1) \cdots 2 \cdot 1$. For other values of $x$, except negative integers, $x!$ is often defined as
$$x! = \Gamma(x+1).$$

$\Gamma(\alpha)$  The (complete) gamma function. For $\alpha$ not equal to a nonpositive integer,
$$\Gamma(\alpha) = \int_0^\infty t^{\alpha-1} e^{-t}\, dt.$$

We have the useful relationship $\Gamma(\alpha) = (\alpha - 1)!$. An important argument is $\frac{1}{2}$, and $\Gamma(\frac{1}{2}) = \sqrt{\pi}$.

$\Gamma_x(\alpha)$        The incomplete gamma function:

$$\Gamma_x(\alpha) = \int_0^x t^{\alpha-1} e^{-t}\, dt.$$

$\mathrm{B}(\alpha,\beta)$        The (complete) beta function:

$$\mathrm{B}(\alpha,\beta) = \int_0^1 t^{\alpha-1}(1-t)^{\beta-1}\, dt,$$

where $\alpha > 0$ and $\beta > 0$. A useful relationship is

$$\mathrm{B}(\alpha,\beta) = \frac{\Gamma(\alpha)\Gamma(\beta)}{\Gamma(\alpha+\beta)}.$$

$\mathrm{B}_x(\alpha,\beta)$        The incomplete beta function:

$$\mathrm{B}_x(\alpha,\beta) = \int_0^x t^{\alpha-1}(1-t)^{\beta-1}\, dt.$$

# Appendix B

# Solutions and Hints for Selected Exercises

**1.5.** With $a = 17$, the correlation of pairs of successive numbers should be about 0.09, and the plot should show 17 lines. With $a = 85$, the correlation of lag 1 is about 0.03, but the correlation of lag 2 is about $-0.09$.

**1.8.** 35 planes for 65 541 and 15 planes for 65 533.

**1.10.** 950 706 376, 129 027 171, 1 728 259 899, 365 181 143, 1 966 843 080, 1 045 174 992, 636 176 783, 1 602 900 997, 640 853 092, 429 916 489.

**1.13.** We seek $x_0$ such that

$$16\,807 x_0 - (2^{31} - 1)c_1 = 2^{31} - 2$$

for some integer $c_1$. First, observe that $2^{31} - 2$ is equivalent to $-1$, so we use Euler's method (see, e.g., Ireland and Rosen, 1991, or Fang and Wang, 1994) with that simpler value and write

$$x_0 = \frac{((2^{31} - 1)c_1 - 1)}{16\,807}$$

$$= 127\,773 c_1 + \frac{(2836 c_1 - 1)}{16\,807}.$$

Because the latter term must also be an integer, we write

$$16\,807 c_2 = 2836 c_1 - 1$$

or

$$c_1 = 5 c_2 + \frac{(2627 c_2 + 1)}{2836}.$$

323

for some integer $c_2$. Continuing,

$$c_2 = c_3 + \frac{(209c_3 - 1)}{2627},$$

$$c_3 = 12c_4 + \frac{(119c_4 + 1)}{209},$$

$$c_4 = c_5 + \frac{(90c_5 - 1)}{119},$$

$$c_5 = c_6 + \frac{(29c_6 + 1)}{90},$$

$$c_6 = 3c_7 + \frac{(3c_7 - 1)}{29},$$

$$c_7 = 9c_8 + \frac{(3c_8 + 1)}{3},$$

$$c_8 = c_9 + \frac{(c_9 - 1)}{2}.$$

Letting $c_9 = 1$, we can backsolve to get $x_0 = 739\,806\,647$.

**1.14.**    Using Maple, for example,

```
> pr := 0:
> while pr < 8191 do
>     pr := primroot(pr, 8191)
> od;
```

yields the 1728 primitive roots, starting with the smallest one, 17, and going through the largest, 8180. To use `primroot`, you may first have to attach the number theory package: `with(numtheory):`.

**1.15.**    0.5.

**2.2c.**    The distribution is degenerate with probability 1 for $r = \min(n, m)$; that is, the matrix is of full rank with probability 1.

**2.3.**    Out of the 100 trials, 97 times the maximum element is in position 1311. The test is not really valid because the seeds are all relatively small and are very close together. Try the same test but with 100 randomly generated seeds.

**4.1a.**    $X$ is a random variable with an absolutely continuous distribution function $P$. Let $Y$ be the random variable $P(X)$. Then, for $0 \leq t \leq 1$, using the existence of $P^{-1}$,

$$
\begin{aligned}
\Pr(Y \leq t) &= \Pr(P(X) \leq t) \\
&= \Pr(X \leq P^{-1}(t)) \\
&= P^{-1}(P(t)) \\
&= t.
\end{aligned}
$$

Hence, $Y$ has a $U(0,1)$ distribution.

**4.2.** Let $Z$ be the random variable delivered. For any $x$, because $Y$ (from the density $g$) and $U$ are independent, we have

$$
\begin{aligned}
\Pr(Z \le x) &= \Pr\left(Y \le x \,\middle|\, U \le \frac{p(Y)}{cg(Y)}\right) \\
&= \frac{\int_{-\infty}^{x} \int_0^{p(t)/cg(t)} g(t)\,ds\,dt}{\int_{-\infty}^{\infty} \int_0^{p(t)/cg(t)} g(t)\,ds\,dt} \\
&= \int_{-\infty}^{x} p(t)\,dt,
\end{aligned}
$$

the distribution function corresponding to $p$. Differentiating this quantity with respect to $x$ yields $p(x)$.

**4.4c.** Using the relationship

$$
\frac{1}{\sqrt{2\pi}} e^{-\frac{x^2}{2}} \le \frac{1}{\sqrt{2\pi}} e^{\frac{1}{2}-|x|}
$$

(see Devroye, 1986a), we have the following algorithm, after simplification.

1. Generate $x$ from the double exponential, and generate $u$ from $U(0,1)$.
2. If $x^2 + 1 - 2|x| \le -2\log u$, then deliver $x$; otherwise, go to step 1.

**4.5.** As $x \to \infty$, there is no $c$ such that $cg(x) \ge p(x)$, where $g$ is the normal density and $p$ is the exponential density.

**4.7a.** $E(T) = c$; $V(T) = c^2 - c$. (Note that $c \ge 1$.)

**4.8.** For any $t$, we have

$$
\begin{aligned}
\Pr(X \le t) &= \Pr(X \le s + rh) \quad (\text{for } 0 \le r \le 1) \\
&= \Pr(U \le r \mid V \le U + p(s + hU)/b) \\
&= \frac{\int_0^r \int_u^{u+p(s+hu)/b} 2\,dv\,du}{\int_0^1 \int_u^{u+p(s+hu)/b} 2\,dv\,du} \\
&= \frac{\int_0^r (p(s+hu)/b)\,du}{\int_0^1 (p(s+hu)/b)\,du} \\
&= \int_s^t p(x)\,dx,
\end{aligned}
$$

where all of the symbols correspond to those in Algorithm 4.7 with the usual convention of uppercase representing random variables and lowercase representing constants or realizations of random variables.

**5.2b.**   We can consider only the case in which $\tau \geq 0$; otherwise, we could make use of the symmetry of the normal distribution and split the algorithm into two regions. Also, for simplicity, we can generate truncated normals with $\mu = 0$ and $\sigma^2 = 1$ and then shift and scale just as we do for the full normal distribution. The probability of acceptance is the ratio of the area under the truncated exponential (the majorizing function) to the area under the truncated normal density. For an exponential with parameter $\lambda$ and truncated at $\tau$, the density is

$$g(x) = \lambda e^{-\lambda(x-\tau)}.$$

To scale this so that it majorizes the truncated normal density requires a constant $c$ that *does not depend on* $\lambda$. We can write the probability of acceptance as

$$c\lambda e^{-\lambda\tau - \lambda^2/2}.$$

Maximizing this quantity (by taking the derivative and equating to 0) yields the equation $\lambda^2 - \lambda\tau - 1 = 0$.

**5.3.**   Use the fact that $U$ and $1 - U$ have the same distribution.

**5.6b.**   A simple program using the IMSL routine bnrdf can be used to compute $r$. Here is a fragment of code that will work:

```
10 pl = bnrdf(z1,z2,rl)
   ph = bnrdf(z1,z2,rh)
   if (abs(ph-pl) .le. eps) go to 99
   rt = rl + (rh-rl)/2.
   pt = bnrdf(z1,z2,rt)
   if (pt .gt. prob) then
      rh = rt
   else
      rl = rt
   endif
   go to 10
99 continue
   print *, rl
```

**5.7b**   We need the partitions of $\Sigma^{-1}$:

$$\Sigma^{-1} = \begin{bmatrix} \Sigma_{11} & \Sigma_{12} \\ \Sigma_{21} & \Sigma_{22} \end{bmatrix}^{-1} = \begin{bmatrix} T_{11} & T_{12} \\ T_{21} & T_{22} \end{bmatrix}.$$

Now,

$$T_{11} = \left(\Sigma_{11} - \Sigma_{12}\Sigma_{22}^{-1}\Sigma_{11}^{T}\right)^{-1}$$

(see Gentle, 1998, page 61).

The conditional distribution of $X_1$ given $X_2 = x_2$ is $N_{d_1}(\mu_1 + \Sigma_{12}\Sigma_{22}^{-1}(x_2 - \mu_2), T_{11}^{-1})$ (see any book on multivariate distributions, such as Kotz, Balakrishnan, and Johnson, 2000). Hence, first, take $y_1$ as rnorm($d_1$). Then,

$$x_1 = T_{11}^{-1/2}y_1 + \mu_1 + \Sigma_{12}\Sigma_{22}^{-1}(x_2 - \mu_2),$$

where $T_{11}^{-1/2}$ is the Cholesky factor of $T_{11}^{-1}$, that is, of $\Sigma_{11} - \Sigma_{12}\Sigma_{22}^{-1}\Sigma_{11}^{T}$.

If $Y_2$ is a $d_2$-variate random variate with a standard circular normal distribution, and $X_1$ has the given relationship, then

$$E(X_1|X_2 = x_2) = \mu_1 + \Sigma_{12}\Sigma_{22}^{-1}(x_2 - \mu_2),$$

and

$$V(X_1|X_2 = x_2) = T_{11}^{-1/2}V(Y_1)T_{11}^{-1/2} = T_{11}^{-1}.$$

**6.2a.** The problem with random sampling using a pseudorandom number generator is that the fixed relationships in the generator must be ignored – else there can be no simple random sample larger than 1. On the other hand, if these fixed relationships are ignored, then it does not make sense to speak of a period.

**7.2b.** The set is a random sample from the distribution with density $f$.

**7.3b.** The integral can be reduced to

$$\int_0^2 \sqrt{\frac{\pi}{y}} \cos \pi y \, dy.$$

Generate $y_i$ as $2u_i$, where $u_i$ are from $U(0,1)$, and estimate the integral as

$$2\sqrt{\pi} \sum \frac{\cos \pi y_i}{\sqrt{y_i}}.$$

**7.6.** The order of the variance is $O(n^{-2})$. The order is obviously dependent on the dimension of the integral, however, and, in higher dimensions, it is not competitive with the crude Monte Carlo method.

**7.9a.** Generate $x_i$ from a gamma(3, 2) distribution, and take your estimator as

$$16 \frac{\sum \sin(\pi x_i)}{n}.$$

**7.10b.** The optimum is $l = d$.

**7.10d.** An unbiased estimator for $\theta$ is

$$\frac{d^2(n_1 + n_2)}{(dl - l^2)n}.$$

The optimum is $l = d$.

**7.14a.**

$$E_{\widehat{P}}(\bar{x}^*) = E_{\widehat{P}}\left(\frac{1}{n}\sum_i x_i^*\right)$$

$$= \frac{1}{n}\sum_i E_{\widehat{P}}(x_i^*)$$

$$= \frac{1}{n}\sum_i \bar{x}$$

$$= \bar{x}.$$

Note that the empirical distribution is a conditional distribution, given the sample. With the sample fixed, $\bar{x}$ is a "parameter" rather than a "statistic".

**7.14b.**

$$E_P(\bar{x}^*) = E_P\left(\frac{1}{n}\sum_i x_i^*\right)$$

$$= \frac{1}{n}\sum_i E_P(x_i^*)$$

$$= \frac{1}{n}\sum_i \mu$$

$$= \mu.$$

Alternatively,

$$E_P(\bar{x}^*) = E_P\left(E_{\widehat{P}}(\bar{x}^*)\right)$$

$$= E_P(\bar{x})$$

$$= \mu.$$

**7.14c.** First, note that

$$E_{\widehat{P}}(\overline{\bar{x}_j^*}) = \bar{x},$$

$$V_{\widehat{P}}(\bar{x}_j^*) = \frac{1}{n}\frac{1}{n}\sum(x_i - \bar{x})^2,$$

and

$$V_{\widehat{P}}(\overline{\bar{x}_j^*}) = \frac{1}{mn^2}\sum(x_i - \bar{x})^2.$$

Now,

$$E_{\widehat{P}}(V) = \frac{1}{m-1}E_{\widehat{P}}\left(\sum_j (\bar{x}_j^* - \overline{\bar{x}_j^*})^2\right)$$

$$
= \frac{1}{m-1} E_{\widehat{P}} \left( \sum_j \bar{x}_j^{*2} - m \overline{\bar{x}_j^*}^2 \right)
$$

$$
= \frac{1}{m-1} \left( m\bar{x}^2 + \frac{m}{n} \sum (x_i - \bar{x})^2 - m\bar{x}^2 - \frac{m}{mn^2} \sum (x_i - \bar{x})^2 \right)
$$

$$
= \frac{1}{m-1} \left( \frac{m}{n^2} \sum (x_i - \bar{x})^2 - \frac{1}{n^2} \sum (x_i - \bar{x})^2 \right)
$$

$$
= \frac{1}{n^2} \sum (x_i - \bar{x})^2
$$

$$
= \frac{1}{n} \sigma_{\widehat{P}}^2.
$$

**7.14d.**

$$
E_P(V) = E_P(E_{\widehat{P}}(V))
$$

$$
= E_P \left( \frac{1}{n} \sum (x_i - \bar{x})^2 / n \right)
$$

$$
= \frac{1}{n} \frac{n-1}{n} \sigma_P^2.
$$

# Bibliography

As might be expected, the literature in the interface of computer science, numerical analysis, and statistics is quite diverse, and articles on random number generation and Monte Carlo methods are likely to appear in journals devoted to quite different disciplines. There are at least ten journals and serials with titles that contain some variants of both "computing" and "statistics", but there are far more journals in numerical analysis and in areas such as "computational physics", "computational biology", and so on that publish articles relevant to the fields of statistical computing and computational statistics. Many of the methods of computational statistics involve random number generation and Monte Carlo methods. The journals in the mainstream of statistics also have a large proportion of articles in the fields of statistical computing and computational statistics because, as we suggested in the preface, recent developments in statistics and in the computational sciences have paralleled each other to a large extent.

There are two well-known learned societies with a primary focus in statistical computing: the International Association for Statistical Computing (IASC), which is an affiliated society of the International Statistical Institute, and the Statistical Computing Section of the American Statistical Association (ASA). The Statistical Computing Section of the ASA has a regular newsletter carrying news and notices as well as articles on practicum. Also, the activities of the Society for Industrial and Applied Mathematics (SIAM) are often relevant to computational statistics.

There are two regular conferences in the area of computational statistics: COMPSTAT, held biennially in Europe and sponsored by the IASC, and the Interface Symposium, generally held annually in North America and sponsored by the Interface Foundation of North America with cooperation from the Statistical Computing Section of the ASA.

In addition to literature and learned societies in the traditional forms, an important source of communication and a repository of information are computer databases and forums. In some cases, the databases duplicate what is available in some other form, but often the material and the communications facilities provided by the computer are not available elsewhere.

# Literature in Computational Statistics

In the Library of Congress classification scheme, most books on statistics, including statistical computing, are in the QA276 section, although some are classified under H, HA, and HG. Numerical analysis is generally in QA279 and computer science in QA76. Many of the books in the interface of these disciplines are classified in these or other places within QA.

*Current Index to Statistics*, published annually by the American Statistical Association and the Institute for Mathematical Statistics, contains both author and subject indexes that are useful in finding journal articles or books in statistics. The *Index* is available in hard copy and on CD-ROM.

The Association for Computing Machinery (ACM) publishes an annual index, by author, title, and keyword, of the literature in the computing sciences.

*Mathematical Reviews*, published by the American Mathematical Society (AMS), contains brief reviews of articles in all areas of mathematics. The areas of "Statistics", "Numerical Analysis", and "Computer Science" contain reviews of articles relevant to computational statistics. The papers reviewed in *Mathematical Reviews* are categorized according to a standard system that has slowly evolved over the years. In this taxonomy, called the AMS MR classification system, "Statistics" is 62Xyy; "Numerical Analysis", including random number generation, is 65Xyy; and "Computer Science" is 68Xyy. ("X" represents a letter and "yy" represents a two-digit number.) *Mathematical Reviews* is available to subscribers via the World Wide Web at MathSciNet:

    http://www.ams.org/mathscinet/

There are various handbooks of mathematical functions and formulas that are useful in numerical computations. Three that should be mentioned are Abramowitz and Stegun (1964), Spanier and Oldham (1987), and Thompson (1997). Anyone doing serious scientific computations should have ready access to at least one of these volumes.

Almost all journals in statistics have occasional articles on computational statistics and statistical computing. The following is a list of journals, proceedings, and newsletters that emphasize this field.

*ACM Transactions on Mathematical Software*, published quarterly by the ACM (Association for Computing Machinery). This journal publishes algorithms in Fortran and C. The ACM collection of algorithms is sometimes called *CALGO*. The algorithms published during the period 1975 through 1999 are available on a CD-ROM from ACM. Most of the algorithms are available through netlib at

    http://www.netlib.org/liblist.html

*ACM Transactions on Modeling and Computer Simulation*, published quarterly by the ACM.

*Applied Statistics*, published quarterly by the Royal Statistical Society. (Until 1998, it included algorithms in Fortran. Some of these algorithms, with cor-

rections, were collected by Griffiths and Hill, 1985. Most of the algorithms
are available through statlib at Carnegie Mellon University.)

*Communications in Statistics — Simulation and Computation*, published quarterly by Marcel Dekker. (Until 1996, it included algorithms in Fortran.
Until 1982, this journal was designated as *Series B*.)

*Computational Statistics*, published quarterly by Physica-Verlag (formerly called
*Computational Statistics Quarterly*).

*Computational Statistics. Proceedings of the xx$^{th}$ Symposium on Computa-
tional Statistics* (COMPSTAT), published biennially by Physica-Verlag. (It
is not refereed.)

*Computational Statistics & Data Analysis*, published by North Holland. The
number of issues per year varies. (This is also the official journal of the International Association for Statistical Computing, and as such incorporates
the *Statistical Software Newsletter*.)

*Computing Science and Statistics*. This is an annual publication containing
papers presented at the Interface Symposium. Until 1992, these proceedings were named *Computer Science and Statistics: Proceedings of the xx$^{th}$
Symposium on the Interface*. (The 24$^{th}$ symposium was held in 1992.) In
1997, Volume 29 was published in two issues: Number 1, which contains the
papers of the regular Interface Symposium, and Number 2, which contains
papers from another conference. The two numbers are not sequentially pag-
inated. Since 1999, the proceedings have been published only in CD-ROM
form, by the Interface Foundation of North America. (It is not refereed.)

*Journal of Computational and Graphical Statistics*, published quarterly by the
American Statistical Association.

*Journal of Statistical Computation and Simulation*, published irregularly in four
numbers per volume by Gordon and Breach.

*Proceedings of the Statistical Computing Section*, published annually by the
American Statistical Association. (It is not refereed.)

*SIAM Journal on Scientific Computing*, published bimonthly by SIAM. This
journal was formerly *SIAM Journal on Scientific and Statistical Computing*.
(Is this a step backward?)

*Statistical Computing & Graphics Newsletter*, published quarterly by the Statistical Computing and the Statistical Graphics Sections of the American
Statistical Association. (It is not refereed and is not generally available in
university libraries.)

*Statistics and Computing*, published quarterly by Chapman & Hall.

There are two journals whose contents are primarily in the subject area
of random number generation, simulation, and Monte Carlo methods: *ACM
Transactions on Modeling and Computer Simulation* (Volume 1 appeared in
1992) and *Monte Carlo Methods and Applications* (Volume 1 appeared in 1995).

There has been a series of conferences concentrating on this area (with
an emphasis on quasirandom methods). The first International Conference
on Monte Carlo and Quasi-Monte Carlo Methods in Scientific Computing was
held in Las Vegas, Nevada, in 1994. The fifth was held in Singapore in 2002.

The proceedings of the conferences have been published in the Lecture Notes in Statistics series of Springer-Verlag. The proceedings of the first conference were published as Niederreiter and Shiue (1995); those of the second as Niederreiter et al. (1998), the third as Niederreiter and Spanier (1999), and of the fourth as Fang, Hickernell, and Niederreiter (2002).

The proceedings of the CRYPTO conferences often contain interesting articles on uniform random number generation, with an emphasis on the cryptographic applications. These proceedings are published in the Lecture Notes in Computer Science series of Springer-Verlag under the name *Proceedings of CRYPTO XX*, where *XX* is a two digit number representing the year of the conference.

There are a number of textbooks, monographs, and survey articles on random number generation and Monte Carlo methods. Some of particular note (listed alphabetically) are Bratley, Fox, and Schrage (1987), Dagpunar (1988), Deák (1990), Devroye (1986a), Fishman (1996), Knuth (1998), L'Ecuyer (1990), L'Ecuyer and Hellekalek (1998), Lewis and Orav (1989), Liu (2001), Morgan (1984), Niederreiter (1992, 1995c), Ripley (1987), Robert and Casella (1999), and Tezuka (1995).

# World Wide Web, News Groups, List Servers, and Bulletin Boards

The best way of storing information is in a digital format that can be accessed by computers. In some cases, the best way for people to access information is by computers. In other cases, the best way is via hard copy, which means that the information stored on the computer must go through a printing process resulting in books, journals, or loose pages.

The references that I have cited in this text are generally traditional books, journal articles, or compact discs. This usually means that the material has been reviewed by someone other than the author. It also means that the author possibly has newer thoughts on the same material. The Internet provides a mechanism for the dissemination of large volumes of information that can be updated readily. The ease of providing material electronically is also the source of the major problem with the material: it is often half-baked and has not been reviewed critically. Another reason that I have refrained from making frequent reference to material available over the Internet is the unreliability of some sites. The average life of a Web site is measured in weeks.

For statistics, one of the most useful sites on the Internet is the electronic repository `statlib`, maintained at Carnegie Mellon University, which contains programs, datasets, and other items of interest. The URL is

`http://lib.stat.cmu.edu.`

The collection of algorithms published in *Applied Statistics* is available in `statlib`. These algorithms are sometimes called the *ApStat* algorithms.

Another very useful site for scientific computing is `netlib`, which was established by research workers at AT&T (now Lucent) Bell Laboratories and national laboratories, primarily Oak Ridge National Laboratory. The URL is

`http://www.netlib.org`

The *Collected Algorithms of the ACM (CALGO)*, which are the Fortran, C, and Algol programs published in *ACM Transactions on Mathematical Software* (or in *Communications of the ACM* prior to 1975), are available in `netlib` under the TOMS link.

The *Guide to Available Mathematical Software* (GAMS) can be accessed at

`http://gams.nist.gov`

A different interface, using Java, is available at

`http://math.nist.gov/HotGAMS/`

A good set of links for software are the Econometric Links of the *Econometrics Journal* (which are not just limited to econometrics):

`http://www.eur.nl/few/ei/links/software.html`

There are two major problems in using the WWW to gather information. One is the sheer quantity of information and the number of sites providing information. The other is the "kiosk problem"; anyone can put up material. Sadly, the average quality is affected by a very large denominator. The kiosk problem may be even worse than a random selection of material; the "fools in public places" syndrome is much in evidence.

There is not much that can be done about the second problem. It was not solved for traditional postings on uncontrolled kiosks, and it will not be solved on the WWW.

For the first problem, there are remarkable programs that automatically crawl through WWW links to build a database that can be searched for logical combinations of terms and phrases. Such systems and databases have been built by several people and companies. One of the most useful is Google at

`http://google.stanford.edu`

A very widely used search program is Yahoo at

`http://www.yahoo.com`

A neophyte can be quickly disabused of an exaggerated sense of the value of such search engines by doing a search on "Monte Carlo".

It is not clear at this time what will be the media for the scientific literature within a few years. Many of the traditional journals will be converted to an electronic version of some kind. Journals will become Web sites. That is for certain; the details, however, are much less certain. Many bulletin boards and discussion groups have already evolved into "electronic journals". A publisher of a standard commercial journal has stated that "we reject 80% of the articles submitted to our journal; those are the ones you can find on the Web".

# References for Software Packages

There is a wide range of software used in the computational sciences. Some of the software is produced by a single individual who is happy to share the software, sometimes for a fee, but who has no interest in maintaining the software. At the other extreme is software produced by large commercial companies whose continued existence depends on a process of production, distribution, and maintenance of the software. Information on much of the software can be obtained from GAMS. Some of the free software can be obtained from `statlib` or `netlib`.

The names of many software packages are trade names or trademarks. In this book, the use of names, even if the name is not especially identified, is not to be taken as a sign that such names, as understood by the Trade Marks and Merchandise Marks Act, may accordingly be used freely by anyone.

# References to the Literature

The following bibliography obviously covers a wide range of topics in random number generation and Monte Carlo methods. Except for a few of the general references, all of these entries have been cited in the text. The purpose of this bibliography is to help the reader get more information; hence, I eschew "personal communications" and references to technical reports that may or may not exist. Those kinds of references are generally for the author rather than for the reader.

In some cases, important original papers have been reprinted in special collections, such as Samuel Kotz and Norman L. Johnson (Editors) (1997), *Breakthroughs in Statistics, Volume III*, Springer-Verlag, New York. In most such cases, because the special collection may be more readily available, I list both sources.

## A Note on the Names of Authors

In these references, I have generally used the names of authors as they appear in the original sources. This may mean that the same author will appear with different forms of names, sometimes with given names spelled out and sometimes abbreviated. In the author index, beginning on page 371, I use a single name for the same author. The name is generally the most unique (i.e., least abbreviated) of any of the names of that author in any of the references. This convention may occasionally result in an entry in the author index that does not occur exactly in any references. For example, a reference to J. Paul Jones together with one to John P. Jones, if I know that the two names refer to the same person, would result in an author index entry for John Paul Jones.

Abramowitz, Milton, and Irene A. Stegun (Editors) (1964), *Handbook of Mathematical Functions with Formulas, Graphs, and Mathematical Tables*, National Bureau of Standards (NIST), Washington. (Reprinted by Dover Publications, New York, 1974. Work on an updated version is occurring at NIST; see http://dlmf.nist.gov/ for the current status.)

Afflerbach, L., and H. Grothe (1985), Calculation of Minkowski-reduced lattice bases, *Computing* **35**, 269–276.

Afflerbach, Lothar, and Holger Grothe (1988), The lattice structure of pseudorandom vectors generated by matrix generators, *Journal of Computational and Applied Mathematics* **23**, 127–131.

Afflerbach, L., and W. Hörmann (1992), Nonuniform random numbers: A sensitivity analysis for transformation methods, *International Workshop on Computationally Intensive Methods in Simulation and Optimization* (edited by U. Dieter and G. C. Pflug), Springer-Verlag, Berlin, 374.

Agarwal, Satish K., and Jamal A. Al-Saleh (2001), Generalized gamma type distribution and its hazard rate function, *Communications in Statistics — Theory and Methods* **30**, 309–318.

Agresti, Alan (1992), A survey of exact inference for contingency tables (with discussion), *Statistical Science* **7**, 131–177.

Ahn, Hongshik, and James J. Chen (1995), Generation of over-dispersed and under-dispersed binomial variates, *Journal of Computational and Graphical Statistics* **4**, 55–64.

Ahrens, J. H. (1995), A one-sample method for sampling from continuous and discrete distributions, *Computing* **52**, 127–146.

Ahrens, J. H., and U. Dieter (1972), Computer methods for sampling from the exponential and normal distributions, *Communications of the ACM* **15**, 873–882.

Ahrens, J. H., and U. Dieter (1974), Computer methods for sampling from gamma, beta, Poisson, and binomial distributions, *Computing* **12**, 223–246.

Ahrens, J. H., and U. Dieter (1980), Sampling from binomial and Poisson distributions: A method with bounded computation times, *Computing* **25**, 193–208.

Ahrens, J. H., and U. Dieter (1985), Sequential random sampling, *ACM Transactions on Mathematical Software* **11**, 157–169.

Ahrens, Joachim H., and Ulrich Dieter (1988), Efficient, table-free sampling methods for the exponential, Cauchy and normal distributions, *Communications of the ACM* **31**, 1330–1337. (See also Hamilton, 1998.)

Ahrens, J. H., and U. Dieter (1991), A convenient sampling method with bounded computation times for Poisson distributions, *The Frontiers of Statistical Computation, Simulation & Modeling* (edited by P. R. Nelson, E. J. Dudewicz, A. Öztürk, and E. C. van der Meulen), American Sciences Press, Columbus, Ohio, 137–149.

Akima, Hirosha (1970), A new method of interpolation and smooth curve fitting based on local procedures, *Journal of the ACM* **17**, 589–602.

Albert, James; Mohan Delampady; and Wolfgang Polasek (1991), A class of distributions for robustness studies, *Journal of Statistical Planning and Inference* **28**, 291–304.

Alonso, Laurent, and René Schott (1995), *Random Generation of Trees: Random Generators in Science*, Kluwer Academic Publishers, Boston.

Altman, N. S. (1989), Bit-wise behavior of random number generators, *SIAM Journal on Scientific and Statistical Computing* **9**, 941–949.

Aluru, S.; G. M. Prabhu; and John Gustafson (1992), A random number generator for parallel computers, *Parallel Computing* **18**, 839–847.

Anderson, N. H., and D. M. Titterington (1993), Cross-correlation between simultaneously generated sequences of pseudo-random uniform deviates, *Statistics and Computing* **3**, 61–65.

Anderson, T. W.; I. Olkin; and L. G. Underhill (1987), Generation of random orthogonal matrices, *SIAM Journal on Scientific and Statistical Computing* **8**, 625–629.

Andrews, D. F.; P. J. Bickel; F. R. Hampel; P. J. Huber; W. H. Rogers; and J. W. Tukey (1972), *Robust Estimation of Location: Survey and Advances*, Princeton University Press, Princeton, New Jersey.

Antonov, I. A., and V. M. Saleev (1979), An economic method of computing $LP_\tau$-sequences, *USSR Computational Mathematics and Mathematical Physics* **19**, 252–256.

Arnason, A. N., and L. Baniuk (1978), A computer generation of Dirichlet variates, *Proceedings of the Eighth Manitoba Conference on Numerical Mathematics and Computing*, Utilitas Mathematica Publishing, Winnipeg, 97–105.

Arnold, Barry C. (1983), *Pareto Distributions*, International Co-operative Publishing House, Fairland, Maryland.

Arnold, Barry C., and Robert J. Beaver (2000), The skew-Cauchy distribution, *Statistics and Probability Letters* **49**, 285–290.

Arnold, Barry C., and Robert J. Beaver (2002), Skewed multivariate models related to hidden truncation and/or selective reporting (with discussion), *Test* **11**, 7–54.

Arnold, Barry C.; Robert J. Beaver; Richard A. Groeneveld; and William Q. Meeker (1993), The nontruncated marginal of a truncated bivariate normal distribution, *Psychometrika* **58**, 471–488.

Atkinson, A. C. (1979), A family of switching algorithms for the computer generation of beta random variates, *Biometrika* **66**, 141–145.

Atkinson, A. C. (1980), Tests of pseudo-random numbers, *Applied Statistics* **29**, 164–171.

Atkinson, A. C. (1982), The simulation of generalized inverse Gaussian and hyperbolic random variables, *SIAM Journal on Scientific and Statistical Computing* **3**, 502–515.

Atkinson, A. C., and M. C. Pearce (1976), The computer generation of beta, gamma and normal random variables (with discussion), *Journal of the Royal Statistical Society, Series A* **139**, 431–460.

Avramidis, Athanassios N., and James R. Wilson (1995), Correlation-induction techniques for estimating quantiles in simulation experiments, *Proceedings of the 1995 Winter Simulation Conference*, Association for Computing Machinery, New York, 268–277.

Azzalini, A., and A. Dalla Valle (1996), The multivariate skew-normal distribution, *Biometrika* **83**, 715–726.

Bacon-Shone, J. (1985), Algorithm AS210: Fitting five parameter Johnson $S_B$ curves by moments, *Applied Statistics* **34**, 95–100.

Bailey, David H., and Richard E. Crandall (2001), On the random character of fundamental constant expressions, *Experimental Mathematics* **10**, 175–190.

Bailey, Ralph W. (1994), Polar generation of random variates with the *t*-distribution, *Mathematics of Computation* **62**, 779–781.

Balakrishnan, N., and R. A. Sandhu (1995), A simple simulation algorithm for generating progressive Type-II censored samples, *The American Statistician* **49**, 229–230.

Banerjia, Sanjeev, and Rex A. Dwyer (1993), Generating random points in a ball, *Communications in Statistics — Simulation and Computation* **22**, 1205–1209.

Banks, David L. (1998), Testing random number generators, *Proceedings of the Statistical Computing Section, ASA*, 102–107.

Barnard, G. A. (1963), Discussion of Bartlett, "The spectral analysis of point processes", *Journal of the Royal Statistical Society, Series B* **25**, 264–296.

Barndorff-Nielsen, Ole E., and Neil Shephard (2001), Non-Gaussian Ornstein-Uhlenbeck-based models and some of their uses in financial economics (with discussion), *Journal of the Royal Statistical Society, Series B* **63**, 167–241.

Bays, Carter, and S. D. Durham (1976), Improving a poor random number generator, *ACM Transactions on Mathematical Software* **2**, 59–64.

Beck, J., and W. W. L. Chen (1987), *Irregularities of Distribution*, Cambridge University Press, Cambridge, United Kingdom.

Becker, P. J., and J. J. J. Roux (1981), A bivariate extension of the gamma distribution, *South African Statistical Journal* **15**, 1–12.

Becker, Richard A.; John M. Chambers; and Allan R. Wilks (1988), *The New S Language*, Wadsworth & Brooks/Cole, Pacific Grove, California.

Beckman, Richard J., and Michael D. McKay (1987), Monte Carlo estimation under different distributions using the same simulation, *Technometrics* **29**, 153–160.

Bélisle, Claude J. P.; H. Edwin Romeijn; and Robert L. Smith (1993), Hit-and-run algorithms for generating multivariate distributions, *Mathematics of Operations Research* **18**, 255–266.

Bendel, R. B., and M. R. Mickey (1978), Population correlation matrices for sampling experiments, *Communications in Statistics — Simulation and Computation* **B7**, 163–182.

Berbee, H. C. P.; C. G. E. Boender; A. H. G. Rinnooy Kan; C. L. Scheffer; R. L. Smith; and J. Telgen (1987), Hit-and-run algorithms for the identification of nonredundant linear inequalities, *Mathematical Programming* **37**, 184–207.

Best, D. J. (1983), A note on gamma variate generators with shape parameter less than unity, *Computing* **30**, 185–188.

Best, D. J., and N. I. Fisher (1979), Efficient simulation of the von Mises distribution, *Applied Statistics* **28**, 152–157.

Beyer, W. A. (1972), Lattice structure and reduced bases of random vectors generated by linear recurrences, *Applications of Number Theory to Numerical Analysis* (edited by S. K. Zaremba), Academic Press, New York, 361–370.

Beyer, W. A.; R. B. Roof; and D. Williamson (1971), The lattice structure of multiplicative congruential pseudo-random vectors, *Mathematics of Computation* **25**, 345–363.

Bhanot, Gyan (1988), The Metropolis algorithm, *Reports on Progress in Physics* **51**, 429–457.

Birkes, David, and Yadolah Dodge (1993), *Alternative Methods of Regression*, John Wiley & Sons, New York.

Blum, L.; M. Blum; and M. Shub (1986), A simple unpredictable pseudorandom number generator, *SIAM Journal of Computing* **15**, 364–383.

Bouleau, Nicolas, and Dominique Lépingle (1994), *Numerical Methods for Stochastic Processes*, John Wiley & Sons, New York.

Boyar, J. (1989), Inferring sequences produced by pseudo-random number generators, *Journal of the ACM* **36**, 129–141.

Boyett, J. M. (1979), Random $R \times C$ tables with given row and column totals, *Applied Statistics* **28**, 329–332.

Braaten, E., and G. Weller (1979), An improved low-discrepancy sequence for multidimensional quasi-Monte Carlo integration, *Journal of Computational Physics* **33**, 249–258.

Bratley, Paul, and Bennett L. Fox (1988), Algorithm 659: Implementing Sobol's quasirandom sequence generator, *ACM Transactions on Mathematical Software* **14**, 88–100.

Bratley, Paul; Bennett L. Fox; and Harald Niederreiter (1992), Implementation and tests of low-discrepancy sequences, *ACM Transactions on Modeling and Computer Simulation* **2**, 195–213.

Bratley, Paul; Bennett L. Fox; and Harald Niederreiter (1994), Algorithm 738: Programs to generate Niederreiter's low-discrepancy sequences, *ACM Transactions on Mathematical Software* **20**, 494–495.

Bratley, Paul; Bennett L. Fox; and Linus E. Schrage (1987), *A Guide to Simulation*, second edition, Springer-Verlag, New York.

Brooks, S. P., and G. O. Roberts (1999) Assessing convergence of iterative simulations, *Statistics and Computing* **8**, 319–335.

Brophy, John F.; James E. Gentle; Jing Li; and Philip W. Smith (1989), Software for advanced architecture computers, *Computer Science and Statistics: Proceedings of the Twenty-first Symposium on the Interface* (edited by Kenneth Berk and Linda Malone), American Statistical Association, Alexandria, Virginia, 116–120.

Brown, Morton B., and Judith Bromberg (1984), An efficient two-stage procedure for generating random variates from the multinomial distribution, *The American Statistician* **38**, 216–219.

Buckheit, Jonathan B., and David L. Donoho (1995), WaveLab and reproducible research, *Wavelets and Statistics* (edited by Anestis Antoniadis and Georges Oppenheim), Springer-Verlag, New York, 55–81.

Buckle, D. J. (1995), Bayesian inference for stable distributions, *Journal of the American Statistical Association* **90**, 605–613.

Burr, I. W. (1942), Cumulative frequency functions, *Annals of Mathematical Statistics* **13**, 215–232.

Burr, Irving W., and Peter J. Cislak (1968), On a general system of distributions. I. Its curve-shape characteristics. II. The sample median, *Journal of the American Statistical Association* **63**, 627–635.

Cabrera, Javier, and Dianne Cook (1992), Projection pursuit indices based on fractal dimension, *Computing Science and Statistics* **24**, 474–477.

Caflisch, Russel E., and Bradley Moskowitz (1995), Modified Monte Carlo methods using quasi-random sequences, *Monte Carlo and Quasi-Monte Carlo Methods in Scientific Computing* (edited by Harald Niederreiter and Peter Jau-Shyong Shiue), Springer-Verlag, New York, 1–16.

Carlin, Bradley P., and Thomas A. Louis (1996), *Bayes and Empirical Bayes Methods for Data Analysis*, Chapman & Hall, New York.

Carta, David G. (1990), Two fast implementations of the "minimal standard" random number generator, *Communications of the ACM* **33**, Number 1 (January), 87–88.

Casella, George, and Edward I. George (1992), Explaining the Gibbs sampler, *The American Statistician* **46**, 167–174.

Chalmers, C. P. (1975), Generation of correlation matrices with given eigenstructure, *Journal of Statistical Computation and Simulation* **4**, 133–139.

Chamayou, J.-F. (2001), Pseudo random numbers for the Landau and Vavilov distributions, *Computational Statistics* **19**, 131–152.

Chambers, John M. (1997), The evolution of the S language, *Computing Science and Statistics* **28**, 331–337.

Chambers, J. M.; C. L. Mallows; and B. W. Stuck (1976), A method for simulating stable random variables, *Journal of the American Statistical Association* **71**, 340–344 (Corrections, 1987, *ibid.* **82**, 704, and 1988, *ibid.* **83**, 581).

Chen, H. C., and Y. Asau (1974), On generating random variates from an empirical distribution, *AIIE Transactions* **6**, 163–166.

Chen, Huifen, and Bruce W. Schmeiser (1992), Simulation of Poisson processes with trigonometric rates, *Proceedings of the 1992 Winter Simulation Conference*, Association for Computing Machinery, New York, 609–617.

Chen, Ming-Hui, and Bruce Schmeiser (1993), Performance of the Gibbs, hit-and-run, and Metropolis samplers, *Journal of Computational and Graphical Statistics* **3**, 251–272.

Chen, Ming-Hui, and Bruce W. Schmeiser (1996), General hit-and-run Monte Carlo sampling for evaluating multidimensional integrals, *Operations Research Letters* **19**, 161–169.

Chen, Ming-Hui; Qi-Man Shao; and Joseph G. Ibrahim (2000), *Monte Carlo Methods in Bayesian Computation*, Springer-Verlag, New York.

Cheng, R. C. H. (1978), Generating beta variates with nonintegral shape parameters, *Communications of the ACM* **21**, 317–322.

Cheng, R. C. H. (1984), Generation of inverse Gaussian variates with given sample mean and dispersion, *Applied Statistics* **33**, 309–316.

Cheng, R. C. H. (1985), Generation of multivariate normal samples with given mean and covariance matrix, *Journal of Statistical Computation and Simulation* **21**, 39–49.

Cheng, R. C. H., and G. M. Feast (1979), Some simple gamma variate generators, *Applied Statistics* **28**, 290–295.

Cheng, R. C. H., and G. M. Feast (1980), Gamma variate generators with increased shape parameter range, *Communications of the ACM* **23**, 389–393.

Chernick, Michael R. (1999), *Bootstrap Methods: A Practitioner's Guide*, John Wiley & Sons, New York.

Chib, Siddhartha, and Edward Greenberg (1995), Understanding the Metropolis–Hastings algorithm, *The American Statistician* **49**, 327–335.

Chou, Wun-Seng, and Harald Niederreiter (1995), On the lattice test for inversive congruential pseudorandom numbers, *Monte Carlo and Quasi-Monte Carlo Methods in Scientific Computing* (edited by Harald Niederreiter and Peter Jau-Shyong Shiue), Springer-Verlag, New York, 186–197.

Chou, Youn-Min; S. Turner; S. Henson; D. Meyer; and K. S. Chen (1994), On using percentiles to fit data by a Johnson distribution, *Communications in Statistics — Simulation and Computation* **23**, 341–354.

Cipra, Barry A. (1987), An introduction to the Ising model, *The American Mathematical Monthly* **94**, 937–959.

Coldwell, R. L. (1974), Correlational defects in the standard IBM 360 random number generator and the classical ideal gas correlational function, *Journal of Computational Physics* **14**, 223–226.

Collings, Bruce Jay (1987), Compound random number generators, *Journal of the American Statistical Association* **82**, 525–527.

Compagner, Aaldert (1991), Definitions of randomness, *American Journal of Physics* **59**, 700–705.

Compagner, A. (1995), Operational conditions for random-number generation, *Physical Review E* **52**, 5634–5645.

Cook, R. Dennis, and Mark E. Johnson (1981), A family of distributions for modelling non-elliptically symmetric multivariate data, *Journal of the Royal Statistical Society, Series B* **43**, 210–218.

Cook, R. Dennis, and Mark E. Johnson (1986), Generalized Burr–Pareto-logistic distributions with applications to a uranium exploration data set, *Technometrics* **28**, 123–131.

Couture, R., and Pierre L'Ecuyer (1994), On the lattice structure of certain linear congruential sequences related to AWC/SWB generators, *Mathematics of Computation* **62**, 799–808.

Couture, Raymond, and Pierre L'Ecuyer (1995), Linear recurrences with carry as uniform random number generators, *Proceedings of the 1995 Winter Sim-*

*ulation Conference*, Association for Computing Machinery, New York, 263–267.

Couture, Raymond, and Pierre L'Ecuyer (1997), Distribution properties of multiply-with-carry random number generators, *Mathematics of Computation* **66**, 591–607.

Coveyou, R. R., and R. D. MacPherson (1967), Fourier analysis of uniform random number generators, *Journal of the ACM* **14**, 100–119.

Cowles, Mary Kathryn, and Bradley P. Carlin (1996), Markov chain Monte Carlo convergence diagnostics: A comparative review, *Journal of the American Statistical Association* **91**, 883–904.

Cowles, Mary Kathryn; Gareth O. Roberts; and Jeffrey S. Rosenthal (1999), Possible biases induced by MCMC convergence diagnostics, *Journal of Statistical Computation and Simulation* **64**, 87–104.

Cowles, Mary Kathryn, and Jeffrey S. Rosenthal (1998), A simulation approach to convergence rates for Markov chain Monte Carlo algorithms, *Statistics and Computing* **8**, 115–124.

Cuccaro, Steven A.; Michael Mascagni; and Daniel V. Pryor (1994), Techniques for testing the quality of parallel pseudorandom number generators, *Proceedings of the Seventh SIAM Conference on Parallel Processing for Scientific Computing*, Society for Industrial and Applied Mathematics, Philadelphia, 279–284.

Currin, Carla; Toby J. Mitchell; Max Morris; and Don Ylvisaker (1991), Bayesian prediction of deterministic functions, with applications to the design and analysis of computer experiments, *Journal of the American Statistical Association* **86**, 953–963.

D'Agostino, Ralph B. (1986), Tests for the normal distribution, *Goodness-of-Fit Techniques* (edited by Ralph B. D'Agostino and Michael A. Stephens), Marcel Dekker, New York, 367–419.

Dagpunar, J. S. (1978), Sampling of variates from a truncated gamma distribution, *Journal of Statistical Computation and Simulation* **8**, 59–64.

Dagpunar, John (1988), *Principles of Random Variate Generation*, Clarendon Press, Oxford, United Kingdom.

Dagpunar, J. (1990), Sampling from the von Mises distribution via a comparison of random numbers, *Journal of Applied Statistics* **17**, 165–168.

Damien, Paul; Purushottam W. Laud; and Adrian F. M. Smith (1995), Approximate random variate generation from infinitely divisible distributions with applications to Bayesian inference, *Journal of the Royal Statistical Society, Series B* **57**, 547–563.

Damien, Paul, and Stephen G. Walker (2001), Sampling truncated normal, beta, and gamma densities, *Journal of Computational and Graphical Statistics* **10**, 206–215.

David, Herbert A. (1981), *Order Statistics*, second edition, John Wiley & Sons, New York.

Davis, Charles S. (1993), The computer generation of multinomial random variates, *Computational Statistics & Data Analysis* **16**, 205–217.

Davis, Don; Ross Ihaka; and Philip Fenstermacher (1994), Cryptographic randomness from air turbulence in disk drives, *Advances in Cryptology —
CRYPTO '94*, edited by Yvo G. Desmedt, Springer-Verlag, New York, 114–
120.

Davison, A. C., and D. V. Hinkley (1997), *Bootstrap Methods and Their Application*, Cambridge University Press, Cambridge, United Kingdom.

Deák, I. (1981), An economical method for random number generation and a
normal generator, *Computing* **27**, 113–121.

Deák, I. (1986), The economical method for generating random samples from
discrete distributions, *ACM Transactions on Mathematical Software* **12**, 34–
36.

Deák, István (1990), *Random Number Generators and Simulation*, Akadémiai
Kiadó, Budapest.

Dellaportas, Petros (1995), Random variate transformations in the Gibbs sampler: Issues of efficiency and convergence, *Statistics and Computing* **5**, 133–
140.

Dellaportas, P., and A. F. M. Smith (1993), Bayesian inference for generalized linear and proportional hazards models via Gibbs sampling, *Applied
Statistics* **42**, 443–459.

De Matteis, A., and S. Pagnutti (1990), Long-range correlations in linear and
non-linear random number generators, *Parallel Computing* **14**, 207–210.

De Matteis, A., and S. Pagnutti (1993), Long-range correlation analysis of the
Wichmann–Hill random number generator, *Statistics and Computing* **3**, 67–
70.

Deng, Lih-Yuan; Kwok Hung Chan; and Yilian Yuan (1994), Random number
generators for multiprocessor systems, *International Journal of Modelling
and Simulation* **14**, 185–191.

Deng, Lih-Yuan, and E. Olusegun George (1990), Generation of uniform variates from several nearly uniformly distributed variables, *Communications
in Statistics — Simulation and Computation* **19**, 145–154.

Deng, Lih-Yuan, and E. Olusegun George (1992), Some characterizations of
the uniform distribution with applications to random number generation,
*Annals of the Institute of Statistical Mathematics* **44**, 379–385.

Deng, Lih-Yuan, and Dennis K. J. Lin (2000), Random number generation for
the new century, *The American Statistician* **54**, 145–150.

Deng, L.-Y.; D. K. J. Lin; J. Wang; and Y. Yuan (1997), Statistical justification
of combination generators, *Statistica Sinica* **7** 993–1003.

Devroye, L. (1984a), Random variate generation for unimodal and monotone
densities, *Computing* **32**, 43–68.

Devroye, L. (1984b), A simple algorithm for generating random variates with a
log-concave density, *Computing* **33**, 247–257.

Devroye, Luc (1986a), *Non-Uniform Random Variate Generation*, Springer-
Verlag, New York.

Devroye, Luc (1986b), An automatic method for generating random variates with a given characteristic function, *SIAM Journal on Applied Mathematics* **46**, 698–719.

Devroye, Luc (1987), A simple generator for discrete log-concave distributions, *Computing* **39**, 87–91.

Devroye, Luc (1989), On random variate generation when only moments or Fourier coefficients are known, *Mathematics and Computers in Simulation* **31**, 71–79.

Devroye, Luc (1991), Algorithms for generating discrete random variables with a given generating function or a given moment sequence, *SIAM Journal on Scientific and Statistical Computing* **12**, 107–126.

Devroye, Luc (1997), Random variate generation for multivariate unimodal densities, *ACM Transactions on Modeling and Computer Simulation* **7**, 447–477.

Devroye, Luc; Peter Epstein; and Jörg-Rüdiger Sack (1993), On generating random intervals and hyperrectangles, *Journal of Computational and Graphical Statistics* **2**, 291–308.

Dieter, U. (1975), How to calculate shortest vectors in a lattice, *Mathematics of Computation* **29**, 827–833.

Do, Kim-Anh (1991), Quasi-random resampling for the bootstrap, *Computer Science and Statistics: Proceedings of the Twenty-third Symposium on the Interface* (edited by Elaine M. Keramidas), Interface Foundation of North America, Fairfax, Virginia, 297–300.

Dodge, Yadolah (1996), A natural random number generator, *International Statistical Review* **64**, 329–344.

Doucet, Arnaud; Nando de Freitas; and Neil Gordon (Editors) (2001), *Sequential Monte Carlo Methods in Practice*, Springer-Verlag, New York.

Efron, Bradley, and Robert J. Tibshirani (1993), *An Introduction to the Bootstrap*, Chapman & Hall, New York.

Eichenauer, J.; H. Grothe; and J. Lehn (1988), Marsaglia's lattice test and non-linear congruential pseudo random number generators, *Metrika* **35**, 241–250.

Eichenauer, J., and H. Niederreiter (1988), On Marsaglia's lattice test for pseudorandom numbers, *Manuscripta Mathematica* **62**, 245–248.

Eichenauer, Jürgen, and Jürgen Lehn (1986), A non-linear congruential pseudo random number generator, *Statistische Hefte* **27**, 315–326.

Eichenauer-Herrmann, Jürgen (1995), Pseudorandom number generation by nonlinear methods, *International Statistical Review* **63**, 247–255.

Eichenauer-Herrmann, Jürgen (1996), Modified explicit inversive congruential pseudorandom numbers with power of 2 modulus, *Statistics and Computing* **6**, 31–36.

Eichenauer-Herrmann, J., and H. Grothe (1989), A remark on long-range correlations in multiplicative congruential pseudorandom number generators, *Numerische Mathematik* **56**, 609–611.

Eichenauer-Herrmann, J., and H. Grothe (1990), Upper bounds for the Beyer ratios of linear congruential generators, *Journal of Computational and Applied Mathematics* **31**, 73–80.

Eichenauer-Herrmann, Jürgen; Eva Herrmann; and Stefan Wegenkittl (1998),
    A survey of quadratic and inversive congruential pseudorandom numbers,
    *Monte Carlo and Quasi-Monte Carlo Methods 1996* (edited by Harald Nieder-
    reiter, Peter Hellekalek, Gerhard Larcher, and Peter Zinterhof), Springer-
    Verlag, New York, 66–97.

Eichenauer-Herrmann, J., and K. Ickstadt (1994), Explicit inversive congru-
    ential pseudorandom numbers with power of 2 modulus, *Mathematics of
    Computation* **62**, 787–797.

Emrich, Lawrence J., and Marion R. Piedmonte (1991), A method for gener-
    ating high-dimensional multivariate binary variates, *The American Statis-
    tician* **45**, 302–304.

Erber, T.; P. Everett; and P. W. Johnson (1979), The simulation of random
    processes on digital computers with Chebyshev mixing transformations,
    *Journal of Computational Physics* **32**, 168–211.

Ernst, Michael D. (1998), A multivariate generalized Laplace distribution, *Com-
    putational Statistics* **13**, 227–232.

Evans, Michael, and Tim Swartz (2000), *Approximating Integrals via Monte
    Carlo and Deterministic Methods*, Oxford University Press, Oxford, United
    Kingdom.

Everitt, B. S. (1998), *The Cambridge Dictionary of Statistics*, Cambridge Uni-
    versity Press, Cambridge, United Kingdom.

Everson, Philip J., and Carl N. Morris (2000), Simulation from Wishart distrib-
    utions with eigenvalue constraints, *Journal of Computational and Graphical
    Statistics* **9**, 380–389.

Falk, Michael (1999), A simple approach to the generation of uniformly dis-
    tributed random variables with prescribed correlations, *Communications in
    Statistics — Simulation and Computation* **28**, 785–791.

Fang, Kai-Tai, and T. W. Anderson (Editors) (1990), *Statistical Inference in
    Elliptically Contoured and Related Distributions*, Allerton Press, New York.

Fang, K.-T.; F. J. Hickernell; and H. Niederreiter (Editors) (2002), *Monte Carlo
    and Quasi-Monte Carlo Methods 2000*, Springer-Verlag, New York.

Fang, Kai-Tai, and Run-Ze Li (1997), Some methods for generating both an NT-
    net and the uniform distribution on a Stiefel manifold and their applications,
    *Computational Statistics & Data Analysis* **24**, 29–46.

Fang, Kai-Tai, and Yuan Wang (1994), *Number Theoretic Methods in Statistics*,
    Chapman & Hall, New York.

Faure, H. (1986), On the star discrepancy of generalised Hammersley sequences
    in two dimensions. *Monatshefte für Mathematik* **101**, 291–300.

Ferrenberg, Alan M.; D. P. Landau; and Y. Joanna Wong (1992), Monte Carlo
    simulations: Hidden errors from "good" random number generators, *Physi-
    cal Review Letters* **69**, 3382–3384.

Fill, James A. (1998), An interruptible algorithm for perfect sampling via
    Markov Chains, *Annals of Applied Probability* **8**, 131–162.

Fill, James Allen; Motoya Machida; Duncan J. Murdoch; and Jeffrey S. Rosenthal (2000), Extensions of Fill's perfect rejection sampling algorithm to general chains, *Random Structures and Algorithms* **17**, 290–316.

Fishman, George S. (1996), *Monte Carlo Concepts, Algorithms, and Applications*, Springer-Verlag, New York.

Fishman, George S., and Louis R. Moore, III (1982), A statistical evaluation of multiplicative random number generators with modulus $2^{31} - 1$, *Journal of the American Statistical Association* **77**, 129–136.

Fishman, George S., and Louis R. Moore, III (1986), An exhaustive analysis of multiplicative congruential random number generators with modulus $2^{31} - 1$, *SIAM Journal on Scientific and Statistical Computing* **7**, 24–45.

Fleishman, Allen I. (1978), A method for simulating non-normal distributions, *Psychometrika* **43**, 521–532.

Flournoy, Nancy, and Robert K. Tsutakawa (Editors) (1991), *Statistical Multiple Integration*, American Mathematical Society (Contemporary Mathematics, Volume 115), Providence, Rhode Island.

Forster, Jonathan J.; John W. McDonald; and Peter W. F. Smith (1996), Monte Carlo exact conditional tests for log-linear and logistic models, *Journal of the Royal Statistical Society, Series B* **55**, 3–24.

Fouque, Jean-Pierre; George Papanicolaou; and K. Ronnie Sircar (2000), *Derivatives in Financial Markets with Stochastic Volatility*, Cambridge University Press, Cambridge, United Kingdom.

Fox, Bennett L. (1986), Implementation and relative efficiency of quasirandom sequence generators, *ACM Transactions on Mathematical Software* **12**, 362–376.

Frederickson, P.; R. Hiromoto; T. L. Jordan; B. Smith; and T. Warnock (1984), Pseudo-random trees in Monte Carlo, *Parallel Computing* **1**, 175–180.

Freimer, Marshall; Govind S. Mudholkar; Georgia Kollia; and Thomas C. Lin (1988), A study of the generalized Tukey lambda family, *Communications in Statistics — Theory and Methods* **17**, 3547–3567.

Freund, John E. (1961), A bivariate extension of the exponential distribution, *Journal of the American Statistical Association* **56**, 971–977.

Friedman, Jerome H.; Jon Louis Bentley; and Raphael Ari Finkel (1977), An algorithm for finding best matches in logarithmic expected time, *ACM Transactions on Mathematical Software* **3**, 209–226.

Frigessi, Arnoldo; Fabio Martinelli; and Julian Stander (1997), Computational complexity of Markov chain Monte Carlo methods for finite Markov random fields, *Biometrika* **84**, 1–18.

Fuller, A. T. (1976), The period of pseudo-random numbers generated by Lehmer's congruential method, *Computer Journal* **19**, 173–177.

Fushimi, Masanori (1990), Random number generation with the recursion $X_t = X_{t-3p} \oplus X_{t-3q}$, *Journal of Computational and Applied Mathematics* **31**, 105–118.

Gamerman, Dani (1997), *Markov Chain Monte Carlo*, Chapman & Hall, London.

Gange, Stephen J. (1995), Generating multivariate categorical variates using the iterative proportional fitting algorithm, *The American Statistician* **49**, 134–138.

Gelfand, Alan E., and Adrian F. M. Smith (1990), Sampling-based approaches to calculating marginal densities, *Journal of the American Statistical Association* **85**, 398–409. (Reprinted in Samuel Kotz and Norman L. Johnson (Editors) (1997), *Breakthroughs in Statistics, Volume III*, Springer-Verlag, New York, 526–550.)

Gelfand, Alan E., and Sujit K. Sahu (1994), On Markov chain Monte Carlo acceleration, *Journal of Computational and Graphical Statistics* **3**, 261–276.

Gelman, Andrew (1992), Iterative and non-iterative simulation algorithms, *Computing Science and Statistics* **24**, 433–438.

Gelman, Andrew, and Xiao-Li Meng (1998), Simulating normalizing constants: From importance sampling to bridge sampling to path sampling, *Statistical Science* **13**, 163–185.

Gelman, Andrew, and Donald B. Rubin (1992a), Inference from iterative simulation using multiple sequences (with discussion), *Statistical Science* **7**, 457–511.

Gelman, Andrew, and Donald B. Rubin (1992b), A single series from the Gibbs sampler provides a false sense of security, *Bayesian Statistics 4* (edited by J. M. Bernardo, J. O. Berger, A. P. Dawid, and A. F. M. Smith), Oxford University Press, Oxford, United Kingdom, 625–631.

Gelman, Andrew; John B. Carlin; Hal S. Stern; and Donald B. Rubin (1995), *Bayesian Data Analysis*, Chapman & Hall, London.

Geman, Stuart, and Donald Geman (1984), Stochastic relaxation, Gibbs distributions, and the Bayesian restoration of images, *IEEE Transactions on Pattern Analysis and Machine Intelligence* **6**, 721–741.

Gentle, James E. (1981), Portability considerations for random number generators, *Computer Science and Statistics: Proceedings of the 13th Symposium on the Interface* (edited by William F. Eddy), Springer-Verlag, New York, 158–164.

Gentle, James E. (1990), Computer implementation of random number generators, *Journal of Computational and Applied Mathematics* **31**, 119–125.

Gentleman, Robert, and Ross Ihaka (1997), The R language, *Computing Science and Statistics* **28**, 326–330.

Gerontidis, I., and R. L. Smith (1982), Monte Carlo generation of order statistics from general distributions, *Applied Statistics* **31**, 238–243.

Geweke, John (1991a), Efficient simulation from the multivariate normal and Student-$t$ distributions subject to linear constraints, *Computer Science and Statistics: Proceedings of the Twenty-third Symposium on the Interface* (edited by Elaine M. Keramidas), Interface Foundation of North America, Fairfax, Virginia, 571–578.

Geweke, John (1991b), Generic, algorithmic approaches to Monte Carlo integration in Bayesian inference, *Statistical Multiple Integration* (edited by

Nancy Flournoy and Robert K. Tsutakawa), American Mathematical Society, Rhode Island, 117–135.

Geyer, Charles J. (1992), Practical Markov chain Monte Carlo (with discussion), *Statistical Science* **7**, 473–511.

Geyer, Charles J., and Elizabeth A. Thompson (1995), Annealing Markov chain Monte Carlo with applications to ancestral inference, *Journal of the American Statistical Association* **90**, 909–920.

Ghitany, M. E. (1998), On a recent generalization of gamma distribution, *Communications in Statistics — Theory and Methods* **27**, 223–233.

Gilks, W. R. (1992), Derivative-free adaptive rejection sampling for Gibbs sampling, *Bayesian Statistics 4* (edited by J. M. Bernardo, J. O. Berger, A. P. Dawid, and A. F. M. Smith), Oxford University Press, Oxford, United Kingdom, 641–649.

Gilks, W. R.; N. G. Best; and K. K. C. Tan (1995), Adaptive rejection Metropolis sampling within Gibbs sampling, *Applied Statistics* **44**, 455–472 (Corrections, Gilks, et al., 1997, *ibid.* **46**, 541–542).

Gilks, W. R.; S. Richardson; and D. J. Spiegelhalter (Editors) (1996), *Markov Chain Monte Carlo in Practice*, Chapman & Hall, London.

Gilks, Walter R., and Gareth O. Roberts (1996), Strategies for improving MCMC, *Practical Markov Chain Monte Carlo* (edited by W. R. Gilks, S. Richardson, and D. J. Spiegelhalter), Chapman & Hall, London, 89–114.

Gilks, W. R.; G. O. Roberts; and E. I. George (1994), Adaptive direction sampling, *The Statistician* **43**, 179–189.

Gilks, W. R.; A. Thomas; and D. J. Spiegelhalter (1992), Software for the Gibbs sampler, *Computing Science and Statistics* **24**, 439–448.

Gilks, W. R.; A. Thomas; and D. J. Spiegelhalter (1994), A language and program for complex Bayesian modelling, *The Statistician* **43**, 169–178.

Gilks, W. R., and P. Wild (1992), Adaptive rejection sampling for Gibbs sampling, *Applied Statistics* **41**, 337–348.

Gleser, Leon Jay (1976), A canonical representation for the noncentral Wishart distribution useful for simulation, *Journal of the American Statistical Association* **71**, 690–695.

Golder, E. R., and J. G. Settle (1976), The Box–Muller method for generating pseudo-random normal deviates, *Applied Statistics* **25**, 12–20.

Golomb, S. W. (1982), *Shift Register Sequences*, second edition, Aegean Part Press, Laguna Hills, California.

Gordon, J. (1989), Fast multiplicative inverse in modular arithmetic, *Cryptography and Coding* (edited by H. J. Beker and F. C. Piper), Clarendon Press, Oxford, United Kingdom, 269–279.

Gordon, N. J.; D. J. Salmond; and A. F. M. Smith (1993), Novel approach to nonlinear/non-Gaussian Bayesian state estimation, *IEE Proceedings F, Communications, Radar, and Signal Processing* **140**, 107–113.

Grafton, R. G. T. (1981), The runs-up and runs-down tests, *Applied Statistics* **30**, 81–85.

Greenwood, J. Arthur (1976a), The demands of trivial combinatorial problems on random number generators, *Proceedings of the Ninth Interface Symposium on Computer Science and Statistics* (edited by David C. Hoaglin and Roy E. Welsch), Prindle, Weber, and Schmidt, Boston, 222–227.

Greenwood, J. A. (1976b), A fast machine-independent long-period generator for 31-bit pseudo-random numbers, *Compstat 1976: Proceedings in Computational Statistics* (edited by J. Gordesch and P. Naeve), Physica-Verlag, Vienna, 30–36.

Greenwood, J. Arthur (1976c), Moments of time to generate random variables by rejection, *Annals of the Institute for Statistical Mathematics* **28**, 399–401.

Griffiths, P., and I. D. Hill (Editors) (1985), *Applied Statistics Algorithms*, Ellis Horwood Limited, Chichester, United Kingdom.

Grothe, H. (1987), Matrix generators for pseudo-random vector generation, *Statistische Hefte* **28**, 233–238.

Guerra, Victor O.; Richard A. Tapia; and James R. Thompson (1976), A random number generator for continuous random variables based on an interpolation procedure of Akima, *Computer Science and Statistics: 9th Annual Symposium on the Interface* (edited by David C. Hoaglin and Roy E. Welsch), Prindle, Weber, and Schmidt, Boston, 228–230.

Halton, J. H. (1960), On the efficiency of certain quasi-random sequences of points in evaluating multi-dimensional integrals, *Numerische Mathematik* **2**, 84–90 (Corrections, 1960, *ibid.* **2**, 190).

Hamilton, Kenneth G. (1998), Algorithm 780: Exponential pseudorandom distribution, *ACM Transactions on Mathematical Software* **24**, 102–106.

Hammersley, J. M., and D. C. Handscomb (1964), *Monte Carlo Methods*, Methuen & Co., London.

Hartley, H. O., and D. L. Harris (1963), Monte Carlo computations in normal correlation procedures, *Journal of the ACM* **10**, 302–306.

Hastings, W. K. (1970), Monte Carlo sampling methods using Markov chains and their applications. *Biometrika* **57**, 97–109. (Reprinted in Samuel Kotz and Norman L. Johnson (Editors) (1997), *Breakthroughs in Statistics, Volume III*, Springer-Verlag, New York, 240–256.)

Heiberger, Richard M. (1978), Algorithm AS127: Generation of random orthogonal matrices, *Applied Statistics* **27**, 199–205. (See Tanner and Thisted, 1982.)

Hellekalek, P. (1984), Regularities of special sequences, *Journal of Number Theory* **18**, 41–55.

Hesterberg, Tim (1995), Weighted average importance sampling and defensive mixture distributions, *Technometrics* **37**, 185–194.

Hesterberg, Timothy C., and Barry L. Nelson (1998), Control variates for probability and quantile estimation, *Management Science* **44**, 1295–1312.

Hickernell, Fred J. (1995), A comparison of random and quasirandom points for multidimensional quadrature, *Monte Carlo and Quasi-Monte Carlo Meth-*

ods in Scientific Computing (edited by Harald Niederreiter and Peter Jau-Shyong Shiue), Springer-Verlag, New York, 212–227.

Hill, I. D.; R. Hill, R.; and R. L. Holder (1976), Algorithm AS99: Fitting Johnson curves by moments, Applied Statistics 25, 180–189 (Remark, 1981, ibid. 30, 106).

Hoaglin, David C., and David F. Andrews (1975), The reporting of computation-based results in statistics, The American Statistician 29, 122–126.

Hope, A. C. A. (1968), A simplified Monte Carlo significance test procedure, Journal of the Royal Statistical Society, Series B 30, 582–598.

Hopkins, T. R. (1983), A revised algorithm for the spectral test, Applied Statistics 32, 328–335. (See http://www.cs.ukc.ac.uk/pubs/1997 for an updated version.)

Hörmann, W. (1994a), A universal generator for discrete log-concave distributions, Computing 52, 89–96.

Hörmann, Wolfgang (1994b), A note on the quality of random variates generated by the ratio of uniforms method, ACM Transactions on Modeling and Computer Simulation 4, 96–106.

Hörmann, Wolfgang (1995), A rejection technique for sampling from T-concave distributions, ACM Transactions on Mathematical Software 21, 182–193.

Hörmann, Wolfgang (2000), Algorithm 802: An automatic generator for bivariate log-concave distributions, ACM Transactions on Mathematical Software 26, 201–219.

Hörmann, Wolfgang, and Gerhard Derflinger (1993), A portable random number generator well suited for the rejection method, ACM Transactions on Mathematical Software 19, 489–495.

Hörmann, Wolfgang, and Gerhard Derflinger (1994), The transformed rejection method for generating random variables, an alternative to the ratio of uniforms method, Communications in Statistics — Simulation and Computation 23, 847–860.

Hosack, J. M. (1986), The use of Chebyshev mixing to generate pseudo-random numbers, Journal of Computational Physics 67, 482–486.

Huber, Peter J. (1985), Projection pursuit (with discussion), The Annals of Statistics 13, 435–525.

Hull, John C. (2000), Options, Futures, & Other Derivatives, Prentice–Hall, Englewood Cliffs, New Jersey.

Ireland, Kenneth, and Michael Rosen (1991), A Classical Introduction to Modern Number Theory, Springer-Verlag, New York.

Jäckel, Peter (2002), Monte Carlo Methods in Finance, John Wiley & Sons Ltd., Chichester.

Jaditz, Ted (2000), Are the digits of $\pi$ an independent and identically distributed sequence?, The American Statistician 54, 12–16.

James, F. (1990), A review of pseudorandom number generators, Computer Physics Communications 60, 329–344.

James, F. (1994), RANLUX: A Fortran implementation of the high-quality pseudorandom number generator of Lüscher, *Computer Physics Communications* **79**, 111–114.

Jöhnk, M. D. (1964), Erzeugung von Betaverteilter und Gammaverteilter Zufallszahlen, *Metrika* **8**, 5–15.

Johnson, Mark E. (1987), *Multivariate Statistical Simulation*, John Wiley & Sons, New York.

Johnson, Valen E. (1996), Studying convergence of Markov chain Monte Carlo algorithms using coupled sample paths, *Journal of the American Statistical Association* **91**, 154–166.

Jones, G.; C. D. Lai; and J. C. W. Rayner (2000), A bivariate gamma mixture distribution, *Communications in Statistics — Theory and Methods* **29**, 2775–2790.

Joy, Corwin; Phelim P. Boyle; and Ken Seng Tan (1996), Quasi-Monte Carlo methods in numerical finance, *Management Science* **42**, 926–938.

Juneja, Sandeep, and Perwez Shahabudding (2001), Fast simulation of Markov chains with small transition probabilities, *Management Science* **47**, 547–562.

Kachitvichyanukul, Voratas (1982), *Computer Generation of Poisson, Binomial, and Hypergeometric Random Variables*, unpublished Ph.D. dissertation, Purdue University, West Lafayette, Indiana.

Kachitvichyanukul, Voratas; Shiow-Wen Cheng; and Bruce Schmeiser (1988), Fast Poisson and binomial algorithms for correlation induction, *Journal of Statistical Computation and Simulation* **29**, 17–33.

Kachitvichyanukul, Voratas, and Bruce Schmeiser (1985), Computer generation of hypergeometric random variates, *Journal of Statistical Computation and Simulation* **22**, 127–145.

Kachitvichyanukul, Voratas, and Bruce W. Schmeiser (1988a), Binomial random variate generation, *Communications of the ACM* **31**, 216–223.

Kachitvichyanukul, Voratas, and Bruce W. Schmeiser (1988b), Algorithm 668: H2PEC: Sampling from the hypergeometric distribution, *ACM Transactions on Mathematical Software* **14**, 397–398.

Kachitvichyanukul, Voratas, and Bruce W. Schmeiser (1990), BTPEC: Sampling from the binomial distribution, *ACM Transactions on Mathematical Software* **16**, 394–397.

Kahn, H., and A. W. Marshall (1953), Methods of reducing sample size in Monte Carlo computations, *Journal of the Operations Research Society of America* **1**, 263–278.

Kankaala, K.; T. Ala-Nissila; and I. Vattulainen (1993), Bit-level correlations in some pseudorandom number generators, *Physical Review E* **48**, R4211–R4214.

Kao, Chiang, and H. C. Tang (1997a), Upper bounds in spectral test for multiple recursive random number generators with missing terms, *Computers and Mathematical Applications* **33**, 113–118.

Kao, Chiang, and H. C. Tang (1997b), Systematic searches for good multiple recursive random number generators, *Computers and Operations Research* **24**, 899–905.

Karian, Zaven A., and Edward J. Dudewicz (1999), Fitting the generalized lambda distribution to data: A method based on percentiles, *Communications in Statistics — Simulation and Computation* **28**, 793–819.

Karian, Zaven A., and Edward J. Dudewicz (2000), *Fitting Statistical Distributions*, CRC Press, Boca Raton, Florida.

Karian, Zaven A.; Edward J. Dudewicz; and Patrick McDonald (1996), The extended generalized lambda distribution system for fitting distributions to data: History, completion of theory, tables, applications, the "final word" on moment fits, *Communications in Statistics — Simulation and Computation* **25**, 611–642.

Kato, Takashi; Li-ming Wu; and Niro Yanagihara (1996a), On a nonlinear congruential pseudorandom number generator, *Mathematics of Computation* **65**, 227–233.

Kato, Takashi; Li-ming Wu; and Niro Yanagihara (1996b), The serial test for a nonlinear pseudorandom number generator, *Mathematics of Computation* **65**, 761–769.

Kemp, A. W. (1981), Efficient generation of logarithmically distributed pseudorandom variables, *Applied Statistics* **30**, 249–253.

Kemp, A. W. (1990), Patchwork rejection algorithms, *Journal of Computational and Applied Mathematics* **31**, 127–131.

Kemp, C. D. (1986), A modal method for generating binomial variables, *Communications in Statistics — Theory and Methods* **15**, 805–813.

Kemp, C. D., and Adrienne W. Kemp (1987), Rapid generation of frequency tables, *Applied Statistics* **36**, 277–282.

Kemp, C. D., and Adrienne W. Kemp (1991), Poisson random variate generation, *Applied Statistics* **40**, 143–158.

Kinderman, A. J., and J. F. Monahan (1977), Computer generation of random variables using the ratio of uniform deviates, *ACM Transactions on Mathematical Software* **3**, 257–260.

Kinderman, A. J., and J. F. Monahan (1980), New methods for generating Student's $t$ and gamma variables, *Computing* **25**, 369–377.

Kinderman, A. J., and J. G. Ramage (1976), Computer generation of normal random variables, *Journal of the American Statistical Association* **71**, 893–896.

Kirkpatrick, S.; C. D. Gelatt; and M. P. Vecchi (1983), Optimization by simulated annealing, *Science* **220**, 671–679.

Kirkpatrick, Scott, and Erich P. Stoll (1981), A very fast shift-register sequence random number generator, *Journal of Computational Physics* **40**, 517–526.

Kleijnen, Jack P. C. (1977), Robustness of a multiple ranking procedure: A Monte Carlo experiment illustrating design and analysis techniques, *Communications in Statistics — Simulation and Computation* **B6**, 235–262.

Knuth, Donald E. (1975), Estimating the efficiency of backtrack programs, *Mathematics of Computation* **29**, 121–136.

Knuth, Donald E. (1998), *The Art of Computer Programming, Volume 2, Semi-*

numerical *Algorithms*, third edition, Addison–Wesley Publishing Company, Reading, Massachusetts.

Kobayashi, K. (1991), On generalized gamma functions occurring in diffraction theory, *Journal of the Physical Society of Japan* **60**, 1501–1512.

Kocis, Ladislav, and William J. Whiten (1997), Computational investigations of low-discrepancy sequences, *ACM Transactions on Mathematical Software* **23**, 266–294.

Koehler, J. R., and A. B. Owen (1996), Computer experiments, *Handbook of Statistics, Volume 13* (edited by S. Ghosh and C. R. Rao), Elsevier Science Publishers, Amsterdam, 261–308.

Kotz, Samuel; N. Balakrishnan; and Norman L. Johnson (2000), *Continuous Multivariate Distributions*, second edition, John Wiley & Sons, New York.

Kovalenko, I. N. (1972), Distribution of the linear rank of a random matrix, *Theory of Probability and Its Applications* **17**, 342–346.

Kozubowski, Tomasz J., and Krzysztof Podgórski (2000), A multivariate and asymmetric generalization of Laplace distribution, *Computational Statistics* **15**, 531–540.

Krawczyk, Hugo (1992), How to predict congruential generators, *Journal of Algorithms* **13**, 527–545.

Krommer, Arnold R., and Christoph W. Ueberhuber (1994), *Numerical Integration on Advanced Computer Systems*, Springer-Verlag, New York.

Kronmal, Richard A., and Arthur V. Peterson (1979a), On the alias method for generating random variables from a discrete distribution, *The American Statistician* **33**, 214–218.

Kronmal, R. A., and A. V. Peterson (1979b), The alias and alias-rejection-mixture methods for generating random variables from probability distributions, *Proceedings of the 1979 Winter Simulation Conference*, Institute of Electrical and Electronics Engineers, New York, 269–280.

Kronmal, Richard A., and Arthur V. Peterson (1981), A variant of the acceptance-rejection method for computer generation of random variables, *Journal of the American Statistical Association* **76**, 446–451 (Corrections, 1982, *ibid.* **77**, 954).

Kronmal, Richard A., and Arthur V. Peterson (1984), An acceptance-complement analogue of the mixture-plus-acceptance-rejection method for generating random variables, *ACM Transactions on Mathematical Software* **10**, 271–281.

Kumada, Toshihiro; Hannes Leeb; Yoshiharu Kurita; and Makoto Matsumoto (2000), New primitive $t$-nomials ($t = 3, 5$) over $GF(2)$ whose degree is a Mersenne exponent, *Mathematics of Computation* **69**, 811–814.

Lagarias, Jeffrey C. (1993), Pseudorandom numbers, *Statistical Science* **8**, 31–39.

Laud, Purushottam W.; Paul Ramgopal; and Adrian F. M. Smith (1993), Random variate generation from $D$-distributions, *Statistics and Computing* **3**, 109–112.

Lawrance, A. J. (1992), Uniformly distributed first-order autoregressive time series models and multiplicative congruential random number generators, *Journal of Applied Probability* **29**, 896–903.

Learmonth, G. P., and P. A. W. Lewis (1973), Statistical tests of some widely used and recently proposed uniform random number generators, *Computer Science and Statistics: 7th Annual Symposium on the Interface* (edited by William J. Kennedy), Statistical Laboratory, Iowa State University, Ames, Iowa, 163–171.

L'Ecuyer, Pierre (1988), Efficient and portable combined random number generators, *Communications of the ACM* **31**, 742–749, 774.

L'Ecuyer, Pierre (1990), Random numbers for simulation, *Communications of the ACM* **33**, Number 10 (October), 85–97.

L'Ecuyer, Pierre (1996), Combined multiple recursive random number generators, *Operations Research* **44**, 816–822.

L'Ecuyer, Pierre (1997), Tests based on sum-functions of spacings for uniform random numbers, *Journal of Statistical Computation and Simulation* **59**, 251–269.

L'Ecuyer, Pierre (1998), Random number generators and empirical tests, *Monte Carlo and Quasi-Monte Carlo Methods 1996* (edited by Harald Niederreiter, Peter Hellekalek, Gerhard Larcher, and Peter Zinterhof), Springer-Verlag, New York, 124–138.

L'Ecuyer, Pierre (1999), Good parameters and implementations for combined multiple recursive random number generators, *Operations Research* **47**, 159–164.

L'Ecuyer, Pierre; François Blouin; and Raymond Couture (1993), A search for good multiple recursive random number generators, *ACM Transactions on Modeling and Computer Simulation* **3**, 87–98.

L'Ecuyer, Pierre; Jean-Françoise Cordeau; and Richard Simard (2000), Close-point spatial tests and their application to random number generators, *Operations Research* **48**, 308–317.

L'Ecuyer, Pierre, and Peter Hellekalek (1998), Random number generators: Selection criteria and testing, *Random and Quasi-Random Point Sets* (edited by Peter Hellekalek and Gerhard Larcher), Springer-Verlag, New York, 223–266.

L'Ecuyer, Pierre, and Richard Simard (1999), Beware of linear congruential generators with multipliers of the form $a = \pm 2^q \pm 2^r$, *ACM Transactions on Mathematical Software* **25**, 367–374.

L'Ecuyer, Pierre, and Shu Tezuka (1991), Structural properties for two classes of combined random number generators, *Mathematics of Computation* **57**, 735–746.

Lee, A. J. (1993), Generating random binary deviates having fixed marginal distributions and specified degrees of association, *The American Statistician* **47**, 209–215.

Leeb, Hannes, and Stefan Wegenkittl (1997), Inversive and linear congruential pseudorandom number generators in empirical tests, *ACM Transactions on Modeling and Computer Simulation* **7**, 272–286.

Lehmer, D. H. (1951), Mathematical methods in large-scale computing units, *Proceedings of the Second Symposium on Large Scale Digital Computing Machinery*, Harvard University Press, Cambridge, Massachusetts. 141–146.

Leva, Joseph L. (1992a), A fast normal random number generator, *ACM Transactions on Mathematical Software* **18**, 449–453.

Leva, Joseph L. (1992b), Algorithm 712: A normal random number generator, *ACM Transactions on Mathematical Software* **18**, 454–455.

Lewis, P. A. W.; A. S. Goodman; and J. M. Miller (1969), A pseudo-random number generator for the System/360, *IBM Systems Journal* **8**, 136–146.

Lewis, P. A. W., and E. J. Orav (1989), *Simulation Methodology for Statisticians, Operations Analysts, and Engineers, Volume I*, Wadsworth & Brooks/Cole, Pacific Grove, California.

Lewis, P. A. W., and G. S. Shedler (1979), Simulation of nonhomogeneous Poisson processes by thinning, *Naval Logistics Quarterly* **26**, 403–413.

Lewis, T. G., and W. H. Payne (1973), Generalized feedback shift register pseudorandom number algorithm, *Journal of the ACM* **20**, 456–468.

Leydold, Josef (1998), A rejection technique for sampling from log-concave multivariate distributions, *ACM Transactions on Modeling and Computer Simulation* **8**, 254–280.

Leydold, Josef (2000), Automatic sampling with the ratio-of-uniforms method, *ACM Transactions on Mathematical Software* **26**, 78–98.

Leydold, Josef (2001), A simple universal generator for continuous and discrete univariate $T$-concave distributions, *ACM Transactions on Mathematical Software* **27**, 66–82.

Li, Kim-Hung (1994), Reservoir-sampling algorithms of time complexity $O(n(1+ \log(N/n)))$, *ACM Transactions on Mathematical Software* **20**, 481–493.

Li, Shing Ted, and Joseph L. Hammond (1975), Generation of pseudo-random numbers with specified univariate distributions and correlation coefficients, *IEEE Transactions on Systems, Man, and Cybernetics* **5**, 557–560.

Liao, J. G., and Ori Rosen (2001), Fast and stable algorithms for computing and sampling from the noncentral hypergeometric distribution, *The American Statistician* **55**, 366–369.

Liu, Jun S. (1996), Metropolized independent sampling with comparisons to rejection sampling and importance sampling, *Statistics and Computing* **6**, 113–119.

Liu, Jun S. (2001), *Monte Carlo Strategies in Scientific Computing*, Springer-Verlag, New York.

Liu, Jun S.; Rong Chen; and Tanya Logvinenko (2001), A theoretical framework for sequential importance sampling with resampling, *Sequential Monte Carlo Methods in Practice* (edited by Arnaud Doucet, Nando de Freitas, and Neil Gordon) Springer-Verlag, New York, 225–246.

Liu, Jun S.; Rong Chen; and Wing Hung Wong (1998), Rejection control and sequential importance sampling *Journal of the American Statistical Association* **93**, 1022–1031.

London, Wendy B., and Chris Gennings (1999), Simulation of multivariate gamma data with exponential marginals for independent clusters, *Communications in Statistics — Simulation and Computation* **28**, 487–500.

Luby, Michael (1996), *Pseudorandomness and Cryptographic Applications*, Princeton University Press, Princeton.

Lurie, D., and H. O. Hartley (1972), Machine generation of order statistics for Monte Carlo computations, *The American Statistician* **26**(1), 26–27.

Lurie, D., and R. L. Mason (1973), Empirical investigation of general techniques for computer generation of order statistics, *Communications in Statistics* **2**, 363–371.

Lurie, Philip M., and Matthew S. Goldberg (1998), An approximate method for sampling correlated random variables from partially specified distributions, *Management Science* **44**, 203–218.

Lüscher, Martin (1994), A portable high-quality random number generator for lattice field theory simulations, *Computer Physics Communications* **79**, 100–110.

MacEachern, Steven N., and L. Mark Berliner (1994), Subsampling the Gibbs sampler, *The American Statistician* **48**, 188–190.

MacLaren, M. D., and G. Marsaglia (1965), Uniform random number generators, *Journal of the ACM* **12**, 83–89.

Manly, Bryan F. J. (1997), *Randomization, Bootstrap and Monte Carlo Methods in Biology*, second edition, Chapman & Hall, London.

Marasinghe, Mervyn G., and William J. Kennedy, Jr. (1982), Direct methods for generating extreme characteristic roots of certain random matrices, *Communications in Statistics — Simulation and Computation* **11**, 527–542.

Marinari, Enzo, and G. Parisi (1992), Simulated tempering: A new Monte Carlo scheme, *Europhysics Letters* **19**, 451–458.

Marriott, F. H. C. (1979), Barnard's Monte Carlo tests: How many simulations?, *Applied Statistics* **28**, 75–78.

Marsaglia, G. (1962), Random variables and computers, *Information Theory, Statistical Decision Functions, and Random Processes* (edited by J. Kozesnik), Czechoslovak Academy of Sciences, Prague, 499–510.

Marsaglia, G. (1963), Generating discrete random variables in a computer, *Communications of the ACM* **6**, 37–38.

Marsaglia, George (1964), Generating a variable from the tail of a normal distribution, *Technometrics* **6**, 101–102.

Marsaglia, G. (1968), Random numbers fall mainly in the planes, *Proceedings of the National Academy of Sciences* **61**, 25–28.

Marsaglia, G. (1972a), The structure of linear congruential sequences, *Applications of Number Theory to Numerical Analysis* (edited by S. K. Zaremba), Academic Press, New York, 249–286.

Marsaglia, George (1972b), Choosing a point from the surface of a sphere, *Annals of Mathematical Statistics* **43**, 645–646.

Marsaglia, G. (1977), The squeeze method for generating gamma variates, *Computers and Mathematics with Applications* **3**, 321–325.

Marsaglia, G. (1980), Generating random variables with a *t*-distribution, *Mathematics of Computation* **34**, 235–236.

Marsaglia, George (1984), The exact-approximation method for generating random variables in a computer, *Journal of the American Statistical Association* **79**, 218–221.

Marsaglia, George (1985), A current view of random number generators, *Computer Science and Statistics: 16th Symposium on the Interface* (edited by L. Billard), North-Holland, Amsterdam, 3–10.

Marsaglia, George (1991), Normal (Gaussian) random variables for supercomputers, *Journal of Supercomputing* **5**, 49–55.

Marsaglia, George (1995), *The Marsaglia Random Number CDROM, including the DIEHARD Battery of Tests of Randomness*, Department of Statistics, Florida State University, Tallahassee, Florida. Available at `http://stat.fsu.edu/~geo/diehard.html`.

Marsaglia, G., and T. A. Bray (1964), A convenient method for generating normal variables, *SIAM Review* **6**, 260–264.

Marsaglia, G.; M. D. MacLaren; and T. A. Bray (1964), A fast method for generating normal random variables, *Communications of the ACM* **7**, 4–10.

Marsaglia, George, and Ingram Olkin (1984), Generating correlation matrices, *SIAM Journal on Scientific and Statistical Computing* **5**, 470–475.

Marsaglia, George, and Wai Wan Tsang (1984), A fast, easily implemented method for sampling from decreasing or symmetric unimodal density functions, *SIAM Journal of Scientific and Statistical Computing* **5**, 349–359.

Marsaglia, George, and Wai Wan Tsang (1998), The Monty Python method for generating random variables, *ACM Transactions on Mathematical Software* **24**, 341–350.

Marsaglia, George, and Liang-Huei Tsay (1985), Matrices and the structure of random number sequences, *Linear Algebra and Its Applications* **67**, 147–156.

Marsaglia, George, and Arif Zaman (1991), A new class of random number generators, *The Annals of Applied Probability* **1**, 462–480.

Marsaglia, George; Arif Zaman; and John C. W. Marsaglia (1994), Rapid evaluation of the inverse normal distribution function, *Statistics and Probability Letters* **19**, 259–266.

Marshall, Albert W., and Ingram Olkin (1967), A multivariate exponential distribution, *Journal of the American Statistical Association* **62**, 30–44.

Marshall, Albert W., and Ingram Olkin (1979), *Inequalities — Theory of Majorization and Its Applications*, Academic Press, New York.

Mascagni, Michael; M. L. Robinson; Daniel V. Pryor; and Steven A. Cuccaro (1995), Parallel pseudorandom number generation using additive lagged-Fibonacci recursions, *Monte Carlo and Quasi-Monte Carlo Methods in Scientific Computing* (edited by Harald Niederreiter and Peter Jau-Shyong Shiue), Springer-Verlag, New York, 262–267.

Mascagni, Michael, and Ashok Srinivasan (2000), SPRNG: A scalable library for pseudorandom number generation, *ACM Transactions on Mathematical Software* **26**, 346–461. (Assigned as Algorithm 806, 2000, *ibid.* **26**, 618–619).

Matsumoto, Makoto, and Yoshiharu Kurita (1992), Twisted GFSR generators, *ACM Transactions on Modeling and Computer Simulation* **2**, 179–194.

Matsumoto, Makoto, and Yoshiharu Kurita (1994), Twisted GFSR generators II, *ACM Transactions on Modeling and Computer Simulation* **4**, 245–266.

Matsumoto, Makoto, and Takuji Nishimura (1998), Mersenne twister: A 623-dimensionally equidistributed uniform pseudo-random generator, *ACM Transactions on Modeling and Computer Simulation* **8**, 3–30.

Maurer, Ueli M. (1992), A universal statistical test for random bit generators, *Journal of Cryptology* **5**, 89–105.

McCullough, B. D. (1999), Assessing the reliability of statistical software: Part II, *The American Statistician* **53**, 149–159.

McKay, Michael D.; William J. Conover; and Richard J. Beckman (1979), A comparison of three methods for selecting values of input variables in the analysis of output from a computer code, *Technometrics* **21**, 239–245.

McLeod, A. I., and D. R. Bellhouse (1983), A convenient algorithm for drawing a simple random sample, *Applied Statistics* **32**, 182–184.

Mendoza-Blanco, José R., and Xin M. Tu (1997), An algorithm for sampling the degrees of freedom in Bayesian analysis of linear regressions with $t$-distributed errors, *Applied Statistics* **46**, 383–413.

Mengersen, Kerrie L.; Christian P. Robert; and Chantal Guihenneuc-Jouyaux (1999), MCMC convergence diagnostics: A reviewwww, *Bayesian Statistics 6* (edited by J. M. Bernardo, J. O. Berger, A. P. Dawid, and A. F. M. Smith), Oxford University Press, Oxford, United Kingdom, 415–440.

Metropolis, N.; A. W. Rosenbluth; M. N. Rosenbluth; A. H. Teller; and E. Teller (1953), Equations of state calculation by fast computing machines, *Journal of Chemical Physics* **21**, 1087–1092. (Reprinted in Samuel Kotz and Norman L. Johnson (Editors) (1997), *Breakthroughs in Statistics, Volume III*, Springer-Verlag, New York, 127–139.)

Meyn, S. P., and R. L. Tweedie (1993), *Markov Chains and Stochastic Stability*, Springer-Verlag, New York.

Michael, John R.; William R. Schucany; and Roy W. Haas (1976), Generating random variates using transformations with multiple roots, *The American Statistician* **30**, 88–90.

Mihram, George A., and Robert A. Hultquist (1967), A bivariate warning-time/failure-time distribution, *Journal of the American Statistical Association* **62**, 589–599.

Modarres, R., and J. P. Nolan (1994), A method for simulating stable random vectors, *Computational Statistics* **9**, 11–19.

Møller, Jesper, and Katja Schladitz (1999), Extensions of Fill's algorithm for perfect simulation, *Journal of the Royal Statistical Society, Series B* **61**, 955–969.

Monahan, John F. (1987), An algorithm for generating chi random variables, *ACM Transactions on Mathematical Software* **13**, 168–171 (Corrections, 1988, *ibid.* **14**, 111).

Morel, Jorge G. (1992), A simple algorithm for generating multinomial random vectors with extravariation, *Communications in Statistics — Simulation and Computation* **21**, 1255–1268.

Morgan, B. J. T. (1984), *Elements of Simulation*, Chapman & Hall, London.

Nagaraja, H. N. (1979), Some relations between order statistics generated by different methods, *Communications in Statistics — Simulation and Computation* **B8**, 369–377.

Neal, Radford M. (1996), Sampling from multimodal distributions using tempered transitions, *Statistics and Computing* **6**, 353–366.

Neave, H. R. (1973), On using the Box–Muller transformation with multiplicative congruential pseudo-random number generators, *Applied Statistics* **22**, 92–97.

Newman, M. E. J., and G. T. Barkema (1999), *Monte Carlo Methods in Statistical Physics*, Oxford University Press, Oxford, United Kingdom.

Niederreiter, H. (1988), Remarks on nonlinear congruential pseudorandom numbers, *Metrika* **35**, 321–328.

Niederreiter, H. (1989), The serial test for congruential pseudorandom numbers generated by inversions, *Mathematics of Computation* **52**, 135–144.

Niederreiter, Harald (1992), *Random Number Generation and Quasi-Monte Carlo Methods*, Society for Industrial and Applied Mathematics, Philadelphia.

Niederreiter, Harald (1993), Factorization of polynomials and some linear-algebra problems over finite fields, *Linear Algebra and Its Applications* **192**, 301–328.

Niederreiter, Harald (1995a), The multiple-recursive matrix method for pseudorandom number generation, *Finite Fields and Their Applications* **1**, 3–30.

Niederreiter, Harald (1995b), Pseudorandom vector generation by the multiple-recursive matrix method, *Mathematics of Computation* **64**, 279–294.

Niederreiter, Harald (1995c), New developments in uniform pseudorandom number and vector generation, *Monte Carlo and Quasi-Monte Carlo Methods in Scientific Computing* (edited by Harald Niederreiter and Peter Jau-Shyong Shiue), Springer-Verlag, New York, 87–120.

Niederreiter, Harald (1995d), Some linear and nonlinear methods for pseudorandom number generation, *Proceedings of the 1995 Winter Simulation Conference*, Association for Computing Machinery, New York, 250–254.

Niederreiter, Harald; Peter Hellekalek; Gerhard Larcher; and Peter Zinterhof (Editors) (1998), *Monte Carlo and Quasi-Monte Carlo Methods 1996*, Springer-Verlag, New York.

Niederreiter, Harald, and Peter Jau-Shyong Shiue (Editors) (1995), *Monte*

Carlo and Quasi-Monte Carlo Methods in Scientific Computing, Springer-Verlag, New York.

Niederreiter, Harald, and Jerome Spanier (Editors) (1999), Monte Carlo and Quasi-Monte Carlo Methods 1998, Springer-Verlag, New York.

NIST (2000), A Statistical Test Suite for Random and Pseudorandom Number Generators for Cryptographic Applications, NIST Special Publication 800-22, National Institute for Standards and Technology, Gaithersburg, Maryland.

Nolan, John P. (1998a), Multivariate stable distributions: Approximation, estimation, simulation and identification, A Practical Guide to Heavy Tails: Statistical Techniques and Applications (edited by Robert J. Adler, Raisa E. Feldman, and Murad S. Taqqu), Birkhäuser, Boston, 509–526.

Nolan, John P. (1998b), Univariate stable distributions: Parameterizations and software, A Practical Guide to Heavy Tails: Statistical Techniques and Applications (edited by Robert J. Adler, Raisa E. Feldman, and Murad S. Taqqu), Birkhäuser, Boston, 527–533.

Norman, J. E., and L. E. Cannon (1972), A computer program for the generation of random variables from any discrete distribution, Journal of Statistical Computation and Simulation 1, 331–348.

Odell, P. L., and A. H. Feiveson (1966), A numerical procedure to generate a sample covariance matrix, Journal of the American Statistical Association 61, 199–203.

Ogata, Yosihiko (1990), A Monte Carlo method for an objective Bayesian procedure, Annals of the Institute for Statistical Mathematics 42, 403–433.

Oh, Man-Suk, and James O. Berger (1993), Integration of multimodal functions by Monte Carlo importance sampling, Journal of the American Statistical Association 88, 450–456.

Øksendal, Bernt (1998), Stochastic Differential Equations. An Introduction with Applications, fifth edition, Springer-Verlag, Berlin.

Ökten, Giray (1998), Error estimates for quasi-Monte Carlo methods, Monte Carlo and Quasi-Monte Carlo Methods 1996 (edited by Harald Niederreiter, Peter Hellekalek, Gerhard Larcher, and Peter Zinterhof), Springer-Verlag, New York, 353–358.

Olken, Frank, and Doron Rotem (1995a), Random sampling from databases: A survey, Statistics and Computing 5, 25–42.

Olken, Frank, and Doron Rotem (1995b), Sampling from spatial databases, Statistics and Computing 5, 43–57.

Owen, A. B. (1992a), A central limit theorem for Latin hypercube sampling, Journal of the Royal Statistical Society, Series B 54, 541–551.

Owen, A. B. (1992b), Orthogonal arrays for computer experiments, integration and visualization, Statistica Sinica 2, 439–452.

Owen, A. B. (1994a), Lattice sampling revisited: Monte Carlo variance of means over randomized orthogonal arrays, Annals of Statistics 22, 930–945.

Owen, Art B. (1994b), Controlling correlations in Latin hypercube samples, Journal of the American Statistical Association 89, 1517–1522.

Owen, Art B. (1995), Randomly permuted $(t, m, s)$-nets and $(t, s)$-sequences, *Monte Carlo and Quasi-Monte Carlo Methods in Scientific Computing* (edited by Harald Niederreiter and Peter Jau-Shyong Shiue), Springer-Verlag, New York, 299–317.

Owen, Art B. (1997), Scrambled net variance for integrals of smooth functions, *Annals of Statistics* **25**, 1541–1562.

Owen, Art B. (1998), Latin supercube sampling for very high-dimensional simulations, *ACM Transactions on Modeling and Computer Simulation* **8**, 71–102.

Papageorgiou, A., and J. F. Traub (1996), Beating Monte Carlo, *Risk* (June), 63–65.

Park, Chul Gyu; Tasung Park; and Dong Wan Shin (1996), A simple method for generating correlated binary variates, *The American Statistician* **50**, 306–310.

Park, Stephen K., and Keith W. Miller (1988), Random number generators: Good ones are hard to find, *Communications of the ACM* **31**, 1192–1201.

Parrish, Rudolph S. (1990), Generating random deviates from multivariate Pearson distributions, *Computational Statistics & Data Analysis* **9**, 283–295.

Patefield, W. M. (1981), An efficient method of generating $r \times c$ tables with given row and column totals, *Applied Statistics* **30**, 91–97.

Pearson, E. S.; N. L. Johnson; and I. W. Burr (1979), Comparisons of the percentage points of distributions with the same first four moments, chosen from eight different systems of frequency curves, *Communications in Statistics — Simulation and Computation* **8**, 191–230.

Perlman, Michael D., and Michael J. Wichura (1975), Sharpening Buffon's needle, *The American Statistician* **29**, 157–163.

Peterson, Arthur V., and Richard A. Kronmal (1982), On mixture methods for the computer generation of random variables, *The American Statistician* **36**, 184–191.

Philippe, Anne (1997), Simulation of right and left truncated gamma distributions by mixtures, *Statistics and Computing* **7**, 173–181.

Pratt, John W. (1981), Concavity of the log likelihood, *Journal of the American Statistical Association* **76**, 103–106.

Press, William H.; Saul A. Teukolsky; William T. Vetterling; and Brian P. Flannery (1992), *Numerical Recipes in Fortran*, second edition, Cambridge University Press, Cambridge, United Kingdom.

Press, William H.; Saul A. Teukolsky; William T. Vetterling; and Brian P. Flannery (2002), *Numerical Recipes in C++*, second edition, Cambridge University Press, Cambridge, United Kingdom.

Propp, James Gary, and David Bruce Wilson (1996), Exact sampling with coupled Markov chains and applications to statistical mechanics, *Random Structures and Algorithms* **9**, 223–252.

Propp, James, and David Wilson (1998), Coupling from the past: A user's guide, *Microsurveys in Discrete Probability* (edited by D. Aldous and J.

Propp), American Mathematical Society, Providence, Rhode Island, 181–192.

Pullin, D. I. (1979), Generation of normal variates with given sample mean and variance, *Journal of Statistical Computation and Simulation* **9**, 303–309.

Rabinowitz, M., and M. L. Berenson (1974), A comparison of various methods of obtaining random order statistics for Monte-Carlo computations. *The American Statistician* **28**, 27–29.

Rajasekaran, Sanguthevar, and Keith W. Ross (1993), Fast algorithms for generating discrete random variates with changing distributions, *ACM Transactions on Modeling and Computer Simulation* **3**, 1–19.

Ramberg, John S., and Bruce W. Schmeiser (1974), An approximate method for generating asymmetric random variables, *Communications of the ACM* **17**, 78–82.

RAND Corporation (1955), *A Million Random Digits with 100,000 Normal Deviates*, Free Press, Glencoe, Illinois.

Ratnaparkhi, M. V. (1981), Some bivariate distributions of $(X, Y)$ where the conditional distribution of $Y$, given $X$, is either beta or unit-gamma, *Statistical Distributions in Scientific Work. Volume 4 – Models, Structures, and Characterizations* (edited by Charles Taillie, Ganapati P. Patil, and Bruno A. Baldessari), D. Reidel Publishing Company, Boston, 389–400.

Reeder, H. A. (1972), Machine generation of order statistics, *The American Statistician* **26**(4), 56–57.

Relles, Daniel A. (1972), A simple algorithm for generating binomial random variables when $N$ is large, *Journal of the American Statistical Association* **67**, 612–613.

Ripley, Brian D. (1987), *Stochastic Simulation*, John Wiley & Sons, New York.

Robert, Christian P. (1995), Simulation of truncated normal variables, *Statistics and Computing* **5**, 121–125.

Robert, Christian P. (1998a), A pathological MCMC algorithm and its use as a benchmark for convergence assessment techniques, *Computational Statistics* **13**, 169–184.

Robert, Christian P. (Editor) (1998b), *Discretization and MCMC Convergence Assessment*, Springer-Verlag, New York.

Robert, Christian P., and George Casella (1999), *Monte Carlo Statistical Methods*, Springer-Verlag, New York.

Roberts, G. O. (1992), Convergence diagnostics of the Gibbs sampler, *Bayesian Statistics 4* (edited by J. M. Bernardo, J. O. Berger, A. P. Dawid, and A. F. M. Smith), Oxford University Press, Oxford, United Kingdom, 775–782.

Roberts, Gareth O. (1996), Markov chain concepts related to sampling algorithms, *Practical Markov Chain Monte Carlo* (edited by W. R. Gilks, S. Richardson, and D. J. Spiegelhalter), Chapman & Hall, London, 45–57.

Robertson, J. M., and G. R. Wood (1998), Information in Buffon experiments, *Journal of Statistical Planning and Inference* **66**, 21–37.

Ronning, Gerd (1977), A simple scheme for generating multivariate gamma distributions with non-negative covariance matrix, *Technometrics* **19**, 179–183.

Rosenbaum, Paul R. (1993), Sampling the leaves of a tree with equal probabilities, *Journal of the American Statistical Association* **88**, 1455–1457.

Rosenthal, Jeffrey S. (1995), Minorization conditions and convergence rates for Markov chain Monte Carlo, *Journal of the American Statistical Association* **90**, 558–566.

Rousseeuw, Peter J., and Annick M. Leroy (1987), *Robust Regression and Outlier Detection*, John Wiley & Sons, New York.

Rubin, Donald B. (1987), Comment on Tanner and Wong, "The calculation of posterior distributions by data augmentation", *Journal of the American Statistical Association* **82**, 543–546.

Rubin, Donald B. (1988), Using the SIR algorithm to simulate posterior distributions (with discussion), *Bayesian Statistics 3* (edited by J. M. Bernardo, M. H. DeGroot, D. V. Lindley, and A. F. M. Smith), Oxford University Press, Oxford, United Kingdom, 395–402.

Ryan, Thomas P. (1980), A new method of generating correlation matrices, *Journal of Statistical Computation and Simulation* **11**, 79–85.

Sacks, Jerome; William J. Welch; Toby J. Mitchell; and Henry P. Wynn (1989), Design and analysis of computer experiments (with discussion), *Statistical Science* **4**, 409–435.

Sarkar, P. K., and M. A. Prasad (1987), A comparative study of pseudo and quasirandom sequences for the solution of integral equations, *Journal of Computational Physics* **68**, 66–88.

Sarkar, Tapas K. (1996), A composition-alias method for generating gamma variates with shape parameter greater than 1, *ACM Transactions on Mathematical Software* **22**, 484–492.

Särndal, Carl-Erik; Bengt Swensson; and Jan Wretman (1992), *Model Assisted Survey Sampling*, Springer-Verlag, New York.

Schafer, J. L. (1997), *Analysis of Incomplete Multivariate Data*, Chapman & Hall, London.

Schervish, Mark J., and Bradley P. Carlin (1992) On the convergence of successive substitution sampling, *Journal of Computational and Graphical Statistics* **1**, 111–127.

Schmeiser, Bruce (1983), Recent advances in generation of observations from discrete random variates, *Computer Science and Statistics: The Interface* (edited by James E. Gentle), North-Holland Publishing Company, Amsterdam, 154–160.

Schmeiser, Bruce, and A. J. G. Babu (1980), Beta variate generation via exponential majorizing functions, *Operations Research* **28**, 917–926.

Schmeiser, Bruce, and Voratas Kachitvichyanukul (1990), Noninverse correlation induction: Guidelines for algorithm development, *Journal of Computational and Applied Mathematics* **31**, 173–180.

Schmeiser, Bruce, and R. Lal (1980), Squeeze methods for generating gamma variates, *Journal of the American Statistical Association* **75**, 679–682.

Schucany, W. R. (1972), Order statistics in simulation, *Journal of Statistical Computation and Simulation* **1**, 281–286.

Selke, W.; A. L. Talapov; and L. N. Shchur (1993), Cluster-flipping Monte Carlo algorithm and correlations in "good" random number generators, *JETP Letters* **58**, 665–668.

Shao, Jun, and Dongsheng Tu (1995), *The Jackknife and Bootstrap*, Springer-Verlag, New York.

Shaw, J. E. H. (1988), A quasi-random approach to integration in Bayesian statistics, *Annals of Statistics* **16**, 895–914.

Shchur, Lev N., and Henk W. J. Blöte (1997), Cluster Monte Carlo: Scaling of systematic errors in the two-dimensional Ising model, *Physical Review E* **55**, R4905–R4908.

Sibuya, M. (1961), Exponential and other variable generators, *Annals of the Institute for Statistical Mathematics* **13**, 231–237.

Sinclair, C. D., and B. D. Spurr (1988), Approximations to the distribution function of the Anderson–Darling test statistic, *Journal of the American Statistical Association* **83**, 1190–1191.

Smith, A. F. M., and G. O. Roberts (1993), Bayesian computation via the Gibbs sampler and related Markov chain Monte Carlo methods, *Journal of the Royal Statistical Society, Series B* **55**, 3–24.

Smith, Robert L. (1984), Efficient Monte Carlo procedures for generating points uniformly distributed over bounded regions, *Operations Research* **32**, 1297–1308.

Smith, W. B., and R. R. Hocking (1972), Algorithm AS53: Wishart variate generator, *Applied Statistics* **21**, 341–345.

Sobol', I. M. (1967), On the distribution of points in a cube and the approximate evaluation of integrals, *USSR Computational Mathematics and Mathematical Physics* **7**, 86–112.

Sobol', I. M. (1976), Uniformly distributed sequences with an additional uniform property, *USSR Computational Mathematics and Mathematical Physics* **16**, 236–242.

Spanier, Jerome, and Keith B. Oldham (1987), *An Atlas of Functions*, Hemisphere Publishing Corporation, Washington (also Springer-Verlag, Berlin).

Srinivasan, Ashok; Michael Mascagni; and David Ceperley (2003), Testing parallel random number generators, *Parallel Computing* **29**, 69–94.

Stacy, E. W. (1962), A generalization of the gamma distribution, *Annals of Mathematical Statistics* **33**, 1187–1191.

Stadlober, Ernst (1990), The ratio of uniforms approach for generating discrete random variates, *Journal of Computational and Applied Mathematics* **31**, 181–189.

Stadlober, Ernst (1991), Binomial variate generation: A method based on ratio of uniforms, *The Frontiers of Statistical Computation, Simulation & Modeling* (edited by P. R. Nelson, E. J. Dudewicz, A. Öztürk, and E. C. van der Meulen), American Sciences Press, Columbus, Ohio, 93–112.

Steel, S. J., and N. J. le Roux (1987), A reparameterisation of a bivariate gamma extension, *Communications in Statistics — Theory and Methods* **16**, 293–305.

Stefănescu, S., and I. Văduva (1987), On computer generation of random vectors by transformations of uniformly distributed vectors, *Computing* **39**, 141–153.

Stein, Michael (1987), Large sample properties of simulations using Latin hypercube sampling, *Technometrics* **29**, 143–151.

Stephens, Michael A. (1986), Tests based on EDF statistics, *Goodness-of-Fit Techniques* (edited by Ralph B. D'Agostino and Michael A. Stephens), Marcel Dekker, New York, 97–193.

Stewart, G. W. (1980), The efficient generation of random orthogonal matrices with an application to condition estimators, *SIAM Journal of Numerical Analysis* **17**, 403–409.

Stigler, Stephen M. (1978), Mathematical statistics in the early states, *Annals of Statistics* **6**, 239–265.

Stigler, Stephen M. (1991), Stochastic simulation in the nineteenth century, *Statistical Science* **6**, 89–97.

Student (1908a), On the probable error of a mean, *Biometrika* **6**, 1–25.

Student (1908b), Probable error of a correlation coefficient, *Biometrika* **6**, 302–310.

Sullivan, Stephen J. (1993), Another test for randomness, *Communications of the ACM* **33**, Number 7 (July), 108.

Tadikamalla, Pandu R. (1980a), Random sampling from the exponential power distribution, *Journal of the American Statistical Association* **75**, 683–686.

Tadikamalla, Pandu R. (1980b), On simulating non-normal distributions, *Psychometrika* **45**, 273–279.

Tadikamalla, Pandu R., and Norman L. Johnson (1982), Systems of frequency curves generated by transformations of logistic variables, *Biometrika* **69**, 461–465.

Takahasi, K. (1965), Note on the multivariate Burr's distribution, *Annals of the Institute of Statistical Mathematics* **17**, 257–260.

Tang, Boxin (1993), Orthogonal array-based Latin hypercubes, *Journal of the American Statistical Association* **88**, 1392–1397.

Tanner, Martin A. (1996), *Tools for Statistical Inference*, third edition, Springer-Verlag, New York.

Tanner, M. A., and R. A. Thisted (1982), A remark on AS127. Generation of random orthogonal matrices, *Applied Statistics* **31**, 190–192.

Tanner, Martin A., and Wing Hung Wong (1987), The calculation of posterior distributions by data augmentation (with discussion), *Journal of the American Statistical Association* **82**, 528–549.

Tausworthe, R. C. (1965), Random numbers generated by linear recurrence modulo two, *Mathematics of Computation* **19**, 201–209.

Taylor, Malcolm S., and James R. Thompson (1986), Data based random number generation for a multivariate distribution via stochastic simulation, *Computational Statistics & Data Analysis* **4**, 93–101.

Tezuka, Shu (1991), Neave effect also occurs with Tausworthe sequences, *Proceedings of the 1991 Winter Simulation Conference*, Association for Computing Machinery, New York, 1030–1034.

Tezuka, Shu (1993), Polynomial arithmetic analogue of Halton sequences, *ACM Transactions on Modeling and Computer Simulation* **3**, 99–107.

Tezuka, Shu (1995), *Uniform Random Numbers: Theory and Practice*, Kluwer Academic Publishers, Boston.

Tezuka, Shu, and Pierre L'Ecuyer (1992), Analysis of add-with-carry and subtract-with-borrow generators, *Proceedings of the 1992 Winter Simulation Conference*, Association for Computing Machinery, New York, 443–447.

Tezuka, Shu; Pierre L'Ecuyer; and R. Couture (1994), On the lattice structure of the add-with-carry and subtract-with-borrow random number generators, *ACM Transactions on Modeling and Computer Simulation* **3**, 315–331.

Thomas, Andrew; David J. Spiegelhalter; and Wally R. Gilks (1992), BUGS: A program to perform Bayesian inference using Gibbs sampling, *Bayesian Statistics 4* (edited by J. M. Bernardo, J. O. Berger, A. P. Dawid, and A. F. M. Smith), Oxford University Press, Oxford, United Kingdom, 837–842.

Thompson, James R. (2000), *Simulation: A Modeler's Approach*, John Wiley & Sons, New York.

Thompson, William J. (1997), *Atlas for Computing Mathematical Functions: An Illustrated Guide for Practitioners with Programs in C and Mathematica*, John Wiley & Sons, New York.

Tierney, Luke (1991), Exploring posterior distributions using Markov chains, *Computer Science and Statistics: Proceedings of the Twenty-third Symposium on the Interface* (edited by Elaine M. Keramidas), Interface Foundation of North America, Fairfax, Virginia, 563–570.

Tierney, Luke (1994), Markov chains for exploring posterior distributions (with discussion), *Annals of Statistics* **22**, 1701–1762.

Tierney, Luke (1996), Introduction to general state-space Markov chain theory, *Practical Markov Chain Monte Carlo* (edited by W. R. Gilks, S. Richardson, and D. J. Spiegelhalter), Chapman & Hall, London, 59–74.

Vale, C. David, and Vincent A. Maurelli (1983), Simulating multivariate nonnormal distributions, *Psychometrika* **48**, 465–471.

Vattulainen, I. (1999), Framework for testing random numbers in parallel calculations, *Physical Review E* **59**, 7200–7204.

Vattulainen, I.; T. Ala-Nissila; and K. Kankaala (1994), Physical tests for random numbers in simulations, *Physical Review Letters* **73**, 2513–2516.

Vattulainen, I.; T. Ala-Nissila; and K. Kankaala (1995), Physical models as tests for randomness, *Physical Review E* **52**, 3205–3214.

Vattulainen, I.; K. Kankaala; J. Saarinen; and T. Ala-Nissila (1995), A comparative study of some pseudorandom number generators, *Computer Physics Communications* **86**, 209–226.

Vitter, J. S. (1984), Faster methods for random sampling, *Communications of the ACM* **27**, 703–717.

Vitter, Jeffrey Scott (1985), Random sampling with a reservoir, *ACM Transactions on Mathematical Software* **11**, 37–57.

Von Neumann, J. (1951), *Various Techniques Used in Connection with Random Digits*, NBS Applied Mathematics Series 12, National Bureau of Standards (now National Institute of Standards and Technology), Washington.

Vose, Michael D. (1991), A linear algorithm for generating random numbers with a given distribution, *IEEE Transactions on Software Engineering* **17**, 972–975.

Wakefield, J. C.; A. E. Gelfand; and A. F. M. Smith (1991), Efficient generation of random variates via the ratio-of-uniforms method, *Statistics and Computing* **1**, 129–133.

Walker, A. J. (1977), An efficient method for generating discrete random variables with general distributions, *ACM Transactions on Mathematical Software* **3**, 253–256..

Wallace, C. S. (1976), Transformed rejection generators for gamma and normal pseudo-random variables, *Australian Computer Journal* **8**, 103–105.

Wallace, C. S. (1996), Fast pseudorandom generators for normal and exponential variates, *ACM Transactions on Mathematical Software* **22**, 119–127.

Wichmann, B. A., and I. D. Hill (1982), Algorithm AS183: An efficient and portable pseudo-random number generator, *Applied Statistics* **31**, 188–190 (Corrections, 1984, *ibid.* **33**, 123).

Wikramaratna, R. S. (1989), ACORN — A new method for generating sequences of uniformly distributed pseudo-random numbers, *Journal of Computational Physics* **83**, 16–31.

Wilson, David Bruce, and James Gary Propp (1996), How to get an exact sample from a generic Markov chain and sample a random spanning tree from a directed graph, both within the cover time, *Proceedings of the Seventh Annual ACM-SIAM Symposium on Discrete Algorithms*, ACM, New York, 448–457.

Wolfram, Stephen (1984), Random sequence generation by cellular automata, *Advances in Applied Mathematics* **7**, 123–169. (Reprinted in Wolfram, 1994.)

Wolfram, Stephen (1994), *Cellular Automata and Complexity. Collected Papers*, Addison–Wesley Publishing Company, Reading, Massachusetts.

Wolfram, Stephen (2002), *A New Kind of Science*, Wolfram Media, Inc., Champaign, Illinois.

Wollan, Peter C. (1992), A portable random number generator for parallel computers, *Communications in Statistics — Simulation and Computation* **21**, 1247–1254.

Wu, Pei-Chi (1997), Multiplicative, congruential random-number generators with multiplier $\pm 2^{k_1} \pm 2^{k_2}$ and modulus $2^p - 1$, *ACM Transactions on Mathematical Software* **23**, 255–265.

Yu, Bin (1995), Comment on Besag et al., "Bayesian computation and stochastic systems": Extracting more diagnostic information from a single run using cusum path plot, *Statistical Science* **10**, 54–58.

Zaremba, S. K. (Editor) (1972), *Applications of Number Theory to Numerical Analysis*, Academic Press, New York.

Zeisel, H. (1986), A remark on Algorithm AS183: An efficient and portable pseudo-random number generator, *Applied Statistics* **35**, 89.

Zierler, Neal, and John Brillhart (1968), On primitive trinomials (mod 2), *Information and Control* **13**, 541–554.

Zierler, Neal, and John Brillhart (1969), On primitive trinomials (mod 2), II, *Information and Control* **14**, 566–569.

Ziff, Robert M. (1998), Four-tap shift-register-sequence random-number generators, *Computers in Physics* **12**, 385–392.

Ziv, J., and A. Lempel (1977), A universal algorithm for sequential data compression, *IEEE Transactions on Information Theory* **23**, 337–343.

# Author Index

# Subject Index

377